# Mathematical Systems: An Introduction

# Mathematical Systems:
# An Introduction

**HARRY D. RUDERMAN**
Hunter College

**ABRAHAM M. GLICKSMAN**
Bronx High School of Science

**Benziger, Inc.**  New York, Beverly Hills
*in Association with*
**John Wiley & Sons, Inc.**  New York, London, Sydney, Toronto

Copyright © 1971 by John Wiley and Sons, Inc.

All rights reserved. Published simultaneously in Canada.

No part of this book may be reproduced by any means, nor transmitted, nor translated into a machine language without the written permission of the publisher.

Library of Congress Catalogue Card Number: 70-153090

ISBN 0-471-74465-4

Printed in the United States of America

10 9 8 7 6 5 4 3 2 1

# Preface

As curriculum reform in mathematics progresses into a second decade, it tends more and more to transcend the traditional divisions of arithmetic, algebra, geometry, analysis, etc. There has emerged a clear trend toward unifying the mathematics curriculum at all levels via general concepts such as sets, operations, mappings, and relations. The result has been an increased emphasis on the study of fundamental structures such as groups, rings, fields, and vector spaces, all of which have significant implications for both algebra and geometry.

In seeking texts that cover these fundamental structures, we found that existing books in this area have been written, for the most part, at a relatively advanced college level. It is true that elementary treatments of these topics are available in the form of individual brief pamphlets or monographs covering particular mathematical systems. Although these are suitable for study of an individual topic, it is, however, difficult to base a unified course on such pamphlets.

The aim of *Mathematical Systems* is to bridge this gap by providing a systematic and unified treatment of mathematical systems through vector spaces, with sound presentations of elementary number theory and matrix algebra woven into the text. The book is designed for either a one or two semester course. The chapters are sufficiently detailed and elementary enough for presentation to secondary school students at either the junior level or senior level, yet they are comprehensive and rigorous enough to meet the demands of college teachers seeking a good first course in mathematical structures that includes elementary number theory and linear algebra. *Mathematical Systems* is also appropriate for courses preparing college students to teach mathematics, especially for the junior level and senior secondary level. The book can also be used as the basis for a college level course for high school students who may be approaching an advanced placement level.

The text is almost self-contained and makes minimal demands on prior mathematical information or skills. It is assumed that the student has the customary intuitive background with respect to the real numbers and has mastered the usual elementary skills of arithmetic and algebra. Ability to follow a logical argument is, of course, important and it is hoped that this book will help strengthen that ability and contribute toward developing mathematical maturity.

There is more than an ample number of exercises accompanying each section of the text. These range in difficulty from routine through truly challenging types. The problems we consider difficult are marked with the symbol ★. A few "research" problems are included for the more ambitious reader. There is also an additional set of review exercises at the end of each chapter for the purpose of reinforcing important ideas in the chapter. Solutions to a great many of the exercises appear at the back of the book. Solutions to all the problems and some suggestions on sequence and course content are available in a separate solution manual.

We acknowledge with gratitude the fine cooperation we have received from the school department of John Wiley & Sons. Our wives, Ruth Ruderman and Frances Glicksman, deserve the usual credit for their patience and forbearance during the throes of creation of a book with its inevitable impact on family life.

HARRY D. RUDERMAN
ABRAHAM GLICKSMAN

# Contents

## 1 │ OPERATIONS IN SETS — 1

    1.1 Unifying Diverse Phenomena   1
    1.2 Sets and Operations   4
    1.3 Binary operations in and on a Set   6
    1.4 Modular Addition—A Binary Operation   10
    1.5 Other Binary Operations of Modular Arithmetic   15
    1.6 Closure and Nonclosure   24
    1.7 Commutativity or Noncommutativity   26
    1.8 Associativity and Nonassociativity   32
    1.9 Existence and Nonexistence of an Identity Element   36
    1.10 Existence and Nonexistence of Inverses   40

## 2 │ ELEMENTARY GROUPS — 53

    2.1 Évariste Galois   53
    2.2 The Group Properties   54
    2.3 The Simplest Noncommutative Group   62
    2.4 Basic Theorems for Groups   68
    2.5 Isomorphic Systems   74

## 3 │ ELEMENTARY NUMBER THEORY — 89

    3.1 The Well-Ordering Axiom and the Division Theorem   89
    3.2 Prime Numbers and Composite Numbers   95

- 3.3 GCD and the Euclidean Algorithm   97
- 3.4 The Fundamental Theorem of Arithmetic   100
- 3.5 Congruence Modulo an Integer   110
- 3.6 Basic Theorems about Congruences   113

# 4 ‖ ELEMENTARY RINGS   125

- 4.1 Examples and Definition of a Ring   125
- 4.2 Rings of Two by Two Matrices   131
- 4.3 Basic Theorems for Rings   142
- 4.4 Isomorphic Rings   152

# 5 ‖ FIELDS   161

- 5.1 Examples and Definition of a Field   161
- 5.2 Basic Field Theorems   168
- 5.3 Some Further Basic Field Theorems   176
- 5.4 Square Roots and the Quadratic Formula   182
- 5.5 Further Application of Field Theorems to Solution of Equations   191

# 6 ‖ ORDERED FIELDS   203

- 6.1 Axioms for an Ordered Field   203
- 6.2 Basic Theorems for Ordered Fields   207
- 6.3 Effect of Operations on Order and Applications to Solution of Inequations   212
- 6.4 Theorems Concerning Reciprocals and Products in Ordered Fields and Further Applications   220
- 6.5 Some More Advanced Theorems for Ordered Fields   227
- 6.6 Absolute Value in Ordered Fields   230
- 6.7 Mathematical Induction—the Integers of an Ordered Field   241
- 6.8 Applications of Mathematical Induction   250

# 7 ‖ COMPLETE ORDERED FIELDS 259

    7.1 Lower Bounds and Upper Bounds 259
    7.2 Least Upper Bound and Greatest Lower Bound 263
    7.3 The Completeness Property 268

# 8 ‖ VECTOR SPACES 278

    8.1 Generalizing the Notion of an Operation 279
    8.2 Definition of a Vector Space 282
    8.3 Elementary Vector Space Theorems 290
    8.4 Linear Dependence, Linear Independence, and Bases 297

# 9 ‖ GAUSS-JORDAN METHOD 307

    9.1 Determination of Linear Dependence or Independence by Gaussian Elimination 307
    9.2 An Abbreviated Algorithm for Determining Linear Dependence or Independence 313
    9.3 The Gauss-Jordan Complete Elimination Procedure 320

# 10 ‖ MATRIX ALGEBRA 333

    10.1 Matrices and Linear Systems 333
    10.2 General Definitions: Matrices and Matrix Products 343
    10.3 Other Operations with Matrices 350
    10.4 Matrix Inverses and Linear Independence 359

Answers and Solutions to Selected Exercises 371

Index 423

*Norbert Wiener (1894-1964)*

# 1 Operations in Sets

> To see what is general in what is particular and what is permanent in what is transitory is the aim of scientific thought.
>
> A. N. WHITEHEAD (1861–1947)

## 1.1 UNIFYING DIVERSE PHENOMENA

The following diagram shows a simple electric circuit and a sketch of a ladder leaning against a vertical wall. What do they have in common?

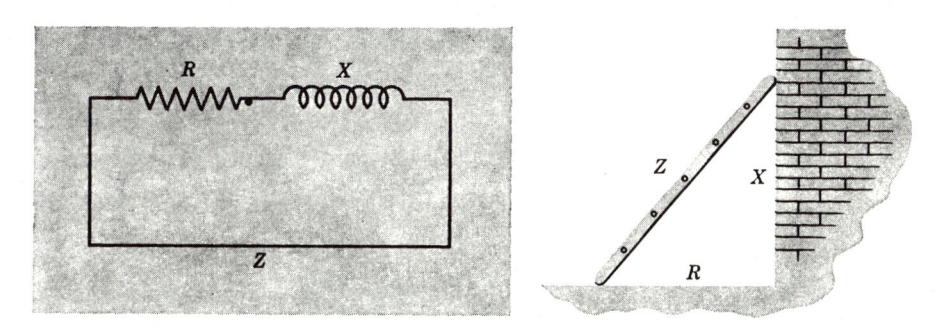

**Figure 1.1**

The average layman would probably see no connection. An electrical engineer, however, would recognize at once that both the length of the ladder and the

1

impedance of the electric circuit can be calculated using the same algebraic formula

$$Z^2 = R^2 + X^2$$

Pythagoras observed this mathematical relationship over 2500 years ago in the course of studying the geometry of right triangles. Today it is precisely what the modern scientist needs to know in studying electrical phenomena discovered within the last century.

Even at the elementary level of ordinary arithmetic, we find that a purely mathematical concept such as the sum of two numbers can be given numerous practical interpretations. If $a$ and $b$ are, respectively, the number of miles traveled each day of a two-day journey, then $a + b$ is the total mileage for the trip. If $a$ and $b$ are, respectively, the number of calories in a glass of milk and a piece of cake, then $a + b$ is the total calorie intake when these items are consumed. Thus the idea of finding the sum of a pair of numbers is an abstraction that embodies the common element in a variety of diverse situations.

The search for common elements in diverse phenomena is probably the most important characteristic which distinguishes the efforts of the scientist from those of most other individuals. This search achieves its purest form in the work of the mathematician.

A mathematician is not content merely to solve a particular problem. He prefers to solve a general problem of which the particular problem is a special case. An example of solving a large class of problems by a single generalization is the quadratic formula

$$x = \frac{-b \pm \sqrt{b^2 - 4ac}}{2a}$$

which solves all quadratic equations of the form

$$ax^2 + bx + c = 0 \quad \text{where } a \neq 0$$

Moreover, in addition to providing mastery over all quadratic equations, this formula has provided mathematicians with new insights into the concept of number and motivated useful extensions of the number system itself.

Here we shall study how diverse mathematical phenomena are unified, how mathematical generalizations are formulated, and how these generalizations are applied to important particular situations, both mathematical and nonmathematical.

# EXERCISES 1.1

1. Find three specific interpretations for the product $a \times b$. (Example: If $a$ is the speed of an airplane in miles per hour and $b$ is the number of hours the plane flies at this speed, then the product $a \times b$ is the distance the plane has flown.)
2. Find solutions among the numbers with which you are familiar for each of these equations:
   (a) $2x - 8 = 0$
   (b) $3x - 9 = 0$
   (c) $3x + 9 = 0$
   (d) $3x + 8 = 0$
   (e) $2x - b = 0$ (where $b$ is a number)
   (f) $ax - 6 = 0$ (where $a$ is a number different from zero)
   (g) $ax + b = 0$ (where $a$ and $b$ are numbers and $a \neq 0$)
   (h) Show that your answer to (g) can be used to obtain each of the answers in (a) to (f). Could you use the answer to solve other equations of the same type?
3. Consider the following computed products:
   $1.5 \times 1.5 = 2.25$, $2.5 \times 2.5 = 6.25$, $3.5 \times 3.5 = ?$, $4.5 \times 4.5 = ?$, etc.
   (a) What seems to be a pattern for these computed products?
   (b) Try to formulate precisely a generalization and then try to prove your generalization.
4. Consider the following computed products:
   $1.6 \times 1.4 = 2.24$, $2.6 \times 2.4 = 6.24$, $3.6 \times 3.4 = ?$, $4.6 \times 4.4 = ?$, etc.
   (a) What seems to be a pattern for these computed products?
   (b) Try to formulate precisely a generalization and then try to prove your generalization.
5. Consider the following computed products:
   $1.7 \times 1.3 = 2.21$, $2.7 \times 2.3 = 6.21$, $3.7 \times 3.3 = ?$, $4.7 \times 4.3 = ?$, etc.
   (a) What seems to be a pattern for these computed products?
   (b) Try to formulate precisely a generalization and then try to prove your generalization.
6. From your investigations in Exercises 3, 4, and 5 try to formulate and prove a more inclusive generalization.
7. Consider the following computed differences:
   $21 - 12 = 9$,   $31 - 13 = 18$,   $41 - 14 = ?$,
   $32 - 23 = ?$,   $42 - 24 = ?$,   $43 - 34 = ?$,
   $51 - 15 = ?$,   $52 - 25 = ?$,   $82 - 28 = ?$

(a) What divisibility property do these differences have in common?

(b) Try to formulate precisely a generalization and then try to prove your generalization.

8. Consider the following computed differences:
$$321 - 123 = ?, \quad 321 - 132 = ?, \quad 312 - 213 = ?,$$
$$874 - 748 = ?, \quad 847 - 478 = ?$$

(a) What divisibility property do these differences have in common?

(b) Try to formulate precisely a generalization and then try to prove your generalization.

9. Find the remainder when each of the following is divided by 3:
$$2^2, \quad 3^2, \quad 4^2, \quad 5^2, \quad 6^2, \quad \ldots, \quad \text{etc.}$$

(a) What seems to be the generalization?

(b) Try to prove your generalization. (*Hint:* Every integer is expressible either as $3n$, $3n + 1$, or $3n - 1$, where $n$ is some integer.)

## 1.2 SETS AND OPERATIONS

When the mathematician seeks to unify diverse mathematical phenomena and to solve large classes of problems, he continually encounters the sort of thing we call a *mathematical system*. The notion of a mathematical system encompasses the essence of mathematics itself. Like many fundamental ideas which are used continually, its importance is sometimes overlooked.

Roughly, a mathematical system consists of a set of objects together with some thing or some things that have to be "done" with these objects. For example, in elementary arithmetic, the objects are numbers; these may be whole numbers, integers, rational numbers, etc. In elementary geometry, the objects might be points, lines, triangles, and other geometric configurations. In probability theory, the objects are usually events made up of outcomes of some experiment. In still more advanced mathematics, the objects may be functions or they may be mappings, groups, topological spaces, etc.

What are some of the things that we "do" with these objects? In arithmetic, we add numbers, subtract numbers, multiply numbers, square numbers, and extract roots. In elementary geometry, we associate with every pair of distinct points a unique line containing them; with certain pairs of distinct lines we associate a unique point of intersection; with each triangle

## 1.2 Sets and Operations

we associate its unique inscribed circle, etc. In probability, we associate with each event a number called the probability of that event. On a still more advanced level, in calculus, we associate with certain functions other functions, for example, their derivatives.

Although these examples could be augmented with countless further illustrations, they are sufficient to indicate the breadth of this basic notion. The things we "do" with the elements of a mathematical system are frequently described as *operations*. For example, in arithmetic there are two basic operations with numbers, addition and multiplication. These are commonly called **binary** operations because each of these operations requires a pair of numbers on which to act. Thus, with the number pair (7, 5), the addition operation associates the sum $7 + 5$ or 12. Mathematicians often indicate this association by writing

$$(7, 5) \rightarrow 7 + 5$$

or also by writing

$$(7, 5) \xrightarrow{+} 12$$

The multiplication operation associates with (7, 5) the product $7 \times 5$ or 35. This association may be indicated by writing

$$(7, 5) \rightarrow 7 \times 5$$

or by writing

$$(7, 5) \xrightarrow{\times} 35$$

Some operations require only a single number on which to act and are therefore called **unary** operations (also singular). For example, the doubling operation assigns to each number $n$ the double of that number, namely $2n$. This doubling operation can be conveniently indicated by writing

$$n \rightarrow 2n$$

We can also picture the doubling operation by a diagram:

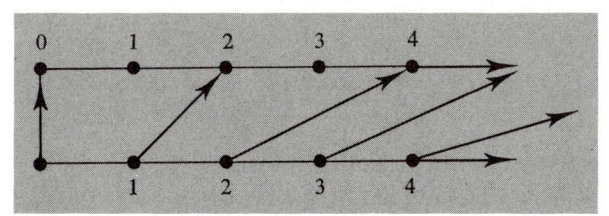

Notice that the symbol for the operation above the arrow is omitted whenever the nature of the operation is clear.

Although there are many other types of operations, they all have an important feature in common.

> Every operation assigns to certain elements of some specified set certain elements of that set or possibly another set.

We shall make this description more precise as we proceed. The arrow symbol → is a very convenient device for exhibiting the assignments for all kinds of operations.

### EXERCISES 1.2

1. Make up three different rules (other than those already mentioned in the text) for assigning a number to
   (a) any given whole number.
   (b) any given pair of whole numbers.
   (c) any given triple of whole numbers.
   Which of these rules defines a binary operation? A unary operation?
2. List three different things that might be assigned to a pair consisting of a line and a point not on the line. Do these assignments define unary or binary operations?
3. List three different things that might be assigned to a pair of distinct, intersecting lines in space. What kind of operations are defined here?
4. List two different rules for assigning a number to
   (a) a square whose side has length $a$.
   (b) a rectangle whose length and width are $a$ and $b$, respectively.
   (c) a circle whose radius is $r$.
   What kind of operation is defined in each case?

## 1.3  BINARY OPERATIONS IN AND ON A SET

Among the most useful mathematical systems are those consisting of a set $S$ together with a **binary operation on S**. Before we discuss this important type of operation, let us make clear another fundamental mathematical idea, the notion of an *ordered pair*.

## 1.3 Binary Operations In and On a Set

By an **ordered pair**, we mean a pair of elements in a definite order. For example, (1, 2) is an ordered pair of natural numbers; (2, 1) is also an ordered pair of natural numbers. These two ordered pairs are not the same. In fact, whenever $a \neq b$, the ordered pair $(a, b)$ is regarded as *distinct* from the ordered pair $(b, a)$. On the other hand, if $a = b$, then the ordered pairs $(a, b)$ and $(b, a)$ are the same, and in this case there is really only *one* ordered pair and this ordered pair may be indicated either as $(a, a)$ or $(b, b)$. In general, if $(a, b)$ and $(c, d)$ are ordered pairs, then

$$(a, b) = (c, d) \quad \text{if and only if} \quad a = c \text{ and } b = d$$

For example, if $(x, y) = (4, 3)$, then $x = 4$ and $y = 3$. For the sake of brevity we shall occasionally use the word "pair" instead of the longer phrase "ordered pair." Unless otherwise stated, these two expressions are to be interpreted as synonymous.

Returning now to our discussion of operations, we define the following.

> **A binary operation on a set** $S$ is an operation which assigns to *every ordered pair* of elements of $S$ a unique element of $S$.

For example, suppose that we choose $S$ to be the set of whole numbers

$$S = \{0, 1, 2, 3, 4, \ldots\}$$

and suppose that we choose addition as our binary operation. In this instance, if $(x, y)$ is any pair of whole numbers, addition assigns to the pair $(x, y)$ a definite whole number $x + y$. Thus to the pair (5, 3) addition assigns the whole number $5 + 3$, namely 8:

$$(5, 3) \xrightarrow{+} 8$$

To the pair (4, 0) addition assigns the whole number $4 + 0$, namely 4:

$$(4, 0) \xrightarrow{+} 4$$

Since a definite whole number of $S$ is assigned to every pair of whole numbers of $S$, we see that addition is an example of a binary operation *on* $S$.

Subtraction, on the other hand, is not a binary operation on $S$ because subtraction is defined only for those whole number pairs $(x, y)$ for which $x \geq y$. Thus to the pair (5, 3) subtraction assigns the definite whole number $5 - 3$, namely 2:

$$(5, 3) \xrightarrow{\phantom{x}} 2$$

Subtraction, however, does not assign a number of $S$ to the pair (3, 5) because $3 - 5$ is not a whole number. We describe this state of affairs by saying that subtraction is a binary operation *in* $S$ (but not on $S$).

> By a **binary operation in a set S** we mean one which assigns to some pairs (that is, at least one pair) of elements of $S$ a unique element of $S$.

From this point of view observe that every binary operation *on* a set $S$ is also a binary operation *in* $S$ (but not the other way around). Notice also that whenever a binary operation in a set $S$ does make an assignment to a pair of elements of $S$ the assignment must be unique. For example, subtraction can and often does assign the same whole number to different pairs of whole numbers:

$$(5, 3) \xrightarrow{\phantom{x}} 2, \quad (6, 4) \xrightarrow{\phantom{x}} 2, \quad (7, 5) \xrightarrow{\phantom{x}} 2, \quad \text{etc.}$$
$$\text{that is,} \quad 5 - 3 = 2, \quad 6 - 4 = 2, \quad 7 - 5 = 2, \quad \text{etc.}$$

In each of these instances, however, there is only *one* number assigned to the given pair. The uniqueness of the assignment is an essential characteristic of every binary operation.

For a nonnumerical illustration let $S$ be the set of all points in a plane and let the operation be choosing the midpoint for each pair of points. For brevity let us call the operation "midpointing." If $A$ and $B$ are distinct points, then clearly midpointing assigns a definite third point $C$ to the pair $(A, B)$, namely, the midpoint of the segment joining $A$ and $B$.

$$(A, B) \xrightarrow{\text{midpointing}} C \text{ (where } C \text{ is the midpoint of segment } AB\text{)}$$

If $A$ and $B$ are the same point, then midpointing simply assigns this same point $A$ to the pair $(A, A)$.

$$A\bullet \qquad (A, A) \xrightarrow{\text{midpointing}} A$$

Midpointing is therefore a binary operation *on* the set $S$ of all points in the plane.

## EXERCISES 1.3

*Each of the following assignments defines an operation. Indicate whether the operation is in or on, and state whether it is unary, binary, or neither.*

1. To a whole number $n$, assign the whole number $n + 1$, that is, $n \to (n + 1)$.
2. To a whole number $n$, assign the whole number $n - 1$, that is, $n \to (n - 1)$.
3. To a circle, assign its center.
4. To a line segment, assign its midpoint.
5. To any pair of distinct points, assign the line containing them.
6. To any whole number $n$, assign the whole number $n^2$, that is, $n \to n^2$.
7. To a pair of whole numbers $a$ and $b$, assign the larger of the two, unless $a = b$, in which case assign the number 0.
8. To any whole number, assign the next greater even whole number.
9. To any real number, assign the next greater whole number.
10. To any pair of distinct points in a plane, assign the perpendicular bisector of the segment joining them.
11. To a pair of distinct intersecting lines in space, assign a line through their intersection, perpendicular to both lines.
12. To a triple of rational numbers $(a, b, c)$, assign the rational number $\dfrac{a + b + c}{3}$, that is,

$$(a, b, c) \to \dfrac{a + b + c}{3}$$

13. To the ordered pair $(x, y)$ of real numbers, assign its second coordinate $y$, that is,

$$(x, y) \to y$$

14. To a pair of rational numbers $(a, b)$, assign a weighted average

$$(a, b) \to \dfrac{2a + 3b}{5}$$

**15.** To any pair of whole numbers $(a, b)$ assign the pair $(a + b, a - b)$.

**16.** To any pair of rational numbers $(a, b)$ assign the pair $(ab, a \div b)$.

## 1.4 MODULAR ADDITION—A BINARY OPERATION

A variety of interesting and useful mathematical systems can be obtained by the use of what is popularly called *clock arithmetic*. Mathematicians prefer the term **modular** arithmetic. A simple example of the use of modular arithmetic is the following:

> A businessman left New York by airplane on a Friday and flew to Europe. If he spent five days in Rome and six days in Paris, on what day of the week did he fly back?

A straightforward way to solve this problem would be to reason as follows: The total trip abroad took $5 + 6 = 11$ days. Starting from Friday, the next day would be Saturday, a second additional day would be Sunday, etc. Continuing this simple counting procedure for a total of 11 days, we find that the return trip occurred on a Tuesday.

This simple counting procedure leaves much to be desired, especially if the counting becomes lengthy. You can undoubtedly figure out a better method and you will then find that you have actually discovered one of the many mathematical systems known as *modular arithmetic*.

If an efficient businessman had to solve problems like this one very often, he would look for a simple procedure to obtain answers accurately and rapidly with a minimum of effort. He could, for example, construct on cardboard a dial with a pointer marked as indicated below:

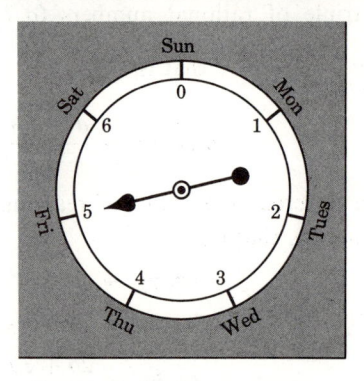

**Figure 1.2**

## 1.4 Modular Addition—A Binary Operation

He could then set the pointer on the numeral corresponding to Friday (this numeral is "5"), move 11 intervals around the circle clockwise, and observe that it stops at "2," which corresponds to Tuesday.

He would eventually discover that there is a still easier way to solve the problem. He would no doubt observe that a rotation of 7 intervals on this 7 interval "clock" brings him back to his starting position. Hence for all practical purposes, instead of moving the pointer through 11 intervals, he could just as well move it only 4 intervals. Instead of counting

$$6 + 5 = 11$$

he could just as well "count"

$$6 + 5 = 4$$

In this new kind of arithmetic, the number 7 counts as zero and any whole number over 7 can be *reduced* to one of the numbers 0, 1, 2, 3, 4, 5, or 6. In this illustration the number 11 is reduced for practical purposes to the number 4.

***Question.*** To which smaller number would 10 be reduced? How about 12, 13, 14, 15, 16, 21?

Now suppose that Sunday is assigned the number 0, Monday the number 1, Tuesday the number 2, etc. (see Figure 1.2), so that each day is assigned exactly one number, which counts the days elapsed since Sunday. Then our businessman need not bother with the dial at all because he can now solve his problem by simple arithmetic. This arithmetic is called **modulo 7 addition**. After figuring that

$$6 + 5 = 4 \quad \text{(instead of 11)}$$

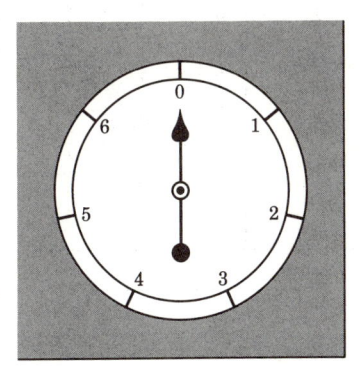

**Figure 1.3**

he could observe that (counting from the zero day Sunday) his day of departure, Friday, corresponds to the number 5. Therefore because

$$5+4=2 \quad \text{(instead of 9)}$$

in this new arithmetic, it then follows that his return day must be Tuesday (because the number 2 corresponds to Tuesday).

As another illustration suppose that he left on Thursday planning to spend 12 days in London and 8 days in Geneva. On what day would he return? We determine this easily as follows. In modulo 7 arithmetic, 12 is the "same" as 5, and 8 is the "same" as 1. Thus we may first calculate

$$12+8 \quad \text{is the "same" as} \quad 5+1=6$$

Now the day of departure, Thursday, corresponds to the number 4 (counting from the zero day Sunday) and

$$6+4=3 \quad \text{(not 10)}$$

Therefore his return home will occur on a Wednesday.

It is important to observe that in modulo 7 addition, each whole number greater than 6 can be replaced by one of the numbers 0, 1, 2, 3, 4, 5, 6. Hence in the computation it is unnecessary to use numbers other than 0, 1, 2, 3, 4, 5, 6. All that is really needed is an addition table for these 7 numbers, as shown below:

*Addition modulo seven*

| +(mod 7) | 0 | 1 | 2 | 3 | 4 | 5 | 6 |
|---|---|---|---|---|---|---|---|
| 0 | 0 | 1 | 2 | 3 | 4 | 5 | 6 |
| 1 | 1 | 2 | 3 | 4 | 5 | 6 | 0 |
| 2 | 2 | 3 | 4 | 5 | 6 | 0 | 1 |
| 3 | 3 | 4 | 5 | 6 | 0 | 1 | 2 |
| 4 | 4 | 5 | 6 | 0 | 1 | 2 | 3 |
| 5 | 5 | 6 | 0 | 1 | 2 | 3 | 4 |
| 6 | 6 | 0 | 1 | 2 | 3 | 4 | 5 |

From this table we can quickly read off all sums of the form

$$a+b \quad \text{(modulo 7)}$$

## 1.4 Modular Addition—A Binary Operation

where $a$ and $b$ can be any pair of numbers in the set $\{0, 1, 2, 3, 4, 5, 6\}$. For example, from the table we immediately see that

$$\left. \begin{array}{r} 4+6=3 \\ 5+2=0 \\ 6+6=5 \end{array} \right\} \quad \text{(modulo 7)}$$

Let us designate the set $\{0, 1, 2, 3, 4, 5, 6\}$ by $W_7$. Observe that addition modulo 7 is a binary operation *on* the set $W_7$.

From a purely mathematical point of view there is no reason to confine ourselves to the modulus 7. We can equally well study addition modulo 3, addition modulo 4, or for that matter addition modulo $n$, where $n$ is any whole number other than 0. For example, the following tables are for addition modulo 3 and addition modulo 4. You can check them by referring to the "3-hour clock" and the "4-hour clock," respectively, below each table.

*Addition modulo three*

| +(mod 3) | 0 | 1 | 2 |
|---|---|---|---|
| 0 | 0 | 1 | 2 |
| 1 | 1 | 2 | 0 |
| 2 | 2 | 0 | 1 |

*Addition modulo four*

| +(mod 4) | 0 | 1 | 2 | 3 |
|---|---|---|---|---|
| 0 | 0 | 1 | 2 | 3 |
| 1 | 1 | 2 | 3 | 0 |
| 2 | 2 | 3 | 0 | 1 |
| 3 | 3 | 0 | 1 | 2 |

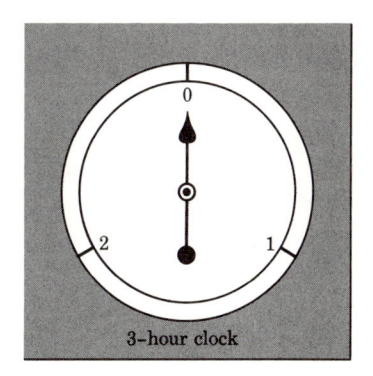

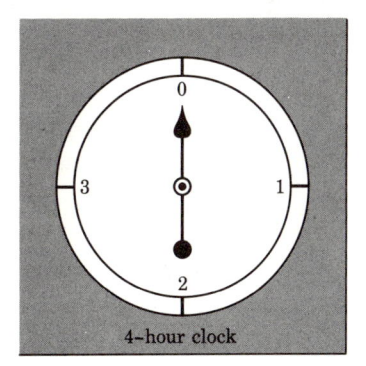

3-hour clock

4-hour clock

Observe that addition modulo 3 is a binary operation on the set $W_3$ where

$$W_3 = \{0, 1, 2\}$$

and addition modulo 4 is a binary operation on the set $W_4$ where

$$W_4 = \{0, 1, 2, 3\}$$

In general, if $n$ is a whole number other than 0, then addition modulo $n$ is a binary operation *on* the set $W_n$ where we define

$$W_n = \{0, 1, 2, \ldots, (n-1)\}$$

Whenever we deal with a set $W_n$ and an integer $x$ not in $W_n$, there will always be a unique number $y$ in $W_n$, such that

$$x = y + (\text{an integer} \times n)$$

In modulo $n$ addition we may always substitute this reduced integer $y$ for the integer $x$. Thus suppose that we are dealing with the set $W_4 = \{0, 1, 2, 3\}$ and the integer $x = 11$. There exists in $W_4$ the number $y = 3$ such that

$$x = y + (\text{an integer} \times 4)$$

namely

$$11 = 3 + (2 \times 4)$$

Hence, in modulo 4 addition, we may always substitute the reduced integer 3 for the integer 11.

## EXERCISES 1.4

1. Construct the addition table modulo 5. Using this table, find a solution in $W_5$ for each of these equations:
   (a) $4 + 3 = x$
   (b) $3 + 4 = x$
   (c) $2 + x = 4$
   (d) $2 + x = 1$
   (e) $2 + x = 2$
   (f) $2 + x = 0$
   (g) $x + 2 = 0$
   (h) $x + 3 = 0$
   (i) $x + 3 = 1$
   (j) $x + 3 = 2$

2. Solve the equation $7 + 8 = x$ (modulo $n$) in each of the following (express your answer as a member of $W_n$):
   (a) mod 3
   (b) mod 4
   (c) mod 5
   (d) mod 6
   (e) mod 7
   (f) mod 8
   (g) mod 9
   (h) mod 10
   (i) mod 15

3. Solve the equation $4 + x = 1$ (modulo $n$) in each of the following (express your answer as a member of $W_n$):
   (a) mod 3
   (b) mod 4
   (c) mod 5
   (d) mod 6
   (e) mod 7
   (f) mod 8
   (g) mod 9
   (h) mod 10
   (i) mod 15

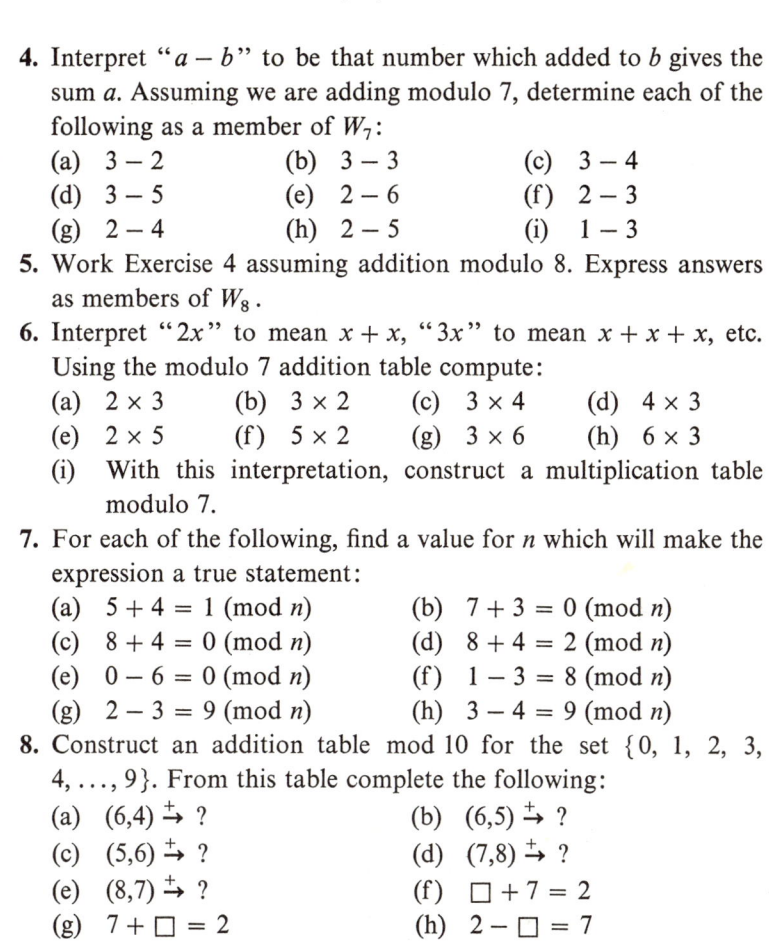

4. Interpret "$a - b$" to be that number which added to $b$ gives the sum $a$. Assuming we are adding modulo 7, determine each of the following as a member of $W_7$:
   (a) $3 - 2$     (b) $3 - 3$     (c) $3 - 4$
   (d) $3 - 5$     (e) $2 - 6$     (f) $2 - 3$
   (g) $2 - 4$     (h) $2 - 5$     (i) $1 - 3$

5. Work Exercise 4 assuming addition modulo 8. Express answers as members of $W_8$.

6. Interpret "$2x$" to mean $x + x$, "$3x$" to mean $x + x + x$, etc. Using the modulo 7 addition table compute:
   (a) $2 \times 3$     (b) $3 \times 2$     (c) $3 \times 4$     (d) $4 \times 3$
   (e) $2 \times 5$     (f) $5 \times 2$     (g) $3 \times 6$     (h) $6 \times 3$
   (i) With this interpretation, construct a multiplication table modulo 7.

7. For each of the following, find a value for $n$ which will make the expression a true statement:
   (a) $5 + 4 = 1 \pmod{n}$     (b) $7 + 3 = 0 \pmod{n}$
   (c) $8 + 4 = 0 \pmod{n}$     (d) $8 + 4 = 2 \pmod{n}$
   (e) $0 - 6 = 0 \pmod{n}$     (f) $1 - 3 = 8 \pmod{n}$
   (g) $2 - 3 = 9 \pmod{n}$     (h) $3 - 4 = 9 \pmod{n}$

8. Construct an addition table mod 10 for the set $\{0, 1, 2, 3, 4, \ldots, 9\}$. From this table complete the following:
   (a) $(6,4) \xrightarrow{+} ?$     (b) $(6,5) \xrightarrow{+} ?$
   (c) $(5,6) \xrightarrow{+} ?$     (d) $(7,8) \xrightarrow{+} ?$
   (e) $(8,7) \xrightarrow{+} ?$     (f) $\square + 7 = 2$
   (g) $7 + \square = 2$     (h) $2 - \square = 7$
   (i) $2 - 7 = \square$     (j) $2 = \square + 7$

9. If Sunday falls on December 4, what day of the week will be
   (a) December 11?     (b) December 12?
   (c) December 19?     (d) December 30?
   (e) 17 days later?     (f) 27 days later?
   (g) 40 days later?     (h) 40 days earlier?

## 1.5 OTHER BINARY OPERATIONS OF MODULAR ARITHMETIC

There is, of course, no reason to confine ourselves only to addition modulo $n$. We can also study subtraction, multiplication, and in some cases (but not all) division modulo $n$. Here are some examples of modular arithmetic showing various modular operations.

***Example 1.*** Compute $11 + 7 \pmod 5$

*Solution:*

METHOD A.  In ordinary arithmetic

$$11 + 7 = 18 = 15 + 3$$

But in modulo 5 arithmetic, $15 = 0$ (because all multiples of 5 reduce to 0 in this system). Hence

$$11 + 7 = 3 \pmod 5$$

METHOD B.  In modulo 5 arithmetic we have

$$11 = 1 \pmod 5 \quad \text{and} \quad 7 = 2 \pmod 5$$

Hence
$$11 + 7 = 1 + 2 \pmod 5$$
$$= 3 \pmod 5$$

***Example 2.*** Compute $11 \times 7 \pmod 5$

*Solution:*

METHOD A.  In ordinary arithmetic

$$11 \times 7 = 77 = 75 + 2$$

but $75 = 0 \pmod 5$ because all multiples of 5 reduce to 0 in modulo 5 arithmetic. Hence

$$11 \times 7 = 2 \pmod 5$$

METHOD B.  In modulo 5 arithmetic we have

$$11 = 1 \pmod 5 \quad \text{and} \quad 7 = 2 \pmod 5$$

Hence
$$11 \times 7 = 1 \times 2 \pmod 5$$
$$= 2 \pmod 5$$

***Example 3.*** Compute $14 \times 6 \pmod 4$

*Solution:*

METHOD A.  In ordinary arithmetic $14 \times 6 = 84$

but $84 = 0 \pmod 4$ because 84 is a multiple of 4 (that is, $84 = 21 \times 4$). Hence

$$14 \times 6 = 0 \pmod 4$$

### 1.5 Other Binary Operations of Modular Arithmetic

METHOD B. In modulo 4 arithmetic we have

$$14 = 2 \pmod 4 \quad \text{and} \quad 6 = 2 \pmod 4$$

Hence
$$14 \times 6 = 2 \times 2 \pmod 4$$
$$= 4 \pmod 4$$
$$= 0 \pmod 4$$

*Example 4.* Compute $11 - 7 \pmod 3$

Solution:

METHOD A. In ordinary arithmetic $11 - 7 = 4 = 3 + 1$. But $3 = 0 \pmod 3$.

Hence $\quad\quad 11 - 7 = 1 \pmod 3$

METHOD B. In modulo 3 arithmetic we have

$$11 = 2 \pmod 3 \quad \text{and} \quad 7 = 1 \pmod 3$$

Hence
$$11 - 7 = 2 - 1 \pmod 3$$
$$= 1 \pmod 3$$

*Example 5.* Compute $7 - 10 \pmod 3$

Solution: We cannot compute $7 - 10$ as ordinary subtraction in the set of whole numbers because there is no such whole number as $7 - 10$. However, we can compute $7 - 10 \pmod 3$ because

$$7 = 1 \pmod 3 \quad \text{and} \quad 10 = 1 \pmod 3.$$

Hence
$$7 - 10 = 1 - 1 \pmod 3$$
$$= 0 \pmod 3$$

In fact, 7 and 10 both represent the same number, namely 1, in the modulo 3 system of arithmetic.

*Example 6.* Compute $7 - 14 \pmod 6$

Solution: If we "reduce" both 7 and 14 modulo 6, we obtain

$$7 = 1 \pmod 6 \quad \text{and} \quad 14 = 2 \pmod 6$$

Hence
$$7 - 14 = 1 - 2 \pmod 6$$

METHOD A. Is there such a number as $1 - 2 \pmod 6$? The answer is Yes because if

$$x \doteq 1 - 2 \pmod 6$$

then
$$x + 2 = 1 \pmod 6$$
Now from the addition table modulo 6 we observe that
$$5 + 2 = 1 \pmod 6$$
Hence we may choose $x = 5$, thus obtaining $1 - 2 = 5 \pmod 6$.

METHOD B. Is there such a number as $1 - 2 \pmod 6$? The answer is Yes because
$$1 = 1 + 6 = 7 \pmod 6$$
Hence
$$1 - 2 = 7 - 2 \pmod 6$$
$$= 5 \pmod 6$$

It appears from these examples that if $a$ and $b$ are whole numbers and if $n$ is a whole number other than 0, it is always possible to compute

$$a + b \pmod n$$
$$a - b \pmod n$$
$$a \times b \pmod n$$

so that the answer will be exactly one of the modulo $n$ numbers

$$0, 1, 2, \ldots, (n-1)$$

Although this assertion can be proved, we shall not do so here. Its significance for us is that it shows that for any whole number $n$ (other than 0) addition modulo $n$, subtraction modulo $n$, and multiplication modulo $n$ are binary operations *on* the set of modulo $n$ numbers, namely the set

$$W_n = \{0, 1, 2, \ldots, (n-1)\}$$

As an illustration let us construct both a multiplication and a subtraction table modulo 5.

*Multiplication modulo five*

| ×(mod 5) | 0 | 1 | 2 | 3 | 4 |
|---|---|---|---|---|---|
| 0 | 0 | 0 | 0 | 0 | 0 |
| 1 | 0 | 1 | 2 | 3 | 4 |
| 2 | 0 | 2 | 4 | 1 | 3 |
| 3 | 0 | 3 | 1 | 4 | 2 |
| 4 | 0 | 4 | 3 | 2 | 1 |

*Subtraction modulo five*

| −(mod 5) | 0 | 1 | 2 | 3 | 4 |
|---|---|---|---|---|---|
| 0 | 0 | 4 | 3 | 2 | 1 |
| 1 | 1 | 0 | 4 | 3 | 2 |
| 2 | 2 | 1 | 0 | 4 | 3 |
| 3 | 3 | 2 | 1 | 0 | 4 |
| 4 | 4 | 3 | 2 | 1 | 0 |

The situation becomes somewhat more complicated when we study division modulo $n$. Suppose, for example, that we want to construct a division table modulo 5. Before we can construct this division table, we must decide what we mean by division modulo 5. In ordinary arithmetic what do we mean by "$6 \div 3$"? $6 \div 3$ is the unique number which, multiplied by 3, gives the product 6. The answer is, of course, 2.

$$6 \div 3 = \boxed{2} \quad \text{because} \quad 6 = \boxed{2} \times 3$$

In general, $a \div b$ is the unique number $c$ (if such exists) which multiplied by $b$ gives the product $a$:

$$a \div b = \boxed{c} \quad \text{whenever there is a unique number } c \text{ such that}$$
$$a = \boxed{c} \times b$$

If $b = 0$, there is no unique number $c$ that satisfies the requirement, hence division by 0 is excluded. In this case, when $a \neq 0$, there is no number at all which satisfies the requirement and when $a = 0$, any number whatsoever satisfies the requirement. Thus, in either case, there is no unique number for $c$ if $b = 0$.

If we carry the above meaning for division over to modular arithmetic, we shall agree that $a \div b = c \pmod{n}$ whenever there is a unique value for $c$ in $W_n$ such that $a = c \times b \pmod{n}$.

Thus in mod 5 arithmetic as in ordinary arithmetic we have

$$4 \div 2 = 2 \pmod{5} \quad \text{because} \quad 4 = 2 \times 2 \pmod{5}$$
$$0 \div 3 = 0 \pmod{5} \quad \text{because} \quad 0 = 0 \times 3 \pmod{5}$$

But "$0 \div 0 \pmod{5}$" does not name a number because there is no *unique* number $c$ in $W_5$ such that $0 = c \times 0 \pmod{5}$. Also "$3 \div 0 \pmod{5}$" names no number because there is simply no number $c$ in $W_5$ such that

$$3 = c \times 0 \pmod{5}$$

Since division by 0 is excluded, we must omit from the modulo 5 division table a column headed by 0. We begin with the following blank table.

| $\div \pmod{5}$ | 1 | 2 | 3 | 4 |
|---|---|---|---|---|
| 0 | | | | |
| 1 | | | | |
| 2 | | | | |
| 3 | | | | |
| 4 | | | | |

Certain entries should be obvious as shown below:

| ÷(mod 5) | 1 | 2 | 3 | 4 |
|---|---|---|---|---|
| 0 | 0 | 0 | 0 | 0 |
| 1 | 1 |   |   |   |
| 2 | 2 | 1 |   |   |
| 3 | 3 |   | 1 |   |
| 4 | 4 | 2 |   | 1 |

The remaining entries, however, require a little more effort. We shall find them by going back to our definition of division. To compute $1 \div 2$ (mod 5) we observe that

$$1 \div 2 = c \text{ (mod 5)}$$

is equivalent to

$$1 = c \times 2 \text{ (mod 5)}$$

We now look at the numbers in $W_5$, namely 0, 1, 2, 3, 4, and see if there is such a unique value for $c$. Referring to the multiplication table modulo 5, we observe that $1 = 3 \times 2$ (mod 5); so 3 certainly satisfies the requirement. Moreover, no other number fits, as may be seen from the multiplication table modulo 5. Hence

$$1 \div 2 = 3 \text{ (mod 5)}$$

In the same manner we find that

$$1 \div 3 = 2 \text{ (mod 5)}$$
$$1 \div 4 = 4 \text{ (mod 5)}$$

The division table modulo 5 now looks like this:

| ÷(mod 5) | 1 | 2 | 3 | 4 |
|---|---|---|---|---|
| 0 | 0 | 0 | 0 | 0 |
| 1 | 1 | 3 | 2 | 4 |
| 2 | 2 | 1 |   |   |
| 3 | 3 |   | 1 |   |
| 4 | 4 | 2 |   | 1 |

## 1.5 Other Binary Operations of Modular Arithmetic

We continue by observing that

$$2 \div 3 = 4 \pmod 5$$
$$2 \div 4 = 3 \pmod 5$$
$$3 \div 2 = 4 \pmod 5$$
$$3 \div 4 = 2 \pmod 5$$
$$4 \div 3 = 3 \pmod 5$$

and the table is completed.

| $\div \pmod 5$ | 1 | 2 | 3 | 4 |
|---|---|---|---|---|
| 0 | 0 | 0 | 0 | 0 |
| 1 | 1 | 3 | 2 | 4 |
| 2 | 2 | 1 | 4 | 3 |
| 3 | 3 | 4 | 1 | 2 |
| 4 | 4 | 2 | 3 | 1 |

Clearly, division modulo 5 is a binary operation *in* the set

$$W_5 = \{0, 1, 2, 3, 4\}$$

but not *on* this set because we cannot compute $a \div b$ for all pairs of numbers $(a, b)$ in $W_5$ but only for those pairs where $b \neq 0$. Notice, however, that if we exclude 0 from the set $W_5$ (let us denote this new set by $W_5^*$):

$$W_5^* = \{1, 2, 3, 4\}$$

then $\div \pmod 5$ is a binary operation *on* the new set $W_5^*$.

| $\div \pmod 5$ | 1 | 2 | 3 | 4 |
|---|---|---|---|---|
| 1 | 1 | 3 | 2 | 4 |
| 2 | 2 | 1 | 4 | 3 |
| 3 | 3 | 4 | 1 | 2 |
| 4 | 4 | 2 | 3 | 1 |

Let us now try to construct a division table modulo 4. We start with the following obvious entries:

| ÷(mod 4) | 1 | 2 | 3 |
|---|---|---|---|
| 0 | 0 |   | 0 |
| 1 | 1 |   |   |
| 2 | 2 | 1 |   |
| 3 | 3 |   | 1 |

The next step is to try to compute $1 \div 2$ (mod 4). Suppose $1 \div 2$ (mod 4) has a value $c$, that is,

$$1 \div 2 = c \text{ (mod 4)}$$

This is equivalent to

$$1 = c \times 2 \text{ (mod 4)}$$

We now examine the numbers in $W_4$, namely 0, 1, 2, 3, to see if there is such a number $c$. Actual trial reveals that no such number $c$ exists. In fact,

$$0 \times 2 = 0, \quad 1 \times 2 = 2, \quad 2 \times 2 = 0, \quad \text{and} \quad 3 \times 2 = 2 \text{ (mod 4)}$$

so our task is impossible. We cannot compute $1 \div 2$ (mod 4) because such a number simply does not exist! Division modulo 4 fails not only when the divisor is 0 but also when the divisor is 2.

Therefore, in order to construct a division table modulo 4, we remove the column headed "2," obtaining

| ÷(mod 4) | 1 | 3 |
|---|---|---|
| 0 | 0 | 0 |
| 1 | 1 | 3 |
| 2 | 2 | 2 |
| 3 | 3 | 1 |

We leave it to the reader to verify these entries. Division modulo 4 is a binary operation *in* each of the sets $W_4^* = \{1, 2, 3\}$ and $W_4 = \{0, 1, 2, 3\}$ but *not on* either of these sets. However, if we remove both 0 and 2 and confine our attention to the set

$$W_4^{**} = \{1, 3\}$$

then the following (reduced) table

| $\div$ (mod 4) | 1 | 3 |
|---|---|---|
| 1 | 1 | 3 |
| 3 | 3 | 1 |

shows that $\div$ (mod 4) is a binary operation on the (reduced) set $W_4^{**}$.

## EXERCISES 1.5

1. Construct an addition table and a multiplication table for
   - (a) mod 2
   - (b) mod 3
   - (c) mod 6
   - (d) mod 7
2. Construct a subtraction table and a division table for
   - (a) mod 2
   - (b) mod 3
   - (c) mod 6
   - (d) mod 7
   
   (Use the tables of Exercise 1, if necessary.)
3. Compute the reduced integer in each of the following (if possible without referring to any tables):
   - (a) $7 \times 3$ (mod 8)
   - (b) $7 \times 8$ (mod 3)
   - (c) $8 \times 3$ (mod 5)
   - (d) $5 \div 3$ (mod 7)
   - (e) $1 \div 5$ (mod 7)
   - (f) $2 \div 5$ (mod 7)
   - (g) $3 \div 5$ (mod 7)
   - (h) $4 \div 5$ (mod 7)
   - (i) $6 \div 5$ (mod 7)
4. Solve each of the following in modulo 7 arithmetic:
   - (a) $2x + 3 = 1$
   - (b) $2x - 3 = 1$
   - (c) $2x + 3 = 0 - 1$
   - (d) $2x - 3 = 0 - 1$
   - (e) $3 - 2x = 1$
   - (f) $4 - 3x = 1$
   - (g) $4 - 3x = 5$
   - (h) $\dfrac{3x}{4} = 5$
5. Do the following computations in modulo 7 arithmetic:
   - (a) $5 + (8 \times 9)$
   - (b) $(5 + 8) \times 9$
   - (c) $8 \times 9 - 5$
   - (d) $8 \times (9 - 5)$
   - (e) $7 \div 5$
   - (f) $2 + (1 \div 5)$
   - (g) $(2 \div 3) + (1 \div 3)$
   - (h) $(2 \div 3) - (3 \div 4)$
   - (i) $(\tfrac{1}{2}) \times (\tfrac{3}{4})$
   - (j) $3 \div 8$
   - (k) $9 \div 8$
   - (l) $1 + (1 \div 8)$
   - (m) $(-1)(-1)$
   - (n) $(-2)(-2)$
   - (o) $(-2)(-3)$
   - (p) $(-3)(-3)$

★6. In general, if $a$, $b$, and $n(\neq 0)$ are integers, then $a = b \pmod{n}$ if and only if for some integer $k$, $a = b + kn$. Prove that whenever $a$, $b$, $c$, $d$, and $n$ are integers (with $n \neq 0$),
(a) $a = a \pmod{n}$
(b) If $a = b \pmod{n}$, then $b = a \pmod{n}$.
(c) If $a = b \pmod{n}$ and $b = c \pmod{n}$, then $a = c \pmod{n}$.
(d) If $a = b \pmod{n}$, then $-a = -b \pmod{n}$.
(e) If $a = b \pmod{n}$ and $c = d \pmod{n}$, then
$a + c = b + d \pmod{n}$
$a - c = b - d \pmod{n}$
$ac = bd \pmod{n}$
It need not be the case that $a \div c = b \div d \pmod{n}$ even when "$a \div c$" and "$b \div d$" are meaningful.

## 1.6 CLOSURE AND NONCLOSURE

We have seen that a binary operation in a set $S$ may or may not be a binary operation on $S$. For example, if $S$ is the set of whole numbers $\{0, 1, 2, \ldots\}$, then subtraction is a binary operation in $S$ but not on $S$. This is due to the fact that if $(x, y)$ is a pair of whole numbers, then $x - y$ is a whole number only when $x \geq y$. If $y$ is larger than $x$, $x - y$ is not a whole number. Thus, although addition is defined for every pair of whole numbers $(x, y)$, subtraction is defined only for certain pairs that satisfy the special restriction $x \geq y$. This idea is expressed by saying that the set $S$ of whole numbers is *closed* with respect to addition, but $S$ is not closed with respect to subtraction.

It is important to note that the property of **closure** or **nonclosure** depends on two things: the set $S$ and the operation in $S$. For example, if we choose $S$ to be the set of integers $\{0, \pm 1, \pm 2, \ldots\}$, then subtraction in $S$ becomes a binary operation on $S$ and $S$ is closed with respect to subtraction. If $S$ is the set of even integers, then $S$ is closed with respect to addition. However, if $S$ is the set of odd integers, then $S$ is not closed with respect to addition. On the other hand, both sets (even as well as odd integers) are closed with respect to multiplication. The reason is that if we multiply any two even numbers, the product is still an even number, and if we multiply any two odd numbers, the product is still an odd number.

Similar remarks apply to unary operations. For example, if $S$ is the set of whole numbers, then $S$ is closed with respect to the unary operation "doubling," but $S$ is not closed with respect to the unary operation "halving."

## EXERCISES 1.6

1. Write Yes or No in the following table to show whether the sets described are or are not closed with respect to each operation listed.

|     | Set | + | − | × | ÷ |
|-----|-----|---|---|---|---|
| (a) | $\{0, \frac{1}{2}, 1, 2\}$ | | | | |
| (b) | $\{0, 1, -1\}$ | | | | |
| (c) | {Odd integers} | | | | |
| (d) | {Even integers} | | | | |
| (e) | {Multiples of 5} | | | | |
| (f) | $\{1, -1\}$ | | | | |
| (g) | {Positive rational numbers} | | | | |
| (h) | {Real numbers} | | | | |

2. (a) State whether the set $W_5 = \{0, 1, 2, 3, 4\}$ is or is not closed with respect to each of the operations: $+$(mod 5), $-$(mod 5), $\times$(mod 5), $\div$(mod 5).
   (b) State whether the set $W_6 = \{0, 1, 2, 3, 4, 5\}$ is or is not closed with respect to each of the operations: $+$(mod 6), $-$(mod 6), $\times$(mod 6), $\div$(mod 6).
   (c) State whether the set $W_7^* = \{1, 2, 3, 4, 5, 6\}$ is or is not closed with respect to each of the operations: $+$(mod 7), $-$(mod 7), $\times$(mod 7), $\div$(mod 7).
3. Write Yes or No in the following table to show whether each of the sets specified is or is not closed with respect to each of the operations specified.

**1** || Operations in Sets

| Operation Defined by | Whole Numbers | Integers | Rational Numbers | Real Numbers | Positive Real Numbers | Negative Real Numbers |
|---|---|---|---|---|---|---|
| (a) $n \to n+2$ | | | | | | |
| (b) $n \to n-2$ | | | | | | |
| (c) $n \to 3n$ | | | | | | |
| (d) $n \to \dfrac{n}{3}$ | | | | | | |
| (e) $n \to n^2$ | | | | | | |
| (f) $n \to \sqrt{n}$ | | | | | | |
| (g) $n \to 2^n$ | | | | | | |
| (h) $(a,b) \to a$ | | | | | | |
| (i) $(a,b) \to 2a+b$ | | | | | | |
| (j) $(a,b) \to 2a-b$ | | | | | | |
| (k) $(a,b) \to a^2+b^2$ | | | | | | |
| (l) $(a,b) \to a^2-b^2$ | | | | | | |
| (m) $(a,b) \to a^2+b^2-2ab$ | | | | | | |
| (n) $(a,b) \to 2^{a+b}$ | | | | | | |
| (o) $(a,b) \to |a+b|$ | | | | | | |
| (p) $(a,b) \to |a|-|b|$ | | | | | | |

## 1.7 COMMUTATIVITY OR NONCOMMUTATIVITY

Addition, subtraction, multiplication, and division are probably the most familiar binary operations with numbers. They are so important that man has developed many ingenious and often expensive devices to perform, rapidly and accurately, computations involving these operations.

Whether we use a machine or just ordinary pencil and paper to com-

## 1.7 Commutativity or Noncommutativity

pute the sum of a pair of numbers, we cannot avoid a feature inherent in all binary operations—they "act" on *ordered* pairs of elements. We must record these elements or feed them to the machine one after the other. The numbers are "operated on" in a definite order.

For some binary operations the order does not matter. For example, suppose we start with the pair of numbers (8, 4) and we set our calculating machine to compute products. We must first enter "8" into the machine and then enter "4," in which case the machine calculates

$$8 \times 4 = 32$$

If we reverse the procedure by entering first "4" and then "8," the machine calculates

$$4 \times 8 = 32$$

Evidently the order does not matter because the same product 32 is obtained in each case. Mathematicians express the fact that

$$8 \times 4 = 4 \times 8$$

by saying that the numbers 8 and 4 **commute** under the operation multiplication. When two numbers commute under an operation, the order in which they are considered does not matter.

If we set our calculator to perform a computation involving the binary operation of division, however, the situation is not so simple. Suppose we enter 8 first and then 4; the machine computes:

$$8 \div 4 = 2$$

If we enter 4 first and then 8, the machine computes:

$$4 \div 8 = 0.5 \quad (\text{or } \tfrac{1}{2})$$

Obviously, the quotient $8 \div 4$ is not the same as $4 \div 8$. For this operation the order of the numbers does make a difference. Mathematicians express the fact that

$$8 \div 4 \neq 4 \div 8$$

by saying that the numbers 8 and 4 do **not commute** under the operation of division.

**Questions.**

    Do 8 and 4 commute under addition?
    Do 8 and 4 commute under subtraction?
    Do 8 and 8 commute under subtraction?
    Do 8 and 8 commute under division?

## 1 ∥ Operations in Sets

Of course, there are many other significant binary operations with numbers besides addition, subtraction, multiplication, and division. For example, the binary operation *exponentiation* occurs very often in algebra and applied mathematics. This operation assigns to an ordered pair of numbers $(a, b)$ the power $a^b$. For example, to the pair of whole numbers $(2, 3)$ exponentiation assigns the power $2^3$, where $2^3$ means $2 \times 2 \times 2$, or 8. Designating the exponentiation operation by "exp," we have

$$a \exp b = a^b$$

so that, for example,

$$2 \exp 3 = 2^3 = 8$$

### Problems.

1. Compute 3 exp 2.
2. Compute 2 exp 4.
3. Compute 4 exp 2.
4. Do 3 and 2 commute under the binary operation exp?
5. Do 2 and 4 commute under the binary operation exp?
6. Compute 0 exp 2.
7. Compute 2 exp 0.
★8. Is exp a binary operation on the set of whole numbers? (*Hint:* Try to calculate 0 exp 0.)
★9. For which pairs of whole numbers $(x, y)$ do we have $x \exp y = y \exp x$?

Observe that 2 and 3 do not commute under the binary operation exponentiation because

$$2 \exp 3 = 2^3 = 8$$

But

$$3 \exp 2 = 3^2 = 9$$

Hence

$$2 \exp 3 \neq 3 \exp 2$$

On the other hand, 2 and 4 do commute under exponentiation because

$$2 \exp 4 = 2^4 = 16$$
$$4 \exp 2 = 4^2 = 16$$

Here, therefore,

$$2 \exp 4 = 4 \exp 2$$

We see from these examples that some pairs of numbers may commute under a given binary operation, whereas other pairs of numbers may not commute under this same operation. Let us generalize this observation.

Suppose we have defined a binary operation ∘ in a set $S$. This means that we have a definite, clear-cut rule for choosing certain ordered pairs of elements of $S$ and then assigning to each of these ordered pairs an element of $S$. If $(x, y)$ is any such chosen pair, there will be an element $z$ in $S$ which our binary operation ∘ assigns to the ordered pair $(x, y)$. This assignment is usually expressed by writing

$$x \circ y = z$$

If the same element $z$ is also assigned to the ordered pair $(y, x)$, that is, if

$$x \circ y = y \circ x$$

then we say that $x$ and $y$ *commute under the binary operation* ∘ *in S*. (If there is only one binary operation under discussion, we shorten this by simply asserting that $x$ and $y$ commute.) For example, 3 and 2 commute under multiplication, which may be shown as follows:

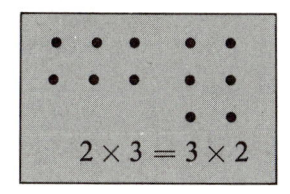

$$2 \times 3 = 3 \times 2$$

Now under what circumstances may $x$ and $y$ fail to commute? This may occur in two ways.

1. The binary operation ∘ may assign one value to $x \circ y$ and a different value to $y \circ x$. (For example, in the set of whole numbers, 2 exp 3 = 8, but 3 exp 2 = 9.)
2. The binary operation ∘ may fail to assign a value to one of the two expressions, "$x \circ y$" or "$y \circ x$." (For example, if $S$ is the set of whole numbers, then $8 \div 2 = 4$, which is a member of $S$, but $2 \div 8$ is not a member of $S$ because $2 \div 8$ is not a whole number.)

If ∘ is a binary operation in $S$ but not on $S$, both possibilities can occur. However, if ∘ is a binary operation on $S$, then the second possibility cannot occur because a binary operation on $S$ assigns an element of $S$ to every ordered pair of elements of $S$. Thus if ∘ is a binary operation on $S$, then $x \circ y$ and $y \circ x$ are defined for every pair of elements, and it makes sense to ask whether *every* pair of elements commute under $\circ$. If it should happen that every pair of elements in $S$ commute under $\circ$, we call $\circ$ a **commutative operation** on $S$, or simply a commutative operation.

**Definition.** A binary operation $\circ$ on a set $S$ is called **commutative** (on $S$) if and only if every pair of elements of $S$ commute under $\circ$, that is, if and only if

$$x \circ y = y \circ x \quad \text{for all } x \in S \text{ and all } y \in S$$

For binary operations in $S$, this definition may not apply because $x \circ y$ may not be defined for every ordered pair $(x, y)$. In these cases, the following "weaker" definition applies.

**Definition.** A binary operation $\circ$ in a set $S$ is called **commutative** *in* $S$ if and only if, whenever $x \circ y$ is defined as an element of $S$, then $y \circ x$ is also defined as an element of $S$ and
$$x \circ y = y \circ x \quad \text{for all such pairs } (x, y)$$

Let us look at some examples.

1. Addition and multiplication are binary operations on the set $S$ of whole numbers. They are both commutative operations on the set of whole numbers because

$$x + y = y + x \quad \text{and} \quad xy = yx$$

for every pair of whole numbers $(x, y)$.

2. Subtraction is not a binary operation on the set $S$ of whole numbers, but it is a binary operation in the set of whole numbers. Subtraction is not commutative in the set of whole numbers because whenever $x - y$ is defined, $y - x$ may or may not be defined (for example, $8 - 2$ is defined, but $2 - 8$ is not defined in the set of whole numbers).

If we extend the set $S$ to include all integers (positive, negative, or zero), then subtraction becomes a binary operation on $S$, that is, $x - y$ is defined for all ordered pairs $(x, y)$. However, even though $x - y$ and $y - x$ are both defined in the set of integers, we find that

$$x - y \neq y - x$$

whenever $x$ and $y$ are different integers. Hence subtraction is not commutative in this set either.

## 1.7 Commutativity or Noncommutativity

3. The operation **averaging** defined by

$$x \diamond y = \frac{x+y}{2}$$

is not a binary operation *on* the set of whole numbers because $x \diamond y$ is a whole number only when $x$ and $y$ are either both odd or both even. However, $\diamond$ is a binary operation *in* the set of whole numbers. It is moreover a commutative operation in this set because whenever $x \diamond y$ is a whole number so is $y \diamond x$, and

$$x \diamond y = \frac{x+y}{2} = \frac{y+x}{2} = y \diamond x$$

for every such pair $(x, y)$.

## EXERCISES 1.7

*Which of the following binary operations are commutative in the sets indicated?*

|     | Operation Defined by | Set |
| --- | --- | --- |
| 1.  | $(x, y) \to x^2 + y^2$ | {Integers} |
| 2.  | $(x, y) \to x^2 - y^2$ | {Integers} |
| 3.  | $(x, y) \to x$ | {Whole numbers} |
| 4.  | $(x, y) \to (x + y)^2$ | {Real numbers} |
| 5.  | $(x, y) \to (x - y)^2$ | {Real numbers} |
| 6.  | $(x, y) \to (x - y)^3$ | {Real numbers} |
| 7.  | $(x, y) \to \|x - y\|$ | {Real numbers} |
| 8.  | $(x, y) \to x^2 - xy + y^2$ | {Integers} |
| 9.  | $(x, y) \to x^2 - xy - y^2$ | {Integers} |
| 10. | $(x, y) \to x + 2y$ | {Whole numbers} |
| 11. | $(x, y) \to \dfrac{x}{y} + \dfrac{y}{x}$ | {Nonzero rational numbers} |
| 12. | $(x, y) \to \dfrac{x}{y^2} + \dfrac{y}{x^2}$ | {Nonzero rational numbers} |
| 13. | $(x, y) \to \dfrac{x}{y^2} + \dfrac{y^2}{x}$ | {Nonzero rational numbers} |
| 14. | $(x, y) \to \dfrac{x}{y} - \dfrac{y}{x}$ | {Nonzero rational numbers} |
| 15. | $(x, y) \to x^y - y^x$ | {Positive integers} |
| 16. | $(x, y) \to 2^{x+y}$ | {Positive integers} |
| 17. | $(x, y) \to 2xy - x - y$ | {Integers} |

18. $(x, y) \rightarrow \dfrac{x}{y}$        {Positive rationals}

19. $(x, y) \rightarrow \dfrac{x}{y}$        $\{1, -1\}$

20. $(x, y) \rightarrow xy \pmod 6$        $\{0, 1, 2, 3, 4, 5\}$

21. $(x, y) \rightarrow \max\{x, y\}$        {Whole numbers}

(*Note:* The operation in exercise 21 means the greater of the pair $(x, y)$ if $x \neq y$ or $x$ if $x = y$.)

## 1.8    ASSOCIATIVITY AND NONASSOCIATIVITY

A binary operation ∘ in or on a set $S$ always operates on pairs of elements of $S$. If $a$, $b$, and $c$ are elements of $S$, it makes sense to ask what elements of $S$ (if any) are represented by

$$a \circ b, \quad b \circ c, \quad a \circ c, \quad \text{etc.}$$

However, an expression such as

$$a \circ b \circ c$$

does not make any sense unless we agree how to apply the binary operation to the triple $(a, b, c)$. There are two obvious ways to do this. We can interpret "$a \circ b \circ c$" to mean

$$(a \circ b) \circ c$$

or we can interpret "$a \circ b \circ c$" to mean

$$a \circ (b \circ c)$$

For many important binary operations these two interpretations yield the same result. For example, let $a$, $b$, $c$, respectively, be the integers 8, 4, 2. If ∘ is the binary operation of addition, we observe that

$$(8 + 4) + 2$$

is the same as

$$8 + (4 + 2)$$

Insertion of parentheses into the expression "$8 + 4 + 2$" so as to "associate" 8 and 4 produces the same result as inserting parentheses so as to associate 4 and 2. We express this fact by saying that $(8, 4, 2)$ *is an associative triple under the operation of addition of integers.*

On the other hand, if ∘ is the binary operation of subtraction of integers, we observe that the two expressions

$$(8 - 4) - 2 \quad \text{and} \quad 8 - (4 - 2)$$

## 1.8 Associativity and Nonassociativity

do not have the same value. We express this fact by saying that (8, 4, 2) *is not an associative triple under the operation of subtraction of integers.*

**Questions.**

1. Is (8, 4, 2) an associative triple under the operation of multiplication of integers?
2. Is (8, 4, 2) an associative triple under the operation of division of integers?
3. Is (8, 4, 0) an associative triple under the operation of subtraction of integers?
4. Is (8, 4, 1) an associative triple under the operation of division integers?
5. Is (3, 2, 1) an associative triple under the operation exponentiation of whole numbers?
6. Is (3, 1, 2) an associative triple under the operation exponentiation of whole numbers?

We see from these examples that certain triples of numbers may be associative under a given binary operation, whereas other triples may not be associative under this same binary operation. Let us generalize these remarks. Suppose that ∘ is a binary operation in a set $S$. Suppose that $(x, y, z)$ is a triple of elements of $S$ and suppose further that both the expressions

$$(x \circ y) \circ z \quad \text{and} \quad x \circ (y \circ z)$$

define definite elements of $S$. If

$$(x \circ y) \circ z = x \circ (y \circ z)$$

then we say that $(x, y, z)$ is an associative triple under ∘.

For example, the following diagram depicts the fact that (6, 4, 2) is an associative triple under addition.

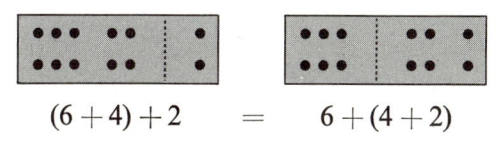

$$(6 + 4) + 2 \quad = \quad 6 + (4 + 2)$$

Now under what circumstances may $(x, y, z)$ fail to be an associative triple under ∘? This may occur in two ways:

1. The binary operation ∘ may assign one value in $S$ to $(x \circ y) \circ z$ and a different value in $S$ to $x \circ (y \circ z)$. For example, in the set of integers

$$(8 \div 4) \div 2 = 1$$

but
$$8 \div (4 \div 2) = 4$$

2. The binary operation ∘ may fail to assign a value in $S$ to one of the expressions
$$(x \circ y) \circ z \quad \text{or} \quad x \circ (y \circ z)$$
(For example, in the set $S$ of integers $6 \div (4 \div 2) = 3$ which is an integer, but $(6 \div 4) \div 2$ is not an integer.)

If ∘ is a binary operation in $S$ but not on $S$, both possibilities can occur. However, when ∘ is a binary operation on $S$, then the second possibility cannot arise, although the first may or may not occur. Thus if ∘ is a binary operation on $S$, then $(x \circ y) \circ z$ and $x \circ (y \circ z)$ are defined for every triple $(x, y, z)$ and it makes sense to inquire whether every such triple is associative under ∘. Should this be the case, we call ∘ an **associative operation** on $S$, or simply an associative operation.

**Definition.** A binary operation ∘ on a set $S$ is called **associative** (on $S$) if and only if every triple $(x, y, z)$ of elements of $S$ is an associative triple under ∘, that is, if and only if
$$x \circ (y \circ z) = (x \circ y) \circ z \quad \text{for all } x, y, z \in S$$
For binary operations in $S$ we use the following "weaker" definition.

**Definition.** A binary operation ∘ in a set $S$ is called **associative** in $S$ if and only if, whenever $(x \circ y) \circ z$ is defined, then $x \circ (y \circ z)$ is also defined and
$$(x \circ y) \circ z = x \circ (y \circ z)$$
for all such triples $(x, y, z)$.

## EXERCISES 1.8

*Indicate whether each of the following is or is not an associative triple with respect to the indicated operation.*

| | Triple | Operation |
|---|---|---|
| 1. | (5, 6, 7) | Multiplication of integers |
| 2. | (8, 4, 1) | Division of integers |
| 3. | (8, 4, 1) | Subtraction of integers |
| 4. | (8, 4, 0) | Subtraction of integers |
| 5. | (8, 0, 4) | Subtraction of integers |

## 1.8 Associativity and Nonassociativity

6. (2, 4, 6)                               Averaging rational numbers
7. (4, 4, 6)                               Averaging rational numbers
8. (6, 4, 6)                               Averaging rational numbers

*Which of the following binary operations are associative in the indicated set? Which of the latter are associative on that set?*

9. $(x, y) \to 2xy$                    {Integers}
10. $(x, y) \to 2(x + y)$            {Integers}
11. $(x, y) \to y$                       {Rational numbers}
12. $(x, y) \to |xy|$                   {Real numbers}
13. $(x, y) \to |x + y|$              {Real numbers}
14. $(x, y) \to y^2$                    {Rational numbers}
15. $(x, y) \to |x - y|$              {Real numbers}
16. $(x, y) \to \sqrt{x^2 + y^2}$      {Real numbers}
17. $(x, y) \to \sqrt{x^2 - y^2}$      {Real numbers}
18. $(x, y) \to x^y$                    {Whole numbers}
19. $(x, y) \to \dfrac{x + y}{2}$            {Rational numbers}
20. $(x, y) \to \dfrac{1}{x} + \dfrac{1}{y}$           {Nonzero rational numbers}
21. $(x, y) \to xy \pmod 7$      {0, 1, 2, 3, 4, 5, 6}
22. $(x, y) \to x \div y \pmod 7$   {0, 1, 2, 3, 4, 5, 6}
23. $(x, y) \to$ ($x$ or $y$, whichever    {Integers}
               is closer to 7.1)
24. $(x, y) \to [x + y]$            {Real numbers}
(*Note:* If $r$ is a real number, $[r]$ is the greatest integer which does not exceed $r$.)
25. $(x, y) \to \text{LCM}\{x, y\}$    {Positive integers}
(*Note:* LCM stands for *least common multiple*.)
26. $(x, y) \to \text{GCD}\{x, y\}$    {Positive integers}
(*Note:* GCD stands for *greatest common divisor*.)
27. $(x, y) \to (x + y + 1)$      {Integers}
28. $(x, y) \to (xy + x + y)$      {Integers}
29. $(x, y) \to$ (see table)

| ∘ | 0 | 1 | 2 | 3 | 4 |
|---|---|---|---|---|---|
| 0 | 0 | 1 | 2 | 3 | 4 |
| 1 | 1 | 2 | 3 | 2 | 1 |
| 2 | 2 | 3 | 2 | 1 | 0 |
| 3 | 3 | 2 | 1 | 0 | 1 |
| 4 | 4 | 1 | 0 | 1 | 2 |

30. $(x, y) \rightarrow xy$ {Integers that are not multiples of 4}

31. $(x, y) \rightarrow xy$ {Integers that are not multiples of 9}

## 1.9 EXISTENCE AND NONEXISTENCE OF AN IDENTITY ELEMENT

Every student knows that computation with numbers (even whole numbers) can often become quite a chore from which relief is possible only by using a calculator, if one is available. However, resorting to a calculating machine is certainly not necessary when a number is to be added to 0 or when a number is to be multiplied by 1. The numbers 0 and 1 play particularly simple (yet important) roles in relation to the binary operations addition and multiplication, respectively,

$$0 + n = n + 0 = n \quad \text{for every number } n$$
$$1 \times n = n \times 1 = n \quad \text{for every number } n$$

Adding 0 to any number or multiplying any number by 1 merely results in that same number. The number 0 plays the role of a "neutral element" for addition and the number 1 plays the role of a "neutral element" for multiplication.

Neutral elements are frequently referred to as **identity elements** and defined mathematically as follows.

**Definition.** Let $S$ by any set of elements and let $\circ$ be a binary operation in $S$. If there is an element $e$ in $S$ having the property that

$$e \circ x = x \circ e = x$$

for every element $x$ in $S$, then $e$ is called an **identity element** for the binary operation $\circ$.

We sometimes say that $e$ is an identity element for the mathematical system $\{S, \circ\}$, that is, for the system composed of the set $S$ together with the binary operation $\circ$ in $S$. We use this language, however, only when there is just one binary operation under consideration. Observe that in order for an element $e$ to be an identity element for a binary operation $\circ$, two requirements must be satisfied:

$$e \circ x = x \quad \text{for every element } x \text{ in } S$$

## 1.9 Existence and Nonexistence of an Identity Element

and
$$x \circ e = x \quad \text{for every element } x \text{ in } S$$
Thus the number 0 is an identity element for addition of integers because
$$0 + x = x \quad \text{for every integer } x$$
and
$$x + 0 = x \quad \text{for every integer } x$$
However, 0 is not an identity element for subtraction of integers because although it is true that
$$x - 0 = x \quad \text{for every integer } x$$
it is not true that
$$0 - x = x \quad \text{for every integer } x$$
In fact, $0 - x = -x$ for every integer $x$, and $-x$ is a different integer from $x$, except in the obvious case where $x = 0$.

Actually, there is no identity element at all for subtraction of integers because the first requirement for an identity element, namely
$$e - x = x \quad \text{for all integers } x$$
requires that
$$e = 2x \quad \text{for all integers } x$$
The second requirement for an identity element, namely,
$$x - e = x \quad \text{for all integers } x$$
requires that
$$e = 0$$
These two requirements imply that
$$2x = 0 \quad \text{for all integers } x$$
which, of course, is not so.

Similarly, although 1 is an identity element for multiplication of integers, 1 is not an identity element for division of integers because, although
$$x \div 1 = x \quad \text{for every integer } x$$
it is not true that
$$1 \div x = x \quad \text{for every integer } x$$
In fact, $1 \div x$ is not even an integer except when $x$ is 1 or $-1$

It should not be assumed that 0 and 1 are the only identity elements when dealing with numbers. Consider, for example, the set (of four numbers)

$$S = \{2, 4, 6, 8\}$$

and let ∘ be the binary operation **max** defined by

$$x \circ y = \max \{x, y\}$$

where max$\{x, y\}$ means the **greater** number of the pair $(x, y)$ if $x \neq y$ and max $\{x, y\} = x$ whenever $x = y$. It is then evident that

$$2 \circ x = x \circ 2 = \max \{2, x\} = x \quad \text{for every } x \text{ in } S$$

showing that 2 is an identity element for this system.

Now let us keep the same set $S = \{2, 4, 6, 8\}$, but instead of the binary operation max let us consider the binary operation **min** defined by

$$x \circ y = \min \{x, y\}$$

where min $\{x, y\}$ means the **smaller** number of the pair $(x, y)$ if $x \neq y$ and min $\{x, y\} = x$ whenever $x = y$. In this case, we see that

$$8 \circ x = x \circ 8 = x \quad \text{for every } x \text{ in } S$$

Thus in this system an identity element is 8.

We have already observed that there is no identity element for some binary operations (for example, subtraction of integers). We now prove that if a binary operation does have an identity element, it cannot have more than one.

**THEOREM 1.1** (*Uniqueness of Identity Element*) In any mathematical system consisting of a set $S$, together with a binary operation ∘ in this set, there can be at most one identity element.

*Proof:* Suppose that $e_1$ and $e_2$ were both identity elements for operation ∘ in $S$. Then, because $e_2$ is an identity element, it follows that

$$e_1 \circ e_2 = e_1$$

But because $e_1$ is also an identity element, it follows that

$$e_1 \circ e_2 = e_2$$

Consequently, $$e_1 = e_2$$

that is, $e_1$ and $e_2$ are the same element.

## 1.9 Existence and Nonexistence of an Identity Element

Observe that in proving $e_1 = e_2$, we have made use of an important axiom which we shall call *replacement*.

**Axiom of Replacement:** *In a mathematical sentence that refers to some object, a name of that object may be replaced by any other name for that same object.*

In the proof of Theorem 1.1 the first equality,

$$e_1 \circ e_2 = e_1$$

asserts tht $e_1 \circ e_2$ and $e_1$ name the same object. Therefore in the mathematical sentence expressed by the second equality

$$e_1 \circ e_2 = e_2$$

we may replace $e_1 \circ e_2$ by $e_1$ to obtain

$$e_1 = e_2$$

## EXERCISES 1.9

*Determine whether each of these mathematical systems does or does not have an identity. If the system has an identity, name it.*

1. $\{\{2, 4\}, (a, b) \to ab \pmod{6}\}$
2. $\{\{2, 4, 6, 8\}, (a, b) \to ab \pmod{10}\}$
3. $\left\{\{\text{Rational numbers}\}, (a, b) \to \dfrac{a+b}{2}\right\}$
4. $\{\{\text{Points in space}\}, \text{midpointing}\}$
5. $\{\{2, 4, 6, 8, 10, 12\}, (a, b) \to ab \pmod{14}\}$
6. $\{\{\text{Integers (positive)}\}, (a, b) \to \max(a, b)\}$
7. $\{\{\text{Integers (positive)}\}, (a, b) \to \min(a, b)\}$
8. $\{\{3, 6, 9, 12\}, (a, b) \to ab \pmod{15}\}$
9. $\{\{\text{Integers}\}, (a, b) \to a \text{ or } b, \text{ whichever is farther from } 7.1\}$
10. $\{\{\text{Integers}\}, (a, b) \to a \text{ or } b, \text{ whichever is closer to } 7.1\}$
11. $\{\{4, 8, 12, 16\}, (a, b) \to ab \pmod{20}\}$
12. $\{\{5, 6, 7, 8, 9\}, (a, b) \to \max\{a, b\}\}$
13. $\{\{5, 6, 7, 8, 9\}, (a, b) \to \min\{a, b\}\}$
14. $\{\{\text{Integers}\}, (a, b) \to b\}$
15. $\{\{\text{Subsets of a given set}\}, \cup\}$
    (*Note:* $\cup$ is the binary operation of *union* on sets.)
16. $\{\{\text{Subsets of a given set}\}, \cap\}$
    (*Note:* $\cap$ is the binary operation of *intersection* on sets.)

**1 ‖ Operations in Sets**

17. {{Subsets of a given set}, $(A, B) \to (A \cap B') \cup (A' \cap B)$}
    (*Note:* $A'$ is the complement of set $A$, $B'$ is the complement of set $B$.)
18. {{Integers}, $(a, b) \to \max\{(a - b), (b - a)\}$}
19. {{Positive integers}, $(a, b) \to \max\left\{\dfrac{a}{b}, \dfrac{b}{a}\right\}$}
20. {{$a, b, c, d$}, $\circ$} where operation $\circ$ is defined by the following table:

| $\circ$ | $a$ | $b$ | $c$ | $d$ |
|---|---|---|---|---|
| $a$ | $a$ | $b$ | $a$ | $d$ |
| $b$ | $a$ | $b$ | $b$ | $d$ |
| $c$ | $a$ | $b$ | $c$ | $d$ |
| $d$ | $a$ | $b$ | $d$ | $d$ |

21. {{Integers}, $(a, b) \to a + b + 2$}
22. {{Rational numbers}, $(a, b) \to ab + a + b$}

## 1.10 EXISTENCE AND NONEXISTENCE OF INVERSES

An example of how diverse phenomena can be unified very effectively is found in some rather elementary procedures for solving equations. Consider, for example, the problem of solving the two simple equations

$$2 + x = 7 \quad \text{and} \quad 2 \cdot x = 7$$

We seek solutions in the set $Q$ of rational numbers. Let us examine in detail a commonly used step-by-step procedure for solving each of these equations

$$
\begin{array}{ll}
2 + x = 7 & 2 \cdot x = 7 \\
-2 + (2 + x) = -2 + 7 & \tfrac{1}{2} \cdot (2 \cdot x) = \tfrac{1}{2} \cdot (7) \\
(-2 + 2) + x = 5 & (\tfrac{1}{2} \cdot 2) \cdot x = \tfrac{7}{2} \\
0 + x = 5 & 1 \cdot x = 3\tfrac{1}{2} \\
x = 5 & x = 3\tfrac{1}{2}
\end{array}
$$

Although these two examples differ in specific details, they have many features in common. In the equation $2 + x = 7$, we add $-2$ to both members. In the equation $2 \cdot x = 7$, we multiply both members by $\tfrac{1}{2}$. These apparently different procedures actually have very similar objectives. In both cases, an identity element finally appears on the left, namely 0 for the binary operation

## 1.10 Existence and Nonexistence of Inverses

addition and 1 for the binary operation multiplication. The similarity of the two situations is more readily appreciated if we let the symbol "∘" stand for either of the two binary operations, addition or multiplication. Then both equations are seen to have the form

$$2 \circ x = 7$$

Each of the two equations was solved by "operating on" (that is, adding or multiplying) both the left and right members of the equation using a suitably chosen number ($-2$ for the first equation, $\frac{1}{2}$ for the second). The particular number was selected so that when it operated on the coefficient 2, the identity element (0 or 1) for the operation was obtained.

If we let the symbol "$2'$" stand for either of the numbers $-2$ or $\frac{1}{2}$ and if we use the usual symbol "$e$" to denote either of the identity elements 0 or 1, whichever is relevant, then we can express the solution procedure for both equations by this single scheme

$$2 \circ x = 7$$
$$2' \circ (2 \circ x) = 2' \circ 7$$
$$(2' \circ 2) \circ x = 2' \circ 7$$
$$e \circ x = 2' \circ 7$$
$$x = 2' \circ 7$$

Note that the associativity of the operation is needed here as well as the existence of an identity element $e$. Fortunately, addition and multiplication possess both of these properties.

The solution obtained here

$$2' \circ 7$$

may now be given its appropriate interpretation in each of the two cases. Thus, in the first case, where ∘ is the binary operation addition, $2'$ is $-2$, so the solution is

$$2' \circ 7 = -2 + 7 = 5$$

In the second case, where ∘ is the binary operation multiplication, $2'$ is $\frac{1}{2}$, so the solution becomes

$$2' \circ 7 = \tfrac{1}{2} \times 7 = 3\tfrac{1}{2}$$

This scheme for solving the equation $2 \circ x = 7$ depends for its success on being able to determine a number $2'$ such that

$$2' \circ 2 = e$$

where $e$ is the identity element for the binary operation $\circ$.

If the original equation $2 \circ x = 7$ is changed to read

$$x \circ 2 = 7$$

we must then seek a number $2'$ such that

$$2 \circ 2' = e$$

Assuming such a number $2'$ is again available, then the scheme for solving the new equation appears as follows:

$$\begin{aligned} x \circ 2 &= 7 \\ (x \circ 2) \circ 2' &= 7 \circ 2' \\ x \circ (2 \circ 2') &= 7 \circ 2' \\ x \circ e &= 7 \circ 2' \\ x &= 7 \circ 2' \end{aligned}$$

As before we are making the additional assumption that the operation $\circ$ is associative. We leave it to the reader to verify that the answer $x = 7 \circ 2'$ can now be interpreted so as to obtain the correct solutions for each of these equations

$$x + 2 = 7 \quad \text{and} \quad x \cdot 2 = 7$$

The effectiveness of the scheme described here clearly depends on the existence of a number $2'$ having the following property

$$2' \circ 2 = 2 \circ 2' = e$$

This number $2'$ is called an **inverse of the number 2 with respect to the operation** $\circ$.

We have seen that the actual value of $2'$ depends on the operation $\circ$. When $\circ$ is addition, $2'$ is $-2$; when $\circ$ is multiplication, $2'$ is $\frac{1}{2}$. Thus $-2$ is the inverse of 2 with respect to addition and $\frac{1}{2}$ is the inverse of 2 with respect to multiplication. The notion of an inverse is formulated precisely in the following definition.

**Definition.** Let $\circ$ be a binary operation in a set $S$ and let there exist, within set $S$, an identity element $e$ for this binary operation $\circ$. If $x'$ and $x$ are elements of $S$ such that

$$x' \circ x = x \circ x' = e$$

then $x'$ is called an **inverse of** $x$ **with respect to** $\circ$.

Whenever there is but one binary operation under discussion, there will be no likelihood of confusion if we omit the phrase "with respect to ∘" and simply say "$x'$ is an inverse of $x$."

It is also worth observing that the requirement

$$x' \circ x = x \circ x' = e$$

is perfectly symmetrical as far as $x$ and $x'$ are concerned. This means that whenever $x'$ is inverse to $x$ it is equally true that $x$ is inverse to $x'$. Hence we can also refer to $x$ and $x'$ as **a pair of inverse elements**. A pair of inverse elements always commutes according to the above definition even if the binary operation is itself not commutative.

To gain further insight into the notion of inverse elements let us examine a few other mathematical systems.

*Example 1.* Let $S = \{\text{natural numbers}\}$ and let $\circ = $ multiplication. The identity element is $e = 1$. In this system $1' = 1$, but $2', 3', 4',$ ..., etc., do not exist. (Why?)

*Example 2.* Let $S = \{\text{positive rational numbers}\}$ and let $\circ = $ multiplication. In this system the identity element is again $e = 1$, but an inverse exists for each element, not just for 1. For example,

$$1' = 1, \quad 2' = \tfrac{1}{2}, \quad 3' = \tfrac{1}{3}, \quad (\tfrac{2}{3})' = \frac{1}{\tfrac{2}{3}} = \tfrac{3}{2}, \ldots,$$

and in general $r' = \dfrac{1}{r}$ for each rational number $r$ in $S$.

*Example 3.* Let $S = W_4 = \{0, 1, 2, 3\}$ and let $\circ = $ addition modulo 4.

| +(mod 4) | 0 | 1 | 2 | 3 |
|---|---|---|---|---|
| 0 | 0 | 1 | 2 | 3 |
| 1 | 1 | 2 | 3 | 0 |
| 2 | 2 | 3 | 0 | 1 |
| 3 | 3 | 0 | 1 | 2 |

In this system the identity element is 0 and every element has an inverse.

$$0' = 0, \quad 1' = 3, \quad 2' = 2, \quad 3' = 1$$

***Example 4.*** Let $S = W_4 = \{0, 1, 2, 3\}$, but this time let $\circ =$ multiplication modulo 4.

| $\times \pmod 4$ | 0 | 1 | 2 | 3 |
|---|---|---|---|---|
| 0 | 0 | 0 | 0 | 0 |
| 1 | 0 | 1 | 2 | 3 |
| 2 | 0 | 2 | 0 | 2 |
| 3 | 0 | 3 | 2 | 1 |

In this system the identity element is $e = 1$ and inverses exist for 1 and 3:

$$1' = 1 \quad \text{and} \quad 3' = 3$$

However, inverses do not exist for 0 and 2.

***Example 5.*** Let $S = \{1, 3, 5, 7\}$ and let $\circ =$ multiplication modulo 8.

| $\times \pmod 8$ | 1 | 3 | 5 | 7 |
|---|---|---|---|---|
| 1 | 1 | 3 | 5 | 7 |
| 3 | 3 | 1 | 7 | 5 |
| 5 | 5 | 7 | 1 | 3 |
| 7 | 7 | 5 | 3 | 1 |

In this system the identity element is again $e = 1$ and each element is its own inverse

$$1' = 1, \quad 3' = 3, \quad 5' = 5, \quad 7' = 7$$

***Example 6.*** Let $S = \{2, 4, 6, 8\}$ and let $\circ =$ multiplication modulo 10.

| $\times \pmod{10}$ | 2 | 4 | 6 | 8 |
|---|---|---|---|---|
| 2 | 4 | 8 | 2 | 6 |
| 4 | 8 | 6 | 4 | 2 |
| 6 | 2 | 4 | 6 | 8 |
| 8 | 6 | 2 | 8 | 4 |

## 1.10 Existence and Nonexistence of Inverses

In this system the identity element is 6 (not 1)! Every element has an inverse:

$$2' = 8, \quad 4' = 4, \quad 6' = 6, \quad 8' = 2$$

**Example 7.** Let $S = \{a, b, c\}$ and let $x \circ y$ be obtained by referring to the following table:

| $\circ$ | $a$ | $b$ | $c$ |
|---|---|---|---|
| $a$ | $a$ | $b$ | $c$ |
| $b$ | $b$ | $c$ | $a$ |
| $c$ | $c$ | $a$ | $b$ |

In this system the identity element is $a$. Every element has an inverse, namely

$a' = a$ because $a \circ a = a$
$b' = c$ because $b \circ c = c \circ b = a$
$c' = b$ because $c \circ b = b \circ c = a$

**Example 8.** Let $S = \{1, -1, i, -i\}$ and let $\circ =$ multiplication and let $i \times i = i^2 = -1$.

| $\times$ | $1$ | $-1$ | $i$ | $-i$ |
|---|---|---|---|---|
| $1$ | $1$ | $-1$ | $i$ | $-i$ |
| $-1$ | $-1$ | $1$ | $-i$ | $i$ |
| $i$ | $i$ | $-i$ | $-1$ | $1$ |
| $-i$ | $-i$ | $i$ | $1$ | $-1$ |

In this system the identity element is 1. Every element has an inverse:

$$1' = 1, \quad (-1)' = -1, \quad i' = -i, \quad (-i)' = i$$

In these examples every element has either no inverse or it has exactly one inverse. The question naturally arises whether an element can have more than one inverse in a mathematical system. The following theorem proves that this cannot occur in mathematical systems where the binary operation is associative.

**THEOREM 1.2** (*Uniqueness of Inverse*) If "∘" is a binary operation on a set $S$ such that
(a)  ∘ is associative on $S$ and
(b)  there is, in $S$, an identity element $e$ for ∘,
then there can be at most one inverse for any given element of $S$.

*Proof:* Suppose that $a_1'$ is an inverse of $a$ and that $a_2'$ is also an inverse of $a$ so that $a_1' \circ a = e$ and $a \circ a_2' = e$.
Then
$\begin{aligned} a_1' &= a_1' \circ e & & e \text{ is an identity for } \circ \\ a_1' &= a_1' \circ (a \circ a_2') & & \text{Replacement, } a \circ a_2' = e \\ a_1' &= (a_1' \circ a) \circ a_2' & & \text{Associativity of } \circ \\ a_1' &= e \circ a_2' & & \text{Replacement, } a_1' \circ a = e \\ a_1' &= a_2' & & e \text{ is an identity for } \circ \end{aligned}$

This proves that $a_1'$ and $a_2'$ are not distinct elements of $S$.

Note that Theorem 1.2 does not prove that an element of $S$ actually has an inverse in $S$. It merely proves that if an element has an inverse, then it cannot have more than one.

Suppose that we are given the mathematical system of Example 8, namely,

$$S = \{1, -1, i, -i\} \qquad \circ = \text{multiplication}$$

and we wish to solve the equation

$$i \circ x = -1$$

We accomplish this very nicely by multiplying each member of the equation by the inverse of $i$, thus:

$\begin{aligned} i' \cdot (i \cdot x) &= i' \cdot (-1) \\ (i' \cdot i)x &= (-i)(-1) \qquad (\text{Note: } i' = -i) \\ 1 \cdot x &= i \\ x &= i \end{aligned}$

This method is exactly like the procedure used earlier to solve the equations

$$2 + x = 7, \qquad 2 \times x = 7, \qquad \text{and} \qquad 2 \circ x = 7.$$

Clearly, it should work for any equation of the form $a \circ x = b$ provided that ∘ is an associative binary operation for which an identity element exists and for which the coefficient $a$ has an inverse.

## 1.10 Existence and Nonexistence of Inverses

To show that the method can be used with any associative binary operation ∘, an important preliminary theorem is needed (which is actually a direct consequence of the Axiom of Replacement).

**THEOREM 1.3** Let ∘ be a binary operation on a set $S$ and let $a, b, c$ be elements of $S$.

If $a = b$

then (a) $c \circ a = c \circ b$

and (b) $a \circ c = b \circ c$

*Proof:* (a) Because ∘ is a binary operation on $S$, $c \circ a$ names a unique object of $S$. This uniqueness may be expressed by the mathematical sentence

$$c \circ a = c \circ a$$

In fact, whenever a symbol, say "$x$," names a unique object we may write $x = x$. Conversely, whenever we write "$x = x$" we are tacitly assuming that "$x$" names a unique object. Both ideas may be expressed simply by saying "equality is reflexive." Therefore, if $a = b$, it follows (using the Axiom of Replacement) that

$$c \circ a = c \circ b$$

Part (b) is proved in exactly the same manner.

With the aid of Theorem 1.3 we can now develop a general method for solving equations of the form

$$a \circ x = b \quad \text{and} \quad x \circ a = b$$

**THEOREM 1.4** Let ∘ be a binary operation on a set $S$ such that
(1) ∘ is associative and
(2) there is, in $S$, an identity element $e$ for ∘.
If $a$ is any element of $S$ for which an inverse $a'$ exists in $S$ and if $b$ is any element of $S$, then the equation

(a) $a \circ x = b$

has the unique solution $a' \circ b$, and the equation

(b) $\quad x \circ a = b$

has the unique solution $b \circ a'$

*Proof:* (a) If there is a solution to the equation

$$a \circ x = b$$

we may obtain this solution as follows. Using Theorem 1.3, we write

| | |
|---|---|
| $a' \circ (a \circ x) = a' \circ b$ | |
| $(a' \circ a) \circ x = a' \circ b$ | Associativity of $\circ$ |
| $e \circ x = a' \circ b$ | Replacement, $a' \circ a = e$ |
| $x = a' \circ b$ | Replacement, $e \circ x = x$ |

So far we have proved that if there is a solution, then this solution is the element $a' \circ b$. To complete the proof we must show that $a' \circ b$ is indeed a solution.

To show that $a' \circ b$ is actually a solution we need merely observe that

| | | |
|---|---|---|
| if | $x = a' \circ b$ | |
| then | $a \circ x = a \circ (a' \circ b)$ | Using Theorem 1.3 |
| | $a \circ x = (a \circ a') \circ b$ | Associativity of $\circ$ |
| | $a \circ x = e \circ b$ | Replacement, $a \circ a' = e$ |
| | $a \circ x = b$ | Replacement, $e \circ b = b$ |

This completes the proof of part (a) of Theorem 1.4. Part (b) is proved in a similar manner.

## EXERCISES 1.10

*For each of the following systems find the identity and all inverses when they exist.*

1. $\{\{2, 4, 6, 8\}, \times \pmod{10}\}$
2. $\{\{0, 1, 2, 3\}, \times \pmod{4}\}$
3. $\{\{0, 1, 2, 3\}, + \pmod{4}\}$
4. $\{\{2, 4\}, \times \pmod{6}\}$
5. $\{\{0, 1, 2, 3, 4\}, + \pmod{5}\}$
6. $\{\{0, 1, 2, 3, 4\}, \times \pmod{5}\}$
7. $\{\{\text{Rationals}\}, +\}$
8. $\{\{\text{Rationals}\}, \times\}$

9. $\{\{2, 4, 6, 8, 10, 12\}, \times \pmod{14}\}$
10. $\{\{0, 4, 8\}, \times \pmod{12}\}$
11. $\{\{0, 3, 6\}, \times \pmod 9\}$
12. $\{\{0, 5, 10\}, \times \pmod{15}\}$
13. $\{\{\text{Integers}\}, (a, b) \rightarrow \text{number farther from } 7.1\}$
14. $\{\{\text{Positive rationals}\}, \times\}$
15. $\{\{1, -1\}, \div\}$
16. $\{\{4, 8, 12, 16\}, \times \pmod{20}\}$
17. $\{\{a, b, c, d\}, \circ\}$, where $\circ$ is defined by

| $\circ$ | $a$ | $b$ | $c$ | $d$ |
|---|---|---|---|---|
| $a$ | $d$ | $c$ | $a$ | $b$ |
| $b$ | $c$ | $d$ | $b$ | $a$ |
| $c$ | $a$ | $b$ | $c$ | $d$ |
| $d$ | $b$ | $a$ | $d$ | $c$ |

18. $\{\{\text{Integers}\}, (a, b) \rightarrow a + b + 1\}$
19. $\{\{\text{Rational numbers}\}, (a, b) \rightarrow ab + a + b\}$
20. $\{\{0, 1, 2, 3, 4\}, \circ\}$, where $\circ$ is defined by

| $\circ$ | 0 | 1 | 2 | 3 | 4 |
|---|---|---|---|---|---|
| 0 | 0 | 1 | 2 | 3 | 4 |
| 1 | 1 | 2 | 3 | 2 | 1 |
| 2 | 2 | 3 | 2 | 1 | 0 |
| 3 | 3 | 2 | 1 | 0 | 1 |
| 4 | 4 | 1 | 0 | 1 | 2 |

21. Prove part b of Theorem 1.3.
22. Prove part b of Theorem 1.4.
23. Let $\circ$ be a binary operation on a set $S$ and let $a, b, c, d$ be elements of $S$. Prove that if $a = b$ and $c = d$ then $a \circ c = b \circ d$.

# REVIEW EXERCISES

1. From a study of the "Pythagorean triplets" shown in the following table try to fill in the rest of the table. Check your answers.

1 ‖ Operations in Sets

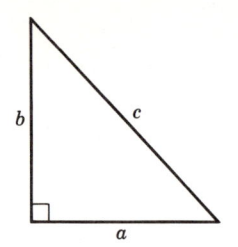

|   | a | b | c |
|---|---|---|---|
|   | 3 | 4 | 5 |
|   | 5 | 12 | 13 |
|   | 7 | 24 | 25 |
|   | 9 | 40 | 41 |
| (a) | 11 | ? | ? |
| (b) | 13 | ? | ? |
| ★(c) | a | ? | ? |

$$\begin{cases} a^2 + b^2 = c^2 \\ b + 1 = c \end{cases}$$

← Express $b$ and $c$ in terms of $a$.

2. Tell whether or not each of the following binary operations *in* the given sets are also *on*.
   (a) $(a, b) \rightarrow a^2 + 2b$     for $a, b$ whole numbers
   (b) $(a, b) \rightarrow a^2 - 2b$     for $a, b$ whole numbers
   (c) $(a, b) \rightarrow \dfrac{a}{b}$     for $a, b$ integers
   (d) $(a, b) \rightarrow \dfrac{a}{b^2 + 1}$     for $a, b$ rational numbers
   (e) $(a, b) \rightarrow \dfrac{2a + b}{a + b}$     for $a, b$ rational numbers

3. Compute each of the following mod 12:
   (a) $5 + 9$            (b) $5 - 9$
   (c) $5 \times 9$            (d) $3 \div 5$

4. For each of the following mathematical systems decide whether or not
   (a) the given set is closed with respect to the operation.
   (b) the operation is commutative.
   (c) the operation is associative.
   (d) there is an identity element among the elements of the given set.
   (e) each element has an inverse.
       (1) $\{\{\text{Even integers}\}, \times\}$
       (2) $\{\{\text{Even integers}\}, +\}$
       (3) $\{\{e, a, b, c, d\}, \circ\}$ where the operation $\circ$ is defined by the table.

| ∘ | e | a | b | c | d |
|---|---|---|---|---|---|
| e | e | a | b | c | d |
| a | a | b | c | b | a |
| b | b | c | b | a | e |
| c | c | b | a | e | a |
| d | d | a | e | a | b |

(4) $\{\{0, 1, 2, 3, r, s, t, u\}, \circ\}$ where the operation $\circ$ is defined by the table.

| ∘ | 0 | 1 | 2 | 3 | r | s | t | u |
|---|---|---|---|---|---|---|---|---|
| 0 | 0 | 1 | 2 | 3 | r | s | t | u |
| 1 | 1 | 2 | 3 | 0 | s | t | u | r |
| 2 | 2 | 3 | 0 | 1 | t | u | r | s |
| 3 | 3 | 0 | 1 | 2 | u | r | s | t |
| r | r | u | t | s | 0 | 3 | 2 | 1 |
| s | s | r | u | t | 1 | 0 | 3 | 2 |
| t | t | s | r | u | 2 | 1 | 0 | 3 |
| u | u | t | s | r | 3 | 2 | 1 | 0 |

(5) $\{\{e, x, y, z, r, s, t, u\}, \circ\}$ where the operation $\circ$ is defined by the following table

| ∘ | e | x | y | z | r | s | t | u |
|---|---|---|---|---|---|---|---|---|
| e | e | x | y | z | r | t | s | u |
| x | x | e | z | y | s | t | u | r |
| y | y | z | e | x | t | u | r | s |
| z | z | y | x | e | u | r | s | t |
| r | r | u | t | s | z | y | x | e |
| s | s | r | u | t | y | x | e | z |
| t | t | s | r | u | x | e | z | y |
| u | u | t | s | r | e | z | y | x |

*Évariste Galois (1811-1832)*

# 2 | Elementary Groups

> Some years ago, I worked out the structure of a group of operators in connection with Dirac's theory of the electron. I afterwards learned that a great deal of what I had written was to be found in a treatise on Kummer's quartic surface .... Perhaps the author of the treatise would have been equally surprised to find that he was dealing with the behavior of the electron.
>
> SIR ARTHUR EDDINGTON

## 2.1 ÉVARISTE GALOIS

The term *group* was first used in the modern, technical, mathematical sense by a 19-year-old French mathematician named Évariste Galois (1811–1832).

Galois was undoubtedly one of the most tragic figures in the history of mathematics. He was born in October of the year 1811 and became a serious student of mathematics at age 13. Nevertheless he was denied admission at age 16 and again at 18 to the École Polytechnique, France's greatest university, although he was already superior to most mathematicians* of his day.

In 1830, at age 19 he solved a problem that had baffled the best of his predecessors and contemporaries. He discovered how group-theoretic notions could be used to prove that there can be no general solution in terms of radicals for equations of degree higher than four.

Beset by frustration, misunderstanding, and lack of recognition, Galois became a political revolutionary. He thereby acquired royalist enemies who plotted to eliminate him. The frameup was successful. On May 30, 1832 Galois' brilliance was extinguished by a bullet in the abdomen, the result of a senseless "duel of honor."

Galois spent most of the night before the duel frantically writing down as many of his mathematical discoveries as he could. "I have not time—I have not time!" he wrote in the margins.

Fourteen years later Galois' writings were published by the French mathematician Liouville. The ideas contained in these sixty odd pages have continued to influence mathematics to the present day.

## 2.2 THE GROUP PROPERTIES

A set of elements, together with at least one operation in (or on) this set, is called a mathematical system. In Chapter 1 we encountered a great many mathematical systems. Some of these were old friends, for example, the set of whole numbers together with such operations as addition, multiplication, and so forth. Other systems were perhaps less familiar, for example, the modulo $n$ systems such as the set of "clock" numbers

$$W_5 = \{0, 1, 2, 3, 4\}$$

together with operations such as modulo 5 addition, and modulo 5 multiplication.

Mathematical systems occur in such endless variety and in such diverse applications that one might be inclined to consider it quite unlikely that they had anything of importance in common. Yet surprising as it may seem there is a basic set of properties that a great many useful mathematical systems do indeed share. These are the **group properties.** To see what they are, let us examine first a few mathematical systems that have these group

---

* Except possibly Cauchy (1789–1857), Gauss (1777–1855), and Abel (1802–1829), who also made contributions to group theory.

## 2.2 The Group Properties

properties. Each of these systems consists of a set together with a binary operation on this set:

1.  Set: $\{1, -1\}$
    Operation: $\times$

| $\times$ | 1  | $-1$ |
|---|---|---|
| 1  | 1  | $-1$ |
| $-1$ | $-1$ | 1  |

2.  Set: $W_4 = \{0, 1, 2, 3\}$
    Operation: $+ \pmod 4$

| $+ \pmod 4$ | 0 | 1 | 2 | 3 |
|---|---|---|---|---|
| 0 | 0 | 1 | 2 | 3 |
| 1 | 1 | 2 | 3 | 0 |
| 2 | 2 | 3 | 0 | 1 |
| 3 | 3 | 0 | 1 | 2 |

3.  Set: $W_5^* = \{1, 2, 3, 4\}$
    Operation: $\times \pmod 5$

| $\times \pmod 5$ | 1 | 2 | 3 | 4 |
|---|---|---|---|---|
| 1 | 1 | 2 | 3 | 4 |
| 2 | 2 | 4 | 1 | 3 |
| 3 | 3 | 1 | 4 | 2 |
| 4 | 4 | 3 | 2 | 1 |

4.  Set: $K = \{1, 3, 5, 7\}$
    Operation: $\times \pmod 8$

| $\times \pmod 8$ | 1 | 3 | 5 | 7 |
|---|---|---|---|---|
| 1 | 1 | 3 | 5 | 7 |
| 3 | 3 | 1 | 7 | 5 |
| 5 | 5 | 7 | 1 | 3 |
| 7 | 7 | 5 | 3 | 1 |

5.   Set: $C = \{S, L, A, R\}$ where $S, L, A, R$ are "commands":

   $L$ = left face,
   $R$ = right face,
   $A$ = about face,
   $S$ = stay (as you are)

Operation: $\circ$ = followed by [for example, "$L \circ A$" means "left face followed by about face." The table asserts that $L \circ A = R$ (right face).]

| $\circ$ | $S$ | $L$ | $A$ | $R$ |
|---|---|---|---|---|
| $S$ | $S$ | $L$ | $A$ | $R$ |
| $L$ | $L$ | $A$ | $R$ | $S$ |
| $A$ | $A$ | $R$ | $S$ | $L$ |
| $R$ | $R$ | $S$ | $L$ | $A$ |

6.   Set: $Z = \{\text{Integers}\}$

Operation: $+$ (addition)

(*Note:* The table extends "infinitely far" in all directions.)

| $+$ | $\cdots$ | $-2$ | $-1$ | $0$ | $1$ | $2$ | $\cdots$ |
|---|---|---|---|---|---|---|---|
| $\vdots$ | | $\vdots$ | $\vdots$ | $\vdots$ | $\vdots$ | $\vdots$ | |
| $-2$ | $\cdots$ | $-4$ | $-3$ | $-2$ | $-1$ | $0$ | $\cdots$ |
| $-1$ | $\cdots$ | $-3$ | $-2$ | $-1$ | $0$ | $1$ | $\cdots$ |
| $0$ | $\cdots$ | $-2$ | $-1$ | $0$ | $1$ | $2$ | $\cdots$ |
| $1$ | $\cdots$ | $-1$ | $0$ | $1$ | $2$ | $3$ | $\cdots$ |
| $2$ | $\cdots$ | $0$ | $1$ | $2$ | $3$ | $4$ | $\cdots$ |
| $\vdots$ | | $\vdots$ | $\vdots$ | $\vdots$ | $\vdots$ | $\vdots$ | |

7.   Set: $Q^* = \{\text{Positive rational numbers}\}$

Operation: $\times$ (multiplication)

(*Note:* This is also an infinite table.)

| $\times$ | $1$ | $\frac{1}{2}$ | $2$ | $3$ | $\frac{1}{3}$ | $\frac{1}{4}$ | $\frac{2}{3}$ | $\frac{3}{2}$ | $\cdots$ |
|---|---|---|---|---|---|---|---|---|---|
| $1$ | $1$ | $\frac{1}{2}$ | $2$ | $3$ | $\frac{1}{3}$ | $\frac{1}{4}$ | $\frac{2}{3}$ | $\frac{3}{2}$ | $\cdots$ |
| $\frac{1}{2}$ | $\frac{1}{2}$ | $\frac{1}{4}$ | $1$ | $\frac{3}{2}$ | $\frac{1}{6}$ | $\frac{1}{8}$ | $\frac{1}{3}$ | $\frac{3}{4}$ | $\cdots$ |
| $2$ | $2$ | $1$ | $4$ | $6$ | $\frac{2}{3}$ | $\frac{1}{2}$ | $\frac{4}{3}$ | $3$ | $\cdots$ |
| $3$ | $3$ | $\frac{3}{2}$ | $6$ | $9$ | $1$ | $\frac{3}{4}$ | $2$ | $\frac{9}{2}$ | $\cdots$ |
| $\frac{1}{3}$ | $\frac{1}{3}$ | $\frac{1}{6}$ | $\frac{2}{3}$ | $1$ | $\frac{1}{9}$ | $\frac{1}{12}$ | $\frac{2}{9}$ | $\frac{1}{2}$ | $\cdots$ |
| $\frac{1}{4}$ | $\frac{1}{4}$ | $\frac{1}{8}$ | $\frac{1}{2}$ | $\frac{3}{4}$ | $\frac{1}{12}$ | $\frac{1}{16}$ | $\frac{1}{6}$ | $\frac{3}{8}$ | $\cdots$ |
| $\frac{2}{3}$ | $\frac{2}{3}$ | $\frac{1}{3}$ | $\frac{4}{3}$ | $2$ | $\frac{2}{9}$ | $\frac{1}{6}$ | $\frac{4}{9}$ | $1$ | $\cdots$ |
| $\frac{3}{2}$ | $\frac{3}{2}$ | $\frac{3}{4}$ | $3$ | $\frac{9}{2}$ | $\frac{1}{2}$ | $\frac{3}{8}$ | $1$ | $\frac{9}{4}$ | $\cdots$ |
| $\vdots$ | $\vdots$ | $\vdots$ | $\vdots$ | $\vdots$ | $\vdots$ | $\vdots$ | $\vdots$ | $\vdots$ | |

These seven mathematical systems certainly look considerably different from each other. Yet in every one of them the binary operation exhibits the following four properties.

1. **Closure**
2. **Associativity**
3. **Existence of an identity**
4. **Existence of inverses**

For example, consider system 7. Does this system have the closure property? Yes, because the product of a positive rational number and a positive rational number is a positive rational number. As another example, consider system 5. Is there an identity element in the system? Yes, because the first row within the table shows that

$$S \circ S = S, \quad S \circ L = L, \quad S \circ A = A, \quad S \circ R = R$$

and the first column within the table shows that

$$S \circ S = S, \quad L \circ S = L, \quad A \circ S = A, \quad R \circ S = R$$

These equalities show that for every element $x$ in the set $C$,

$$S \circ x = x \circ S = x$$

proving that $S$ is an identity element for the binary operation of this system.

We leave it to the reader to check further that each of these systems does indeed possess all four properties.

These systems also possess a fifth property, **commutativity** (that is, in each case the binary operation is commutative on the set $S$). This fact is readily verified by observing that within each table the entries are symmetric about a diagonal line extending from the upper left corner downward toward the lower right corner. If we designate any of the seven binary operations by "$\circ$," the symmetry shows that

$$x \circ y = y \circ x$$

for every pair $x$, $y$ in the set, which means simply that for each system the operation $\circ$ is commutative on the given set.

Although the commutativity property occurs frequently in elementary mathematical systems, it is not included among the requirements for a group. This is partly because there are many interesting and valuable mathematical systems which do not have the commutativity property but that do possess the other four properties. An example of such a system is the following **permutation** group which we discuss in greater detail later.

8.   Set: $D = \{e, p, q, r, s, t\}$
     Operation: ∘ as defined by the table

| ∘ | e | p | q | r | s | t |
|---|---|---|---|---|---|---|
| e | e | p | q | r | s | t |
| p | p | q | e | s | t | r |
| q | q | e | p | t | r | s |
| r | r | t | s | e | q | p |
| s | s | r | t | p | e | q |
| t | t | s | r | q | p | e |

Observe that for this mathematical system, $s \circ t = q$, whereas $t \circ s = p$. Therefore
$$s \circ t \neq t \circ s$$
that is, the elements $s$ and $t$ do not commute under the given operation ∘. (We leave as an exercise the task of discovering which pairs of elements do commute and which do not commute.) Thus the system defined here does not possess the commutative property. It can be verified, however, that it does possess the other four properties. Any mathematical system that has the four properties is called a **group**.

A precise definition of a group can be formulated as follows.

**Definition of a Group.** Let $G = \{S, \circ\}$ be a mathematical system consisting of a set $S$ together with a binary operation ∘ in $S$. The system $G$ is called a **group** if and only if $G$ has the following properties:

1. ∘ is a binary operation on $S$. (Closure)
2. ∘ is an associative operation (on $S$).
3. There is an identity element* for ∘ in $S$.
4. There is an inverse in $S$ for every element of $S$.

The four group properties are often expressed in the following equivalent manner.

**Axioms for a Group.**   1. *For every $x, y$ in $S$ (not necessarily distinct)*
$$x \circ y \text{ is an element of } S.$$
2. *For every $x, y, z$ in $S$ (not necessarily distinct)*
$$(x \circ y) \circ z = x \circ (y \circ z)$$

---

* By virtue of Theorem 1.1 there is at most one such element in $S$.

3. *There is an element e, in S, such that for every x in S*

$$e \circ x = x \circ e = x$$

4. *For every x in S, there is an element x' in S such that*

$$x' \circ x = x \circ x' = e$$

Observe that in order for the closure property (Axiom 1) to hold, the operation ∘ must be a binary operation *on* S, not merely *in* S. For example, the mathematical system $\{W, +\}$ consisting of the set of whole numbers together with the binary operation addition has the closure property, but the system $\{W, -\}$ consisting of the whole numbers, together with the binary operation subtraction does not have the closure property. The reason is that addition is a binary operation *on* W, but subtraction is merely a binary operation *in* W, not *on* W.

Failure of the closure property immediately disqualifies the system $\{W, -\}$ from being a group. But what about the system $\{W, +\}$? Is it a group? It certainly has the first property (closure). Does it have the second property (associativity)? The answer is Yes, because addition of whole numbers is indeed an associative operation, that is,

$$(x + y) + z = x + (y + z)$$

for all whole numbers $x$, $y$, $z$.

We now consider the third property. Is there an identity element in $W$ for the addition operation? Again the answer is Yes, because the whole number zero is precisely such an identity element:

$$0 + x = x + 0 = x$$

for every whole number $x$. What about the fourth property? Is there an "additive inverse" for every whole number in $W$? If $x$ is a whole number, is there another whole number $x'$ such that

$$x' + x = x + x' = 0?$$

Unfortunately, the answer is No for most values of $x$. For example, if $x = 1$, there is no whole number $x'$ such that $x' + 1 = 0$. This one counterexample (although there are, of course, many others) proves that the system $\{W, +\}$ does not have the inverse element property. The mathematical system $\{W, +\}$ is therefore not a group because it does not satisfy all four requirements stipulated by the definition of a group.

## 2 ‖ Elementary Groups

Suppose we enlarge our mathematical system somewhat by using the set $Z$ of integers instead of the set $W$ of whole numbers. Is the system $\{Z, +\}$ a group? The answer is Yes; $\{Z, +\}$ is indeed a group. We leave it to the reader to check that this system possesses all the four requirements.

A few words are in order here regarding the associativity requirement. For many systems this requirement is difficult (or even impossible) to verify by direct examination of cases because there may be too many (or even infinitely many) to consider. Sometimes a general argument can be given that will cover all cases. Such a general argument, however, usually requires further information beyond what may be available at the moment. For example, we have asserted that addition of whole numbers is associative. This means that *every* triple of whole numbers is an associative triple under addition. To prove this assertion would require a deeper study of the whole numbers than is possible within our present scope. Nevertheless, our extensive previous experience with the whole numbers has led us to verify the associativity of addition in so many specific cases that we are inclined to assume that this property holds in all cases. Although this is hardly a proof, we shall have to be content with this sort of incomplete verification of associativity for most of the systems we encounter. When a general argument is possible but not readily available, we verify the associativity in several cases and then go on to examine the other group requirements. In any event some assumptions must be made, and it is reasonable to assume that the operations on the well-known number systems have the usual familiar properties.

## EXERCISES 2.2

1. By referring to the table for system 4,
    (a) determine an inverse for each element of the system. What seems to be special about these inverses?
    (b) find the solution set for
        (1) $x \circ x = 1$
        (2) $x \circ x = 3$
        (3) $x \circ 3 = 7$
2. By referring to the table for system 5, solve each of the following:
    (a) $x \circ x = S$
    (b) $x \circ x = L$
    (c) $x \circ x = A$
    (d) $x \circ x = R$
    (e) $x \circ R = A$
    (f) $A \circ x = L$
    (g) $(x \circ S) \circ x = A$
    (h) $(x \circ L) \circ x = A$
    (i) $(x \circ A) \circ x = A$
    (j) $(x \circ R) \circ x = R$
    (k) $(x \circ L) \circ x = R$
    (l) $x \circ (A \circ x) = L$

## 2.2 The Group Properties

3. In the system consisting of the set of **rational numbers** under the operation addition, find an inverse for each of the following:
   (a) 5    (b) −7    (c) $\frac{2}{3}$    (d) −4.5
4. In the system consisting of the set of rational numbers under the operation multiplication, find an inverse for each of the following:
   (a) $\frac{2}{3}$    (b) $-\frac{6}{5}$    (c) 4.6    (d) −4.6
5. By referring to the table for the permutation group (system 8) find the solution set for each of the following:
   (a) $x \circ x = e$    (b) $x \circ x = p$    (c) $x \circ x = q$
   (d) $x \circ x = r$    (e) $x \circ x = s$    (f) $x \circ x = t$
   (g) $(r \circ x) \circ s = t$    (h) $(s \circ x) \circ r = t$    (i) $(x \circ r) \circ x = e$
   (j) $(x \circ r) \circ x = s$    (k) $(x \circ r) \circ x = t$    (l) $(x \circ r) \circ x = p$
   (m) $x' \circ r = t$    (n) $r \circ x' = t$    (o) $x \circ r = p \circ x$
   (p) $x \circ r = r \circ x$    (q) $(x \circ r) \circ x' = t$    (r) $(x' \circ r) \circ x = t$
6. Determine which group properties each of the following mathematical systems possesses. Which of these systems are groups?
   (a) The set of nonnegative even integers under addition.
   (b) The set of nonnegative even integers under multiplication.
   (c) The set of odd integers under multiplication.
   (d) The set of nonzero rational numbers under multiplication.
   (e) The set of numbers of the form $2^x$, where $x$ is any integer, under multiplication.
   (f) The set of ordered pairs $(x, y)$ where $x, y \in \{0, 1, 2\}$ under the operation addition defined as follows:
   $(x_1, y_1) + (x_2, y_2) = ((x_1 + x_2) \bmod 3, (y_1 + y_2) \bmod 3)$
   (g) The set of integers under the operation defined by the integer $x$ or $y$ whichever is closer to 4.4. $(x, y) \rightarrow$
   (h) The set of integers under the operation defined by the integer $x$ or $y$ whichever is farther from 4.4. $(x, y) \rightarrow$
   (i) The set $\{0, 1, 2, 3, 4, 5, 6, 7\}$ under the operation defined by

| $\circ$ | 0 | 1 | 2 | 3 | 4 | 5 | 6 | 7 |
|---|---|---|---|---|---|---|---|---|
| 0 | 0 | 1 | 2 | 3 | 4 | 5 | 6 | 7 |
| 1 | 1 | 0 | 3 | 2 | 5 | 4 | 7 | 6 |
| 2 | 2 | 3 | 0 | 1 | 6 | 7 | 4 | 5 |
| 3 | 3 | 2 | 1 | 0 | 7 | 6 | 5 | 4 |
| 4 | 4 | 5 | 6 | 7 | 0 | 1 | 2 | 3 |
| 5 | 5 | 4 | 7 | 6 | 1 | 0 | 3 | 2 |
| 6 | 6 | 7 | 4 | 5 | 2 | 3 | 0 | 1 |
| 7 | 7 | 6 | 5 | 4 | 3 | 2 | 1 | 0 |

7. Prove that each of the following mathematical systems is a group:
   (a) Set: $\{2, 4, 6, 8\}$, operation: $\times \pmod{10}$.
   (b) Set: $\{1, 3, 7, 9\}$, operation: $\times \pmod{10}$.
   (c) Set: $\{0, \pm 5, \pm 10, \pm 15, \ldots\}$, operation: $+$.
   (d) Set: {Integers}, operation: $\circ$ defined by $a \circ b = a + b - 1$.
   (e) Set: {Integers}, operation: $\circ$ defined by $a \circ b = a + b - 2$.
   (f) Set: $\{2^x \mid x \text{ is an integer}\}$, operation: $\times$ (multiplication).

8. For each of the following, (a) through (p), construct, if possible, a mathematical system in which the binary operation fulfills all four requirements specified. If such a construction is not possible, state why.

|     | Associativity | Identity | Inverse | Commutativity |
|-----|---------------|----------|---------|---------------|
| (a) | Yes | Yes | Yes | Yes |
| (b) | Yes | Yes | Yes | No  |
| (c) | Yes | Yes | No  | Yes |
| (d) | Yes | No  | Yes | Yes |
| (e) | No  | Yes | Yes | Yes |
| (f) | Yes | Yes | No  | No  |
| (g) | Yes | No  | Yes | No  |
| (h) | No  | Yes | Yes | No  |
| (i) | Yes | No  | No  | Yes |
| (j) | No  | Yes | No  | Yes |
| (k) | No  | No  | Yes | Yes |
| (l) | Yes | No  | No  | No  |
| (m) | No  | Yes | No  | No  |
| (n) | No  | No  | Yes | No  |
| (o) | No  | No  | No  | Yes |
| (p) | No  | No  | No  | No  |

## 2.3 THE SIMPLEST NONCOMMUTATIVE GROUP

Imagine three students occupying specific spots on a gymnasium floor. Let us illustrate their positions as follows:

$$\boxed{1} \quad \boxed{2} \quad \boxed{3}$$

The gym instructor has a simple device for directing them to shift their

## 2.3 The Simplest Noncommutative Group

positions. He holds up a large placard bearing an inscription like this:

$$\begin{pmatrix} 1 & 2 & 3 \\ 3 & 1 & 2 \end{pmatrix}$$

The students readily interpret this instruction and promptly execute it as follows:

> The student in position 1 moves to position 3.
> The student in position 2 moves to position 1.
> The student in position 3 moves to position 2.

The following diagrams illustrate these movements:

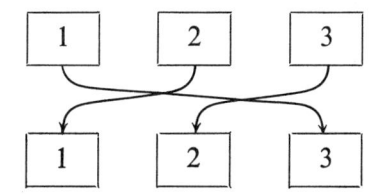

After the above instruction has been executed, suppose that a new one is given:

$$\begin{pmatrix} 1 & 2 & 3 \\ 1 & 3 & 2 \end{pmatrix}$$

This time the student in position 1 remains in position 1, and the students in positions 2 and 3 exchange places. The complete effect of the sequence of two instructions can be pictured as follows:

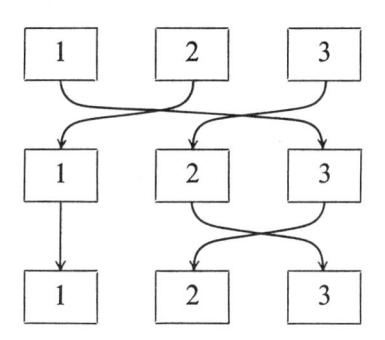

The combined result of both instructions evidently can be achieved by the following single instruction:

$$\begin{pmatrix} 1 & 2 & 3 \\ 2 & 1 & 3 \end{pmatrix}$$

## 2 ‖ Elementary Groups

We indicate this by writing

$$\begin{pmatrix}1 & 2 & 3\\ 3 & 1 & 2\end{pmatrix} \text{ followed by } \begin{pmatrix}1 & 2 & 3\\ 1 & 3 & 2\end{pmatrix} = \begin{pmatrix}1 & 2 & 3\\ 2 & 1 & 3\end{pmatrix}$$

or more briefly

$$\begin{pmatrix}1 & 2 & 3\\ 3 & 1 & 2\end{pmatrix} \circ \begin{pmatrix}1 & 2 & 3\\ 1 & 3 & 2\end{pmatrix} = \begin{pmatrix}1 & 2 & 3\\ 2 & 1 & 3\end{pmatrix}$$

where the symbol "∘" stands for the phrase "followed by." Clearly, ∘ can be interpreted as a binary operation which associates a new instruction with each ordered pair of instructions of this type.

There are altogether six distinct instructions that the gym teacher can give the three students. These are listed below. For convenient reference each instruction is denoted by a single letter.

$$e = \begin{pmatrix}1 & 2 & 3\\ 1 & 2 & 3\end{pmatrix} \qquad r = \begin{pmatrix}1 & 2 & 3\\ 1 & 3 & 2\end{pmatrix}$$

$$p = \begin{pmatrix}1 & 2 & 3\\ 2 & 3 & 1\end{pmatrix} \qquad s = \begin{pmatrix}1 & 2 & 3\\ 3 & 2 & 1\end{pmatrix}$$

$$q = \begin{pmatrix}1 & 2 & 3\\ 3 & 1 & 2\end{pmatrix} \qquad t = \begin{pmatrix}1 & 2 & 3\\ 2 & 1 & 3\end{pmatrix}$$

Observe that in this set of six instructions the instruction $e$ is a "do nothing" instruction. It tells each student to remain where he is and therefore serves as an identity element for the binary operation ∘ (followed by). The set of six instructions, together with the operation ∘, forms a mathematical system which is in fact the permutation group mentioned in Section 2.2 (system 8). The table for this group looks like this.

$$e = \begin{pmatrix}1 & 2 & 3\\ 1 & 2 & 3\end{pmatrix}$$

$$p = \begin{pmatrix}1 & 2 & 3\\ 2 & 3 & 1\end{pmatrix}$$

$$q = \begin{pmatrix}1 & 2 & 3\\ 3 & 1 & 2\end{pmatrix}$$

$$r = \begin{pmatrix}1 & 2 & 3\\ 1 & 3 & 2\end{pmatrix}$$

$$s = \begin{pmatrix}1 & 2 & 3\\ 3 & 2 & 1\end{pmatrix}$$

$$t = \begin{pmatrix}1 & 2 & 3\\ 2 & 1 & 3\end{pmatrix}$$

| ∘ | e | p | q | r | s | t |
|---|---|---|---|---|---|---|
| e | e | p | q | r | s | t |
| p | p | q | e | s | t | r |
| q | q | e | p | t | r | s |
| r | r | t | s | e | q | p |
| s | s | r | t | p | e | q |
| t | t | s | r | q | p | e |

## 2.3 The Simplest Noncommutative Group

To verify the entries in this table it is not really necessary to draw pictorial diagrams as we have done above. For example, suppose we want to "compute" $q \circ r$, namely

$$\begin{pmatrix} 1 & 2 & 3 \\ 3 & 1 & 2 \end{pmatrix} \circ \begin{pmatrix} 1 & 2 & 3 \\ 1 & 3 & 2 \end{pmatrix}$$

We need merely observe that the first instruction $q$ sends a student from position 1 to position 3 and the second instruction then sends this student from position 3 to position 2. The combined instruction $q \circ r$ therefore sends the student from his original position 1 to the final position 2. So far, therefore, we know that

$$q \circ r = \begin{pmatrix} 1 & 2 & 3 \\ 2 & & \end{pmatrix}$$

In a similar manner we can trace the effect of the combined instruction $q \circ r$ on each of the other two students. The first instruction $q$ sends the student who was originally in position 2 to position 1 and then instruction $r$ keeps him in position 1. Therefore the composite instruction $q \circ r$ sends the student from initial position 2 to final position 1:

$$q \circ r = \begin{pmatrix} 1 & 2 & 3 \\ 2 & 1 & \end{pmatrix}$$

The remaining entry here is, of course, obvious, but it is nice to verify it by observing that $q$ sends the student who was originally in position 3 to position 2, and then instruction $r$ sends him back to his original position 3, so we have finally

$$q \circ r = \begin{pmatrix} 1 & 2 & 3 \\ 2 & 1 & 3 \end{pmatrix}$$

or more briefly,

$$q \circ r = t$$

The reader should choose other pairs of instructions and compute their "product," thus verifying other entries in the above table.

We have already observed that the system we are considering is a noncommutative group. Therefore care should be exercised, when computing products, to maintain the order of the factors. For example, suppose we want to compute

$$(q \circ t) \circ p$$

We observe from the table that

$$q \circ t = s$$

and therefore (using the Axiom of Replacement)

$$(q \circ t) \circ p = s \circ p$$

Referring once again to the table, we observe that

$$s \circ p = r$$

and consequently,

$$(q \circ t) \circ p = r$$

If at any point in this procedure we carelessly switched the order of the instructions, we would obtain an incorrect result. For example, if we had $p \circ s$ instead of $s \circ p$, we would obtain the incorrect result $t$ instead of the correct result $r$.

As another illustration, suppose that we want to find an instruction $x$ that will satisfy the equation

$$p \circ x = s$$

We must be careful to choose the instruction $p$ from the left side of the table and then seek the value $x$ from among the instructions listed across the top of the table. We see that $x = r$. On the other hand, if we (incorrectly) choose $p$ from the entries at the top of the table and then seek the value $x$ from among the instructions at the left side of the table, we obtain the (incorrect) solution $t$. Actually, this value $t$ is a correct solution for the equation

$$x \circ p = s$$

but it is not a correct solution for our original equation

$$p \circ x = s$$

When the group is noncommutative, these two equations are not equivalent.

## EXERCISES 2.3

1. For the permutation group (whose elements are $e, p, q, r, s, t$),
   (a) list six pairs of distinct elements that do not commute.
   (b) list six pairs of distinct elements that do commute.
   (c) Verify that each of the following is an associative triple:
   (1) $p, q, r$     (2) $p, r, s$     (3) $r, s, t$
   (4) $s, r, p$     (5) $r, t, r$     (6) $s, s, t$

## 2.3 The Simplest Noncommutative Group

(d) Which element of the system does each of the following represent?
  (1) $e'$        (2) $p'$        (3) $q'$
  (4) $r'$        (5) $s'$        (6) $t'$

(e) Which statements are true and which are false?
  (1) $(p \circ q)' = p' \circ q'$        (2) $(p \circ q)' = q' \circ p'$
  (3) $(p \circ r)' = p' \circ r'$        (4) $(p \circ r)' = r' \circ p'$
  (5) $(p \circ s)' = p' \circ s'$        (6) $(p \circ s)' = s' \circ p'$
  (7) $(r \circ s)' = r' \circ s'$        (8) $(r \circ s)' = s' \circ r'$

(f) Solve:
  (1) $p \circ x \circ t = r$        (2) $p \circ x \circ x = r$
  (3) $x \circ p \circ x = r$        (4) $x \circ p = r \circ x$
  (5) $x \circ r = s \circ x$        (6) $x \circ p = q \circ x$
  (7) $x \circ s = p \circ x$        (8) $x \circ p = p \circ x \circ q$

**2.** Compute:

(a) $\begin{pmatrix} 1 & 2 & 3 & 4 \\ 2 & 3 & 4 & 1 \end{pmatrix} \circ \begin{pmatrix} 1 & 2 & 3 & 4 \\ 2 & 3 & 4 & 1 \end{pmatrix}$

(b) $\begin{pmatrix} 1 & 2 & 3 & 4 \\ 2 & 3 & 4 & 1 \end{pmatrix}'$

(c) $\begin{pmatrix} 1 & 2 & 3 & 4 \\ 2 & 4 & 1 & 3 \end{pmatrix}'$

(d) $\left[\begin{pmatrix} 1 & 2 & 3 & 4 \\ 2 & 3 & 4 & 1 \end{pmatrix} \circ \begin{pmatrix} 1 & 2 & 3 & 4 \\ 4 & 1 & 2 & 3 \end{pmatrix}\right]'$

(e) $\begin{pmatrix} 1 & 2 & 3 & 4 \\ 4 & 1 & 2 & 3 \end{pmatrix}' \circ \begin{pmatrix} 1 & 2 & 3 & 4 \\ 2 & 3 & 4 & 1 \end{pmatrix}'$

(f) Show that the following triple of permutations is an associative triple:

$\begin{pmatrix} 1 & 2 & 3 & 4 \\ 4 & 1 & 2 & 3 \end{pmatrix} \quad \begin{pmatrix} 1 & 2 & 3 & 4 \\ 2 & 3 & 4 & 1 \end{pmatrix} \quad \begin{pmatrix} 1 & 2 & 3 & 4 \\ 3 & 4 & 1 & 2 \end{pmatrix}$

(g) Give an argument to show that every triple of permutations such as in (f) is an associative triple.

**3.** Consider the set of all permutations of four objects together with the operation $\circ$ meaning "followed by". (Several of these permutations are shown in Exercise 2.) In this system,
  (a) find two permutations that form a group under the operation $\circ$.
  (b) find three permutations that form a group under the operation $\circ$.
  (c) find four permutations that form a group under the operation $\circ$.
  (d) find six permutations that form a group under the operation $\circ$.
  (*Note:* Each of these smaller groups is called a *subgroup* of the original group.)

## 2.4 BASIC THEOREMS FOR GROUPS

In our discussion of the noncommutative permutation group (see Section 2.3) we observed that $r$ was a solution for the equation

$$p \circ x = s$$

A glance at the table for this mathematical system quickly reveals that $r$ is the only such solution. In other words, the equation $p \circ x = s$ has a *unique* solution.

Similarly, if we seek inverses for various instructions in the system, we find that each instruction has a unique inverse instruction. For example, the unique inverse of $p$ is $q$, the unique inverse of $r$ is $r$ itself, and so on.

Uniqueness properties such as these are not accidental. They apply to many other mathematical systems in addition to this particular one. In fact, we have already proved several theorems in Chapter 1 that show that uniqueness properties are shared by a great variety of mathematical systems. Included among these is the very broad class of mathematical systems known as **groups**. The fact that every group possesses these uniqueness properties (and other valuable properties) is a logical consequence of the axioms for a group. In this section we deduce some of these logical consequences. The theorems we derive must be applicable to all groups because the axioms from which they originate apply to all groups. It is this generality and wide applicability that make the study of group theory valuable.

For convenient reference let us recall that a group is a mathematical system $\{S, \circ\}$ consisting of a set of elements $S$, together with a binary operation in $S$, which satisfies the following four axioms.

*Axiom G1* (*Closure*) $\circ$ is a binary operation on $S$.

*Axiom G2* (*Associativity*) $\circ$ is an associative operation (on $S$).

*Axiom G3* (*Identity*) There exists an identity element $e$ for $\circ$ in $S$.

*Axiom G4* (*Inverses*) There exists in $S$ an inverse $a'$ for every element $a$ in $S$.

The first few theorems are immediate consequences of these four axioms and the theorems we have already proved in Chapter 1. For example, on page 38 of Chapter 1 we prove the following.

## 2.4 Basic Theorems for Groups

**THEOREM 1.1** In any mathematical system consisting of a set $S$, together with a binary operation ∘ in this set, there can be at most one identity element.

A mathematical system consisting of a set $S$ together with a binary operation on this set is called a **groupoid**. Theorem 1.1 therefore implies that

> If a groupoid has an identity element, then this identity element is unique.

Axiom G1 asserts that every group is, in particular, a groupoid. Axiom G3 asserts that an identity element exists in any group. Hence, applying Theorem 1.1, we may now assert the following.

**THEOREM 2.1** (*Uniqueness of Identity*) In any group there is one and only one identity element.

Another theorem that applies to any groupoid is Theorem 1.3. Applying this theorem to any group, we may assert the following.

**THEOREM 2.2** Let $a$, $b$, and $c$ be elements of a group $\{S, \circ\}$.
If $\quad a = b$
then $\quad c \circ a = c \circ b \quad$ (Left multiplication)
and $\quad a \circ c = b \circ c \quad$ (Right multiplication)

A group, of course, is more than just a groupoid. It has other properties (expressed by Axioms G2, G3, and G4). These include the properties of associativity and existence of an identity element. Because of these additional properties, we may apply Theorem 1.2, to any group. We then obtain the following (slightly "stronger") theorem.

**THEOREM 2.3** (*Uniqueness of Inverses*) In any group $\{S, \circ\}$ each element $a$ has a unique inverse $a'$.

*Proof:* By Axioms G1, G2, G3, ∘ is a binary operation on $S$, ∘ is associative, and there is in $S$ an identity element $e$ for ∘. Hence by Theorem 1.2 there can be at most one inverse for each element of $S$. But by Axiom G4 there exists at least one inverse $a'$ for each element $a$. Consequently, there exists *exactly one* inverse for each element of $S$.

It is only natural that we continue by applying Theorem 1.4 to any group. We immediately obtain Theorem 2.4.

**THEOREM 2.4** (*Unique Solutions*) If $a$ and $b$ are any elements of a group $\{S, \circ\}$, then the equation $a \circ x = b$ has the unique solution $a' \circ b$ and the equation $x \circ a = b$ has the unique solution $b \circ a'$.

As an example of the use of this theorem consider the permutation group of Section 2.3. Suppose we want to solve the equation

$$p \circ x = t$$

According to Theorem 2.4, this equation has the unique solution $x = p' \circ t$. Now by referring to the table for this group we find $p' = q$. Therefore (using the Replacement Axiom) $x = q \circ t$, which, by referring to the table again, becomes $x = s$. Similarly, the equation $x \circ p = t$ has the unique solution

$$x = t \circ p' = t \circ q = r$$

Sometimes it is necessary to apply Theorem 2.4 repeatedly. For example, to solve the equation (because of associativity, parentheses are not needed)

$$q \circ x \circ r = s$$

we may solve first for $x \circ r$

$$x \circ r = q' \circ s$$

and then solve for $x$

$$x = (q' \circ s) \circ r'$$

Referring to the table, we observe that $q' = p$ and $r' = r$. Hence

$$x = (p \circ s) \circ r$$
$$\therefore x = t \circ r$$
$$\therefore x = q$$

Several other useful theorems can be derived from the axioms for a group. Of particular importance is the following *converse* of Theorem 2.2.

**THEOREM 2.5** (*Cancellation Theorem*) Let $a$, $b$, and $c$ be elements of a group $\{S, \circ\}$.
    (a)  If $c \circ a = c \circ b$, then $a = b$    (Left cancellation)
    (b)  If $a \circ c = b \circ c$, then $a = b$    (Right cancellation)

## 2.4 Basic Theorems for Groups

*Proof of 2.5a:*

| | | |
|---|---|---|
| (1) | $c \circ a = c \circ b$ | Given |
| (2) | There exists in $S$ an inverse $c'$ for $c$ | Axiom G4 (inverses) |
| (3) | $c' \circ (c \circ a) = c' \circ (c \circ b)$ | Theorem 2.2 (left multiplication) |
| (4) | $\therefore (c' \circ c) \circ a = (c' \circ c) \circ b$ | Axiom G2 (associativity) |
| (5) | $\therefore e \circ a = e \circ b$ | Axiom G4 (inverses) |
| (6) | $\therefore a = b$ | Axiom G3 (identity) |

(Note that the Axiom of Replacement has been used in the last three steps without explicitly saying so.)

A proof for Theorem 2.5b is very similar and is left for the reader.

**THEOREM 2.6** *(Inverse of an Inverse)* If $a$ is any element of a group $\{S, \circ\}$, then
$$(a')' = a$$

*Proof:* Let $x = (a')'$. Then by Theorem 2.2 (right multiplication),

| | |
|---|---|
| $x \circ a' = (a')' \circ a' = e$ | Axiom G4 (inverses) |
| $(x \circ a') \circ a = e \circ a$ | Theorem 2.2 (right multiplication) |
| $x \circ (a' \circ a) = a$ | Associativity and Axiom G3 |
| $x \circ e = a$ | Axiom G4 |
| $x = a$ | Axiom G3 |
| $(a')' = a$ | Replacement $x = (a')'$ |

**THEOREM 2.7** *(Inverse of a Product)* If $a$ and $b$ are any elements of a group $\{S, \circ\}$, then
$$(a \circ b)' = b' \circ a'$$

*Proof:* Let $x = (a \circ b)'$. Then by Theorem 2.2 (left multiplication),

| | |
|---|---|
| $(a \circ b) \circ x = (a \circ b) \circ (a \circ b)' = e$ | Axiom G4 (inverses) |
| $\therefore \quad a \circ (b \circ x) = e$ | Axiom G2 (associativity) |
| $\therefore \quad b \circ x = a' \circ e$ | Theorem 2.4 |
| that is, $b \circ x = a'$ | Axiom G3 (identity) |
| $\therefore \quad x = b' \circ a'$ | Theorem 2.4 |
| $(a \circ b)' = b' \circ a'$ | Replacement Axiom |

## EXERCISES 2.4

1. Interpret Theorem 2.6, $(a')' = a$, when the group under consideration is the following:
   (a) $\{\{\text{Integers}\}, +\}$
   (b) $\{\{\text{Positive rationals}\}, \times\}$
   (c) the group of permutations of four objects, letting

   $$a = \begin{pmatrix} 1 & 2 & 3 & 4 \\ 4 & 3 & 1 & 2 \end{pmatrix}$$

2. There are various other ways of proving Theorem 2.6, $(a')' = a$. Give another proof of this theorem by
   (a) using the following equality which holds because of associativity

   $$(a')' \circ [a' \circ a] = [(a')' \circ a'] \circ a$$

   (b) using the theorem that every element of the group, in particular the element $a'$, must have a unique inverse.

3. Interpret Theorem 2.7, $(a \circ b)' = b' \circ a'$, when the group under consideration is
   (a) $\{\{\text{Integers}\}, +\}$
   (b) $\{\{\text{Positive rationals}\}, \times\}$
   (c) the group of permutations of four elements, letting

   $$a = \begin{pmatrix} 1 & 2 & 3 & 4 \\ 4 & 3 & 1 & 2 \end{pmatrix} \quad \text{and} \quad b = \begin{pmatrix} 1 & 2 & 3 & 4 \\ 4 & 2 & 1 & 3 \end{pmatrix}$$

4. There are other ways of proving Theorem 2.7, $(a \circ b)' = b' \circ a'$. Give another proof of this theorem
   (a) using the following equality which holds because of associativity

   $$(a \circ b)' \circ [(a \circ b) \circ (b' \circ a')] = [(a \circ b)' \circ (a \circ b)] \circ (b' \circ a')$$

   (b) by showing that $b' \circ a'$ is an inverse of $(a \circ b)$ and using the fact that $(a \circ b)$ must have a unique inverse.

5. Prove that in any group

   $$(a \circ b')' = b \circ a'$$

6. Prove that in any group
   (a) if $a = b$, then $a' = b'$.
   (b) if $a' = b'$, then $a = b$.
   (c) if $a \circ b = e$, then $a = b'$ and $a' = b$.
   (d) if $a \circ b = b \circ a$, then $a \circ b' = b' \circ a$.

## 2.4 Basic Theorems for Groups 73

**7.** Let $\{S, \circ\}$ be a group such that

$$(a \circ b) \circ (a \circ b) = (a \circ a) \circ (b \circ b)$$

for all $a$, $b$ in $S$. Prove that $\circ$ is a commutative operation on $S$.

**8.** Prove that if $x \circ x = e$ for every element $x$ in the group $\{S, \circ\}$, then the operation $\circ$ is commutative on $S$, that is,

$$a \circ b = b \circ a \quad \text{for all } a, b \text{ in } S.$$

(*Hint:* Use the fact that $(a \circ b) \circ (a \circ b) = e$.)

**9.** Prove that if $\{S, \circ\}$ is a group and $\{\bar{S}, \bar{\circ}\}$ is also a group, then the system consisting of the new set $\{(x, y) \mid x \in S \text{ and } y \in \bar{S}\}$ together with the new operation $\cdot$ defined by

$$(x_1, y_1) \cdot (x_2, y_2) = (x_1 \circ x_2, y_1 \bar{\circ} y_2)$$

is also a group (*Note:* This new group is called the *direct product* of the two original groups.)

**10.** Let $c$ be a fixed element in a group $\{S, \circ\}$. Prove that the set of all elements $x$ in $S$, which commute with this fixed element $c$, forms a group under the operation $\circ$. (*Note:* The latter group is an example of a "subgroup" of the given group.)

**11.** If $\{S, \circ\}$ is a group and if $T$ is a subset of $S$ such that the system $\{T, \circ\}$ is also a group, then the system $\{T, \circ\}$ is called a *subgroup* of $\{S, \circ\}$.

Find all subgroups of each of the following groups:

(a) $\{\{1, 3, 5, 7\}, \times \pmod{8}\}$.
(b) $\{\{2, 4, 6, 8\}, \times \pmod{10}\}$.
(c) The permutation group $\{\{e, p, q, r, s, t\}, \circ\}$.

**12.** Prove that if $\{G, \circ\}$ and $\{H, \circ\}$ are subgroups of a given group $\{S, \circ\}$, then the system $\{G \cap H, \circ\}$ is also a subgroup of $\{S, \circ\}$. (This is often stated, "The intersection of two subgroups is a subgroup.")

**★13.** Prove that if a group has an even number of elements, then the equation $x \circ x = e$ is satisfied by an even number of elements in this group.

**★14.** A "weaker" definition for a group is the following: A mathematical system $\{S, \circ\}$ is a group if and only if these conditions are satisfied.

**Axiom (a)** *For every pair $a$, $b$, in $S$, $a \circ b$ is in $S$.*

**Axiom (b)** *For every triple $a$, $b$, $c$ in $S$*

$$(a \circ b) \circ c = a \circ (b \circ c)$$

***Axiom (c)*** There is in S a left identity $e_L$ such that for every $a$ in S

$$e_L \circ a = a$$

***Axiom (d)*** For every $a$ in S, there is an element $a'_L$ in S such that

$$a'_L \circ a = e_L$$

where $e_L$ is a left identity. ($a'_L$ is called a left inverse of $a$.)

Prove that this "weaker" definition is equivalent to the "stronger" definition given in the text. (*Hint:* A suggested sequence of theorems sufficient to prove this is the following.)
(1) If $c \circ a = c \circ b$, then $a = b$.
(2) If $b \circ a = e_L$, then $a \circ b = e_L$.
(3) $a \circ a'_L = e_L$. (This proves that a left inverse is also a right inverse.)
(4) $a \circ e_L = a$. (This proves that a left identity is also a right identity.)

## 2.5  ISOMORPHIC SYSTEMS

We have stressed repeatedly that diverse mathematical systems often resemble each other in that they share certain common properties. For example, although groups occur in a virtually endless variety of "sizes and shapes," all groups must behave in accordance with the group Axioms G1 through G4 and they must obey any theorems that we have derived, or can derive, from these axioms.

Despite these strong resemblances, groups can nevertheless appear very different, and it becomes an interesting problem to investigate whether various particular groups we may encounter are really significantly different or merely superficially different from a mathematical point of view. This investigation leads to a surprisingly deep and often challenging problem with rich mathematical results. In this section we shall take a preliminary glance at this aspect of mathematical systems with particular reference to groups.

Let us start by considering two simple mathematical systems, both of which are groups.

|     System A                          |     System B                       |
|---------------------------------------|------------------------------------|
|  Set:  $\{1, -1\}$                    |  Set:  $\{0, 1\}$                  |
|  Operation:  $\times$ (multiplication)|  Operation:  $+$ (mod 2)           |

## 2.5 Isomorphic Systems

These systems look quite different insofar as they are composed of different sets of elements, $\{1, -1\}$ versus $\{0, 1\}$, and they involve different operations, ordinary multiplication versus addition modulo 2. However, if we look at the operation table for each system, we begin to see some resemblances.

**System A**

| × | 1 | −1 |
|---|---|----|
| 1 | 1 | −1 |
| −1 | −1 | 1 |

**System B**

| +(mod 2) | 0 | 1 |
|---|---|---|
| 0 | 0 | 1 |
| 1 | 1 | 0 |

We observe, for example, that each system has two distinct elements. Moreover, in each system the identity element for that system appears along the main diagonal of the table, and the remaining element of the system appears along the other diagonal of the table. Let us denote the identity element of either system by $e$, the remaining element of that system by $a$, and finally the operation of either system by $\circ$. Thus

In System A
$e = 1$
$a = -1$
$\circ = \times$

In System B
$e = 0$
$a = 1$
$\circ = +\ (\text{mod } 2)$

With this change of notation we can now express systems A and B as follows:

Set: $\{e, a\}$
Operation: $\circ$ (defined by)

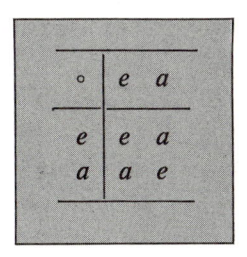

| $\circ$ | $e$ | $a$ |
|---|---|---|
| $e$ | $e$ | $a$ |
| $a$ | $a$ | $e$ |

It becomes clear that the two mathematical systems A and B have essentially the same basic "structure." From an abstract mathematical point of view they are not really different systems. Mathematicians express this idea by stating that the systems A and B are **isomorphic**, which literally means that they have the "same form." We can exhibit this sameness of form by setting up a *one-to-one correspondence*, that is, a matching of the elements and the operations of the two systems as follows.

$$\begin{array}{ccc} \text{System A} & & \text{System B} \\ 1 & \leftrightarrow & 0 \\ -1 & \leftrightarrow & 1 \\ \times & \leftrightarrow & +(\text{mod } 2) \end{array}$$

[We read this as "1 corresponds to 0," "−1 corresponds to 1," "× corresponds to +(mod 2)."]

If we multiply 1 by −1 in system A, we obtain the number −1. This computation corresponds to addition of 0 and 1 in system B:

$$(1) \times (-1) \leftrightarrow 0 + 1$$

In fact, any computation with elements in system A can be "mirrored" by a corresponding computation (using the corresponding elements) in system B. In short, systems A and B do not really represent two distinct groups; they are basically one and the same group but differently expressed.

As a second illustration of isomorphic systems consider the following systems:

$$\begin{array}{ll} \text{System C} & \text{System D} \\ \textit{Set:} \quad \{0, 1, 2, 3\} & \textit{Set:} \quad \{S, L, A, R\} \\ \textit{Operation:} \quad +(\text{mod } 4) & \textit{Operation:} \quad * \text{ (followed by)} \end{array}$$

The operation tables for these systems are

*System C*

| +(mod 4) | 0 | 1 | 2 | 3 |
|---|---|---|---|---|
| 0 | 0 | 1 | 2 | 3 |
| 1 | 1 | 2 | 3 | 0 |
| 2 | 2 | 3 | 0 | 1 |
| 3 | 3 | 0 | 1 | 2 |

*System D*

| * | S | L | A | R |
|---|---|---|---|---|
| S | S | L | A | R |
| L | L | A | R | S |
| A | A | R | S | L |
| R | R | S | L | A |

System D is sometimes called the "Command Group:" S, Stay; L, Left face; A, About face; R, Right face. As explained in Section 2.2, S, L, A, R are commands, and the operation * means "followed by." Here again we see

## 2.5 Isomorphic Systems

systems that are really only superficially different. The two tables really have the same form; it is merely the symbols that are different. If we match the four numbers of system C with the four instructions of system D according to the following scheme:

$$0 \leftrightarrow S$$
$$1 \leftrightarrow L$$
$$2 \leftrightarrow A$$
$$3 \leftrightarrow R$$

we see that either of these tables can be converted into the other by merely replacing each symbol appearing in that table by the corresponding matching symbol. Let us denote the identity element of either system by "$e$" and the remaining elements by "$a$," "$b$," "$c$" according to the following scheme:

| In System C | In System D |
|---|---|
| $e = 0$ | $e = S$ |
| $a = 1$ | $a = L$ |
| $b = 2$ | $b = A$ |
| $c = 3$ | $c = R$ |

Let us also denote the operation of either system by "$\circ$." With this change of notation we can now express either system as follows:

Set: $\{e, a, b, c\}$
Operation: $\circ$ (defined by table)

| $\circ$ | $e$ | $a$ | $b$ | $c$ |
|---|---|---|---|---|
| $e$ | $e$ | $a$ | $b$ | $c$ |
| $a$ | $a$ | $b$ | $c$ | $e$ |
| $b$ | $b$ | $c$ | $e$ | $a$ |
| $c$ | $c$ | $e$ | $a$ | $b$ |

Once again we see that system C and system D have essentially the same underlying structure. They are *isomorphic* groups. (*Note:* The basic structure of both of these systems can be described by the term "Cyclic Group of Order 4." See Exercise 4.)

Although we have described the notion of isomorphism, we have not actually defined this term.

**Definition.** Let $\{S, \circ\}$ and $\{\bar{S}, \bar{\circ}\}$ be mathematical systems such that there is a one-to-one correspondence (also called a one-to-one mapping) between the set $S$ and the set $\bar{S}$. For each $x \in S$ let the corresponding element in $\bar{S}$ (the image of $x$) be $\bar{x}$. (We denote the correspondence by $x \leftrightarrow \bar{x}$.) The systems $\{S, \circ\}$ and $\{\bar{S}, \bar{\circ}\}$ are called **isomorphic** if and only if

Whenever $x \in S$ and $y \in S$, then
$$x \circ y \leftrightarrow \bar{x} \,\bar{\circ}\, \bar{y}.$$

This condition is often expressed by saying that the *image of a product* is the *product of the images*, or symbolically,
$$\overline{x \circ y} = \bar{x} \,\bar{\circ}\, \bar{y}$$
(Observe that the word "product" is used here in a double sense; within each system it refers to the binary operation for that system.)

Whenever a pair of mathematical systems is isomorphic, it can also be said that either system is isomorphic to the other.

The phenomenon of isomorphism applies to infinite groups as well as finite ones. Consider, for example, the following two mathematical systems.

### System I
*Set:* The integers, that is,
$$\{0, \pm 1, \pm 2, \pm 3, \ldots\}$$
*Operation:* $+$ (addition)

### System II
*Set:* The integer powers of 2, that is,
$$\{2^0, 2^1, 2^{-1}, 2^2, 2^{-2}, 2^3, 2^{-3}, \ldots\}$$
*Operation:* $\times$ (multiplication)

Again we apparently have two systems involving different elements (all integers as against powers of 2) and using different operations (addition versus multiplication). We can, however, set up the following one-to-one correspondence between the two systems. We match each integer $n$ of system I with the corresponding power $2^n$ of system II.

$$0 \leftrightarrow 2^0 = 1$$
$$1 \leftrightarrow 2^1 = 2$$
$$2 \leftrightarrow 2^2 = 4$$
$$3 \leftrightarrow 2^3 = 8$$
$$-1 \leftrightarrow 2^{-1} = \tfrac{1}{2}$$
$$-2 \leftrightarrow 2^{-2} = \tfrac{1}{4}$$
$$\ldots \text{etc.} \ldots$$

## 2.5 Isomorphic Systems

In general, we match

$$x \leftrightarrow 2^x \quad \text{for each integer } x$$

Furthermore, whenever we add two integers in system I, say $x + y$, we multiply the corresponding powers $(2^x) \times (2^y)$ in system II. This matching process does indeed produce corresponding elements of the two systems because

$$(2^x) \times (2^y) = 2^{x+y}$$

so that, according to this matching procedure, we have

$$x + y \leftrightarrow 2^{x+y}$$

Comparing the tables for the two systems, we observe a striking similarity of form. Basically, both groups can be built up from a particular element $a$ and its inverse $a'$. (For example, $a = 1$ in system I and $a = 2$ in system II.)

### System I

| + | ... | −2 | −1 | 0 | 1 | 2 | 2 | ... |
|---|---|---|---|---|---|---|---|---|
| ⋮ | | ⋮ | ⋮ | ⋮ | ⋮ | ⋮ | ⋮ | |
| −2 | ... | −4 | −3 | −2 | −1 | 0 | 1 | ... |
| −1 | ... | −3 | −2 | −1 | 0 | 1 | 2 | ... |
| 0 | ... | −2 | −1 | 0 | 1 | 2 | 3 | ... |
| 1 | ... | −1 | 0 | 1 | 2 | 3 | 4 | ... |
| 2 | ... | 0 | 1 | 2 | 3 | 4 | 5 | ... |
| 3 | ... | 1 | 2 | 3 | 4 | 5 | 6 | ... |
| ⋮ | | ⋮ | ⋮ | ⋮ | ⋮ | ⋮ | ⋮ | |

### System II

| × | ... | $\frac{1}{4}$ | $\frac{1}{2}$ | 1 | 2 | 4 | 8 | ... |
|---|---|---|---|---|---|---|---|---|
| ⋮ | | ⋮ | ⋮ | ⋮ | ⋮ | ⋮ | ⋮ | |
| $\frac{1}{4}$ | ... | $\frac{1}{16}$ | $\frac{1}{8}$ | $\frac{1}{4}$ | $\frac{1}{2}$ | 1 | 2 | ... |
| $\frac{1}{2}$ | ... | $\frac{1}{8}$ | $\frac{1}{4}$ | $\frac{1}{2}$ | 1 | 2 | 4 | ... |
| 1 | ... | $\frac{1}{4}$ | $\frac{1}{2}$ | 1 | 2 | 4 | 8 | ... |
| 2 | ... | $\frac{1}{2}$ | 1 | 2 | 4 | 8 | 16 | ... |
| 4 | ... | 1 | 2 | 4 | 8 | 16 | 32 | ... |
| 8 | ... | 2 | 4 | 8 | 16 | 32 | 64 | ... |
| ⋮ | | ⋮ | ⋮ | ⋮ | ⋮ | ⋮ | ⋮ | |

The identity element for each group is $a \circ a'$ and the rest of the elements within each group are "generated" either by the element $a$ or by its inverse $a'$. This is done by operating either with $a$ or with $a'$. For example, in system I,

$$a \circ a = 1 + 1 = 2$$
$$a \circ a \circ a = 1 + 1 + 1 = 3$$
$$a' \circ a' = (-1) + (-1) = -2$$
$$a' \circ a' \circ a' = (-1) + (-1) + (-1) = -3$$
$$\ldots \text{etc.} \ldots$$

Similarly, in system II,

$$a \circ a = 2 \cdot 2 = 4 = 2^2$$
$$a \circ a \circ a = 2 \cdot 2 \cdot 2 = 8 = 2^3$$
$$a' \circ a' = 2^{-1} \cdot 2^{-1} = \tfrac{1}{2} \cdot \tfrac{1}{2} = \tfrac{1}{4} = 2^{-2}$$
$$a' \circ a' \circ a' = 2^{-1} \cdot 2^{-1} \cdot 2^{-1} = \tfrac{1}{2} \cdot \tfrac{1}{2} \cdot \tfrac{1}{2} = \tfrac{1}{8} = 2^{-3}$$
$$\ldots \text{etc.} \ldots$$

Thus the table for either system might be represented as follows:

| $\circ$ | $\ldots$ | $(a')^2$ | $(a')$ | $e$ | $a$ | $a^2$ | $a^3$ | $\ldots$ |
|---|---|---|---|---|---|---|---|---|
| $\vdots$ | | $\vdots$ | $\vdots$ | $\vdots$ | $\vdots$ | $\vdots$ | $\vdots$ | |
| $(a')^2$ | $\ldots$ | $(a')^4$ | $(a')^3$ | $(a')^2$ | $a'$ | $e$ | $a$ | $\ldots$ |
| $(a')$ | $\ldots$ | $(a')^3$ | $(a')^2$ | $a'$ | $e$ | $a$ | $a^2$ | $\ldots$ |
| $e$ | $\ldots$ | $(a')^2$ | $a'$ | $e$ | $a$ | $a^2$ | $a^3$ | $\ldots$ |
| $a$ | $\ldots$ | $a'$ | $e$ | $a$ | $a^2$ | $a^3$ | $a^4$ | $\ldots$ |
| $a^2$ | $\ldots$ | $e$ | $a$ | $a^2$ | $a^3$ | $a^4$ | $a^5$ | $\ldots$ |
| $a^3$ | $\ldots$ | $a$ | $a^2$ | $a^3$ | $a^4$ | $a^5$ | $a^6$ | $\ldots$ |
| $\vdots$ | | $\vdots$ | $\vdots$ | $\vdots$ | $\vdots$ | $\vdots$ | $\vdots$ | |

This table is often rewritten in the following way:

| $\circ$ | $\ldots$ | $a^{-2}$ | $a^{-1}$ | $a^0$ | $a^1$ | $a^2$ | $a^3$ | $\ldots$ |
|---|---|---|---|---|---|---|---|---|
| $\vdots$ | | $\vdots$ | $\vdots$ | $\vdots$ | $\vdots$ | $\vdots$ | $\vdots$ | |
| $a^{-2}$ | $\ldots$ | $a^{-4}$ | $a^{-3}$ | $a^{-2}$ | $a^{-1}$ | $a^0$ | $a^1$ | $\ldots$ |
| $a^{-1}$ | $\ldots$ | $a^{-3}$ | $a^{-2}$ | $a^{-1}$ | $a^0$ | $a^1$ | $a^2$ | $\ldots$ |
| $a^0$ | $\ldots$ | $a^{-2}$ | $a^{-1}$ | $a^0$ | $a^1$ | $a^2$ | $a^3$ | $\ldots$ |
| $a$ | $\ldots$ | $a^{-1}$ | $a^0$ | $a^1$ | $a^2$ | $a^3$ | $a^4$ | $\ldots$ |
| $a^2$ | $\ldots$ | $a^0$ | $a^1$ | $a^2$ | $a^3$ | $a^4$ | $a^5$ | $\ldots$ |
| $a^3$ | $\ldots$ | $a^1$ | $a^2$ | $a^3$ | $a^4$ | $a^5$ | $a^6$ | $\ldots$ |
| $\vdots$ | | $\vdots$ | $\vdots$ | $\vdots$ | $\vdots$ | $\vdots$ | $\vdots$ | |

For convenience, we have denoted $a^0 = e$, $a' = a^{-1}$, $a' \circ a' = a^{-2}$, and so on. This defines an abstract group known as an **infinite cyclic group**. The two isomorphic systems I and II are thus merely different looking examples of this basic cyclic group structure.

Two further observations about isomorphic groups merit specific mention at this time. You may have noticed that in setting up the one-to-one correspondence between the elements of two isomorphic groups.

> The identity element of one system always corresponds to the identity element of the other system.

In Exercise 5a you are asked to prove that this is true for any pair of isomorphic groups.

Closely related to this fact is our second observation: If $x$ and $\bar{x}$ are corresponding elements of isomorphic groups, then their respective inverses $x'$ and $(\bar{x})'$ must also be corresponding elements. This may also be stated as follows:

> Whenever two groups are isomorphic, the image of the inverse of an element is the same as the inverse of its image.

(See Exercise 5b.)

## EXERCISES 2.5

1. Show that $A$ and $\bar{A}$ are isomorphic systems under the indicated mapping. Remember to show two things:
   (1) The mapping is one-to-one between the sets $A$ and $\bar{A}$.
   (2) "Products" are preserved.

| | Mathematical System $A$ | Mapping | Mathematical System $\bar{A}$ |
|---|---|---|---|
| (a) | {{Real numbers}, +} | $n \to 3n$ | Same as System $A$ |
| (b) | {{Integers}, +} | $n \to 3^n$ | {{Powers of 3}, ×} |
| (c) | {{Real numbers}, ×} | $n \to n^3$ | Same as System $A$ |
| (d) | {{Integers}, +} | $n \to -n$ | Same as System $A$ |
| (e) | {{Positive rationals}, ×} | $n \to \dfrac{1}{n}$ | Same as System $A$ |

(f) {{Positive integers}, ×}  $n \to \dfrac{1}{n}$   $\left\{\left\{\left|\dfrac{1}{x}\right| x \text{ is a positive integer}\right\}, \times\right\}$

(g) {{$a + b\sqrt{2}$ | $a, b$ rational}, +}   $a + b\sqrt{2} \to a - b\sqrt{2}$   Same as System $A$

(h) {{$a + b\sqrt{2}$ | $a, b$ rational}, ×}   $a + b\sqrt{2} \to a - b\sqrt{2}$   Same as System $A$

(i) {{Integers}, +}   $n \to 2^{-n}$   {{Powers of 2}, ×}

(j) {{0, 1, 2, 3, 4}, +(mod 5)}   $n \to 3n$   Same as System $A$

2. Construct operation tables for each of the following pairs of mathematical systems and set up a one-to-one correspondence to show that they are isomorphic:
   (a) {{0, 1, 2}, +(mod 3)}    {{0, 2, 4}, +(mod 6)}
   (b) {{1, 2, 3, 4}, ×(mod 5)}    {{0, 1, 2, 3}, +(mod 4)}
   (c) {{1, 3, 5, 7}, ×(mod 8)}    {{1, 5, 7, 11}, ×(mod 12)}
   (d) {{0, 1, 2, 3}, +(mod 4)}    {{2, 4, 6, 8}, ×(mod 10)}
   (e) {{2, 4, 6, 8}, ×(mod 10)}    {{1, 3, 7, 9}, ×(mod 10)}

3. Find an isomorphic mapping between the following pairs of mathematical systems:
   (a) {{Integers}, +}    {{Integer powers of 3}, ×}
   (b) {{Integers}, +}    {{Integer multiples of 3}, +}
   (c) {{Odd integers}, ×}    {{Reciprocals of odd integers}, ×}

4. Show that the group defined by

| ∘ | $e$ | $a$ | $b$ | $c$ |
|---|---|---|---|---|
| $e$ | $e$ | $a$ | $b$ | $c$ |
| $a$ | $a$ | $b$ | $c$ | $e$ |
| $b$ | $b$ | $c$ | $e$ | $a$ |
| $c$ | $c$ | $e$ | $a$ | $b$ |

is a cyclic group that can be generated from the single element $a$ by verifying that
$$a^1 = a, \quad a^2 = b, \quad a^3 = c, \quad a^4 = e = a^0, \quad a^{-1} = c.$$
Does $b$ generate the group?
Does $c$ generate the group?

5. Let $\{S, \circ\}$ and $\{\bar{S}, \bar{\circ}\}$ be isomorphic groups. Prove:
   (a) If $e$ is the identity element for the first system, then $\bar{e}$ (the "image" of $e$) must be the identity element for the second system.

## 2.5 Isomorphic Systems

(b) If $a'$ is the inverse of $a$ in the first system, then $\overline{(a')}$ is the inverse of $\bar{a}$ in the second system (that is, prove that $\overline{(a')} = (\bar{a})'$).

(c) If either group is commutative, then so is the other.

6. Prove the following pairs of systems isomorphic:

   (a) $\{\{a + b\sqrt{2} \mid a, b \text{ integers}\}, +\}$
   $$\{\{a - b\sqrt{2} \mid a, b \text{ integers}\}, +\}$$

   (b) $\{\{a + b\sqrt{2} \mid a, b \text{ integers}\}, \times\}$
   $$\{\{a - b\sqrt{2} \mid a, b \text{ integers}\}, \times\}$$

   (c) $\{\{2^a 3^b \mid a, b \text{ integers}\}, \times\}$
   $$\{\{5^a 7^b \mid a, b \text{ integers}\}, \times\}$$

   (d) $\{\{a + b\sqrt{2} \mid a, b \text{ integers}\}, +\}$
   $$\{\{2^a 3^b \mid a, b \text{ integers}\}, \times\}$$

7. (a) If each element of a set $S$ is assigned to itself, the resulting one-to-one correspondence is called the *identity mapping* on $S$. If $\{G, \circ\}$ is a group, does the identity mapping on $G$ establish that $\{G, \circ\}$ is isomorphic to itself? Explain.

   (b) Find two one-to-one correspondences other than the identity mapping under which the permutation group
   $$\{\{e, p, q, r, s, t\}, \circ\}$$
   is isomorphic to itself.

   (*Note:* Such an isomorphic mapping is called an *automorphism.*)

8. Consider the mathematical system $\{\{\text{Real numbers}\}, +\}$. Find at least two automorphisms of this system (that is, at least two isomorphic mappings of this system with itself).

★9. Let $r + s\sqrt{2}$ be a root of the equation
$$ax^2 + bx + c = 0$$
where $a, b, c, r,$ and $s$ are integers. Prove that $r - s\sqrt{2}$ must also be a root. (*Hint:* There are at least two ways of proving this: (a) by direct substitution and (b) by using the isomorphisms established in Exercises 6a and 6b.)

★10. Extend the result of Exercise 9 to a cubic equation
$$ax^3 + bx^2 + cx + d = 0$$

★11. Extend the results of Exercises 9 and 10 to any $n$th degree equation with integer coefficients.

★12. Prove that

   (a) all groups consisting of exactly two elements are isomorphic.
   
   (b) all groups consisting of exactly three elements are isomorphic.

★13. Prove that any group consisting of exactly four elements must be isomorphic either to the *cyclic* group

$$\{\{0, 1, 2, 3\}, +(\bmod 4)\}$$

or to the so-called *Klein* group

$$\{\{1, 3, 5, 7\}, \times(\bmod 8)\}$$

## REVIEW EXERCISES

1. For each of the following mathematical systems decide whether or not
   (a) the given set is closed with respect to the operation.
   (b) the operation is associative.
   (c) there is an identity element.
   (d) each element has an inverse.
   (e) the operation is commutative.

A.
| ∘ | e | a | b |
|---|---|---|---|
| e | e | a | b |
| a | a | a | a |
| b | b | b | b |

B.
| ∘ | e | a | b |
|---|---|---|---|
| e | e | a | b |
| a | a | e | a |
| b | b | b | e |

C.
| ∘ | e | a | b |
|---|---|---|---|
| e | e | a | b |
| a | a | e | a |
| b | b | a | e |

D.
| ∘ | e | a | b |
|---|---|---|---|
| e | e | a | b |
| a | a | e | a |
| b | b | a | b |

E.
| ∘ | e | a | b |
|---|---|---|---|
| e | e | a | b |
| a | a | e | a |
| b | b | b | b |

F. $\{\{\text{Rationals}\}, (x, y) \to \sqrt{xy}\}$
G. $\{\{\text{Positive real numbers}\}, (x, y) \to \sqrt{xy}\}$
H. $\{\{\text{Even integers}\}, (x, y) \to x + y + 2\}$
I. $\{\{\text{Integers}\}, (x, y) \to x + y + 7\}$
J. $\{\{\text{Reals}\}, (x, y) \to x + y + xy\}$

2. Show that the following mathematical systems
   $\{\{\text{Positive reals}\}, \times\}$ and $\{\{\text{Reals}\}, +\}$
   are isomorphic under the mapping $\begin{pmatrix} x \to \log x \\ \times \to + \end{pmatrix}$

3. Show that the following two groups are isomorphic:
   $\{\{1, -1, i, -i\}, \times\}$ and $\{\{0, 1, 2, 3\}, + \bmod 4\}$

★4. Let $\{G, \circ\}$ be a group with a finite number of elements and $\{H, \circ\}$ be a subgroup of $\{G, \circ\}$ such that
   $$H \neq G, \quad H \neq \{e\}.$$
   Let $Hx = \{h \circ x \mid h \in H\}$ for every $x$ in $G$.
   Prove: (a) If $b \in H$, then $Hb = H$.
   (b) If $b \notin H$, then $Hb$ and $H$ have the same number of elements. (*Hint:* Use the association $x \circ b \leftrightarrow x$ to establish a one-to-one correspondence.)
   (c) If $b$ and $c$ are any elements of $G$, then $Hb$ and $Hc$ are either identical or have no elements in common.
   (d) $G$ may be expressed as a union of mutually disjoint sets
   $$G = H \cup Hb_1 \cup Hb_2 \cup \ldots \cup Hb_n$$
   for some set of elements $b_1, b_2, \ldots, b_n$ in $G$.
   (e) (*Lagrange's Theorem*) The number of elements in $G$ is a multiple of the number of elements in $H$.
   (f) If two groups have five elements each, then the groups must be isomorphic.

5. Compute:
   (a) $\begin{pmatrix} 1 & 2 & 3 & 4 \\ 2 & 4 & 3 & 1 \end{pmatrix} \circ \begin{pmatrix} 1 & 2 & 3 & 4 \\ 4 & 1 & 2 & 3 \end{pmatrix}$
   (b) $\begin{pmatrix} 1 & 2 & 3 & 4 \\ 4 & 1 & 2 & 3 \end{pmatrix}'$

6. Prove that if $\{G, \circ\}$ is a group and if $a$ is an element of $G$ such that $a \circ a = a$, then $a = e$.

7. If $e$ is the identity element of a group, prove that $e' = e$.

8. Let $\{G, \circ\}$ be a group. Let $S$ be the set of those elements of $G$ that commute with every element of $G$, that is,
   $$S = \{x \mid x \circ a = a \circ x \text{ for all } a \in G\}.$$
   Prove that $\{S, \circ\}$ is a group.

9. (a) Let $S = \{0, 1, 2, 3, 4, 5\}$ and define:
   $$x \circ y = \begin{cases} x - y \pmod 6 & \text{if } x \text{ and } y \text{ are both odd} \\ x + y \pmod 6 & \text{in all other cases} \end{cases}$$
   (b) Which of the group properties does the system in (a) have?

(c) If the operation ∘ were taken mod 7 instead of mod 6, which group properties would the new system have?

★10. Let $\{S, \circ\}$ be a commutative group. Show that there is a noncommutative group $\{T, \circ\}$ having a subgroup that is isomorphic to $\{S, \circ\}$.

(*Hint:* Use Exercises 2.4, Exercise 9.)

★11. Let $\{S, \circ\}$ be a commutative group such that not every element in $S$ is its own inverse. For all $x, y$ in $S$ define:
$$(x, 1) \cdot (y, 1) = (x \circ y, 1)$$
$$(x, 1) \cdot (y, -1) = (x \circ y, -1)$$
$$(x, -1) \cdot (y, 1) = (x \circ y', -1)$$
$$(x, -1) \cdot (y, -1) = (x \circ y', 1)$$
Let $T = \{(x, u) \mid x \in S \text{ and } u \in \{1, -1\}\}$.

Prove that $\{T, \cdot\}$ is a noncommutative group. What would happen if every element in $S$ were its own inverse?

12. (a) Let $S$ be the set of all ordered triples $(u, v, w)$ where $u, v, w$ are 0 or 1. Show that $S$ has exactly 8 elements and display these elements.

(b) Define operation ∘ on pairs of elements of $S$ as follows where addition is taken mod 2:

(1) $(a, b, 0) \circ (r, s, t) = (a + r, b + s, t)$

(2) $(a, b, 1) \circ (r, s, t) = (a + s, b + r, 1 + t)$

Show that $\{S, \circ\}$ is a noncommutative group.

(c) Show that the subset $T$ of ordered triples having the third entry 0, $(r, s, 0)$, is a commutative group of order 4 under the operation ∘ and that each element is its own inverse.

*Carl Friedrich Gauss (1777-1855)*

# 3 | Elementary Number Theory

> The integers were created by God, all else is the work of man.
>
> LEOPOLD KRONECKER

## 3.1 THE WELL-ORDERING AXIOM AND THE DIVISION THEOREM

In Chapter 2 we caught a glimpse of the endless variety of groups. Probably the most familiar group among these is the system $\{Z, +\}$ consisting of the set $Z$ of integers,

$$Z = \{0, \pm 1, \pm 2, \pm 3, \ldots\}$$

together with the ordinary addition operation $(+)$. This number system and its important subset, the whole numbers* $W = \{0, 1, 2, 3, \ldots\}$, have probably received almost as much attention as the rest of mathematics.

Problems concerning integers have challenged the minds of the greatest mathematicians from earliest antiquity to the present day. Although

---

* We are identifying the set of *whole numbers* $\{0, 1, 2, 3, \ldots\}$ with the set of *nonnegative* integers $\{0, +1, +2, +3, \ldots\}$, that is, each whole number is identified with the corresponding nonnegative integer. In Chapter 6 we define these sets of integers more precisely.

seemingly innocuous and deceptively simple to formulate, many of these problems often turn out to be exceedingly deep and difficult. Many of the still unsolved problems of mathematics lie within the realm of this particular branch of mathematics known as **number theory**. It is desirable that we become acquainted with a few elementary, number theoretic aspects of the system of integers before proceeding further with the study of more general mathematical systems.

The basic binary operation on integers is, of course, addition. Under this operation the integers form a **commutative group**. Moreover, by using this group structure it becomes possible to define various new binary operations such as subtraction, multiplication, and (for certain pairs of integers) division. For example, a subtraction problem such as

$$8 - 11$$

can be defined as an addition problem

$$8 + (-11)$$

where $-11$ is the additive inverse of $11$. We shall assume that the reader has already acquired a reasonable degree of skill with such computations and that he knows, for example, that $8 + (-11) = -3$.

Multiplication of any integer by a positive integer can be defined as repeated addition of that integer to itself. For example,

$$(4) \times 3 = (4) + (4) + (4) = 12$$
$$(-4) \times 3 = (-4) + (-4) + (-4) = -12$$
$$(4) \times 1 = 4$$
$$(-4) \times 1 = -4$$

The product of any integer by a negative integer can be defined as the additive **inverse** of the corresponding repeated sum defined previously. For example,

$$(4) \times (-3) = -[(4) \times 3] = -12$$
$$(-4) \times (-3) = -[(-4) \times 3] = -[-12] = 12$$
$$(4) \times (-1) = -4$$
$$(-4) \times (-1) = -(-4) = 4$$

The product of any integer by $0$ is defined to be $0$. For example,

$$4 \times 0 = 0, \quad (-4) \times 0 = 0, \quad \text{etc.}$$

Thus, basing the definition on addition of integers, the operation of multiplication is defined for every pair of integers (regardless of whether each is

## 3.1 The Well-Ordering Axiom and the Division Theorem

positive, negative, or zero). We shall assume that computation of products of integers is also quite familiar to the reader.

Several important properties of the multiplication operation are the following:

1. Multiplication of integers is *commutative*, that is,

$$a \times b = b \times a \qquad \text{for all integers } a, b$$

2. Multiplication of integers is associative, that is,

$$a \times (b \times c) = (a \times b) \times c \qquad \text{for all integers } a, b, c$$

3. Multiplication of integers is *distributive* over addition of integers, that is,

$$a \times (b + c) = (a \times b) + (a \times c) \qquad \text{for all integers } a, b, c$$

This last property (distributivity) plays a prominent role in Chapter 4 in connection with the study of mathematical systems known as "rings." At present, however, we focus our attention on the particular "ring" $\{Z, +, \times\}$. This mathematical system consists of the set $Z$ of integers together with the two binary operations addition and multiplication.

Multiplication of integers has been defined as repeated addition. Division of integers can be defined as repeated subtraction. For example, to divide 68 by 12, we note that we can subtract 12 from 68 five times (because five twelves are sixty), leaving a remainder of 8. We can express this procedure as follows:

$$68 - (12 \times 5) = 8$$

or

$$68 = 12 \times 5 + 8$$

In this process the number 12 which was subtracted repeatedly from the dividend 68 is called the **divisor**, the number 5 is called the **quotient**, and the number 8 is called the **remainder**.

More generally, if we start with any nonnegative integer (whole number) $a$ as a dividend and any positive integer $b$ as a divisor, then by subtracting the divisor $b$ from the dividend $a$ a sufficient (whole) number of times, say $q$ times, we obtain eventually a nonnegative integer remainder $r$ which is less than the divisor $b$. We can express this procedure as follows:

$$a - bq = r \qquad \text{where} \qquad 0 \leq r < b$$

or

$$a = bq + r \qquad \text{where} \qquad 0 \leq r < b$$

Although this argument appears quite reasonable, it is hardly a mathematical proof. Can we really be certain that if we subtract $b$ from $a$ repeatedly a sufficient number of times that we will always arrive eventually at a non-negative remainder $r$ such that $r < b$? In order to prove that this is indeed true, we shall resort to a rather obvious, yet very fundamental property of the whole numbers (the integers $\geq 0$). This property is known as the **Well-Ordering Axiom**. For the present we accept this property as an axiom. In Chapter 6 we derive this property assuming certain other properties or axioms.

**The Well-Ordering Axiom.** *In any nonempty set of whole numbers there is always a least number.*

If we grant this very reasonable assertion to be true, we can readily prove the statements we made previously about the division process.

Throughout the remainder of this chapter it will be understood that the only numbers under consideration will be integers.

**THEOREM 3.1** (*Division Theorem for Integers*) If $a$ is a whole number and $b$ is a positive integer, then there exist whole numbers $q$ and $r$ such that

$$a = bq + r \quad \text{where} \quad 0 \leq r < b.$$

*Proof:* Starting with $a$, let us subtract $b$ from $a$ repeatedly. This procedure generates the decreasing sequence of integers

$$a, \quad a-1b, \quad a-2b, \quad a-3b, \ldots \quad \text{etc.}$$

If we select from this sequence only those integers that are *whole numbers*, that is, only those integers that are $\geq 0$, then the set of whole numbers so obtained is not empty since it certainly contains $a$. This set must, according to the Well-Ordering Axiom, have a *least* member. Let this least whole number be $a - qb$, where $q$ is clearly some whole number and let the value of $a - qb$ be $r$:

$$a - qb = r$$

Now certainly $0 \leq r$ because $r$ is a whole number. It remains to prove that $r < b$. Suppose on the contrary that $r \geq b$. Then

$$r - b \geq 0$$

that is,

$$a - qb - b \geq 0$$

## 3.1 The Well-Ordering Axiom and the Division Theorem

or
$$a - (q+1)b \geq 0$$

Our supposition that $r \geq b$ thus implies that $a - (q+1)b$ is a whole number. But $a - (q+1)b$ is less than $a - qb$, which was supposed to be the *least* whole number in the sequence

$$a - 1b, \quad a - 2b, \quad a - 3b, \ldots \quad \text{etc.}$$

This contradiction shows that our supposition $r \geq b$ is false. Hence $r < b$ and we finally obtain

$$a = qb + r \quad \text{where} \quad 0 \leq r < b$$

### EXERCISES 3.1

1. Which of the following sets of real numbers are well-ordered, that is, which of them have the property that every nonempty subset has a least member?
   - (a) {Integers}
   - (b) {Negative integers}
   - (c) {Positive even integers}
   - (d) {Negative even integers}
   - (e) {Positive rationals}
   - (f) $\{2^x \mid x \text{ is an integer}\}$
   - (g) $\{2^x \mid x \text{ is a positive integer}\}$
   - (h) $\{(-2)^x \mid x \text{ is an integer}\}$
   - (i) $\{(-2)^x \mid x \text{ is a positive integer}\}$
   - (j) $\{(-2)^x \mid x \text{ is a positive even integer}\}$

2. Compute the values of $q$ and $r$ so that $a = bq + r$, where $0 \leq r < b$.

(a)

| | $a$ | $b$ | $q$ | $r$ |
|---|---|---|---|---|
| (1) | 65 | 7 | | |
| (2) | 70 | 7 | | |
| (3) | 135 | 7 | | |
| (4) | 15 | 7 | | |
| (5) | 4558 | 7 | | |

(b)

| | $a$ | $b$ | $q$ | $r$ |
|---|---|---|---|---|
| (1) | 65 | 17 | | |
| (2) | 70 | 17 | | |
| (3) | 135 | 17 | | |
| (4) | 15 | 17 | | |
| (5) | 4558 | 17 | | |

3. As derived in the text the Division Theorem (Theorem 3.1) applies only to nonnegative integers. It is also possible to prove the following more general theorem: If $a$ is any integer and $b$ is any integer other than 0, then there exist integers $q$ and $r$ such that

$$a = bq + r, \quad \text{where} \quad 0 \le r < |b|$$

and where $|b|$ is the *absolute value* of $b$, that is, the so-called "numerical value" of $b$. For example,

$$|3| = 3 \quad \text{and} \quad |-3| = 3$$

A more precise definition of absolute value appears in Chapter 6. Compute the values of $q$ and $r$ so that $a = bq + r$, where $0 \le r < |b|$.

(a)

| | $a$ | $b$ | $q$ | $r$ |
|---|---|---|---|---|
| (1) | −65 | 7 | | |
| (2) | −70 | 7 | | |
| (3) | 135 | −7 | | |
| (4) | −15 | −7 | | |
| (5) | −4558 | 7 | | |

(b)

| | $a$ | $b$ | $q$ | $r$ |
|---|---|---|---|---|
| (1) | −65 | 17 | | |
| (2) | −70 | 17 | | |
| (3) | −135 | 17 | | |
| (4) | −15 | 17 | | |
| (5) | −4558 | 17 | | |

(c)

| | $a$ | $b$ | $q$ | $r$ |
|---|---|---|---|---|
| (1) | 58 | 6 | | |
| (2) | 58 | −6 | | |
| (3) | −58 | 6 | | |
| (4) | −58 | −6 | | |
| (5) | 187 | 13 | | |
| (6) | 187 | −13 | | |
| (7) | −187 | 13 | | |
| (8) | −187 | −13 | | |

## 3.2 PRIME NUMBERS AND COMPOSITE NUMBERS

The Division Theorem (Theorem 3.1) asserts that if $a$ is a whole number and $b$ is a positive integer, then there exist whole numbers $q$ and $r$ such that

$$a = bq + r \quad \text{where} \quad 0 \leq r < b$$

In the special case where $r = 0$ this becomes simply

$$a = bq$$

When $a = bq$ we call $b$ a **divisor** or **factor** of $a$. (Of course, $q$ is also a divisor of $a$.) If $d$ is a divisor of $a$, it is customary to include $-d$ also, as a divisor of $a$. For example, if $a = 24$, then the divisors of $a$ are

$$\pm 1, \quad \pm 2, \quad \pm 3, \quad \pm 4, \quad \pm 6, \quad \pm 8, \quad \pm 12, \quad \pm 24$$

Similarly, if $d$ is a divisor of $a$, then we also consider $d$ to be a divisor of $-a$; so it makes sense to talk about the divisors of any integer. For example, the divisors of $-15$ are the same as the divisors of $15$:

$$\pm 1, \quad \pm 3, \quad \pm 5, \quad \pm 15$$

*Question.* What are the divisors of zero?

The correct answer to this question is: Every integer is a divisor of zero. Zero, however, is an exceptional number in this respect. Each integer other than zero has only a finite number of divisors. The simplest case occurs with the integers 1 and $-1$. Each of these integers has exactly two divisors, $\pm 1$. All other integers have at least four divisors. Thus

> The divisors of 2 are $\pm 1, \pm 2$.
> The divisors of 3 are $\pm 1, \pm 3$.
> The divisors of 4 are $\pm 1, \pm 2, \pm 4$.
> The divisors of 5 are $\pm 1, \pm 5$.
> etc.

Integers which have exactly four divisors (neither more nor less) play a most important role in number theory. They are called **prime integers**. Thus $\pm 2, \pm 3, \pm 5, \pm 7, \pm 11, \ldots$, etc., are prime integers. It is customary to refer to the positive prime integers as **prime numbers** or simply **primes**. Thus the first few primes are

$$2, \quad 3, \quad 5, \quad 7, \quad 11, \quad 13, \quad 17, \quad 19, \quad 23, \quad \ldots, \quad \text{etc.}$$

Those integers which have more than four divisors (but not infinitely many) are called **composite integers**. For example,

$$\pm 4, \quad \pm 6, \quad \pm 8, \quad \pm 9, \quad \pm 10, \quad \pm 12, \quad \pm 14, \quad \pm 15$$

are all composite integers. Note that 0, 1, and $-1$ are not included either among the prime integers, or among the composite integers. The positive composite integers are also often called **composite numbers** or even more briefly, **composites**. Thus the first few composites are

$$4, 6, 8, 9, 10, 12, 14, 15, 16, 18, 20, 21, \ldots$$

## EXERCISES 3.2

1. List all the divisors of
   (a) 18   (b) 30   (c) 36   (d) 37   (e) 48
2. What is the greatest positive integer that divides each number in the following pairs?
   (a) (12, 16)                (b) (50, 72)
   (c) (35, 96)                (d) (48, $-48$)
   (e) (100, 124)              (f) (57, 76)
   (g) (69, 115)               (h) (87, 203)
3. Prove that every integer is a divisor of itself.
4. Prove that every even integer other than 0, $\pm 2$, $\pm 4$ has at least 8 divisors.
5. Prove that if $x$ is a divisor of $y$ and $y$ is a divisor of $z$, then $x$ is a divisor of $z$.
6. Prove that if $d$ is a divisor of both $a$ and $b$, then $d$ is also a divisor of $xa + yb$ whenever $x$ and $y$ are integers.
7. Show that if $a$ and $b$ are positive integers such that $a$ is a divisor of $b$, then $a \leq b$. (*Hint:* If $a$ is a divisor of $b$, there must be a whole number $q$ such that $b = aq$ or $b = a + (q-1)a$. This shows that $q \geq 1$.)
8. List all the primes less than 100.
9. Prime pairs such as (11, 13) whose difference is 2 are called *twin primes*. List all the twin primes less than 100.
10. Given any integer, prove that its smallest divisor greater than one must be a prime.
11. (a)  Show that the six consecutive numbers
    $(7! + 2), \quad (7! + 3), \quad (7! + 4), \quad (7! + 5), \quad (7! + 6), \quad (7! + 7)$
    are composites. (*Note:* $7! = 1 \cdot 2 \cdot 3 \cdot 4 \cdot 5 \cdot 6 \cdot 7$. In general, for each positive integer $n$ we define: $n! = 1 \cdot 2 \cdot 3 \cdots n$.)

(b) Show that there exist arbitrarily long sequences of consecutive integers all of whose terms are composite.

12. Prove that there cannot be a greatest prime, that is, there must be infinitely many primes. (*Hint:* Suppose $p$ were the greatest prime. Consider the number $(1 \cdot 2 \cdot 3 \cdots p) + 1$. Since this number is greater than $p$, it would have to be composite. Use Exercise 10 to obtain a contradiction.)

13. Find a whole number that has exactly
    (a) 2 divisors
    (b) 4 divisors
    (c) 6 divisors
    (d) 8 divisors
    (e) 10 divisors.

★14. For each case cited in Exercise 13, try to formulate a generalization.

## 3.3 GCD AND THE EUCLIDEAN ALGORITHM

If an integer $d$ is a divisor of integer $a$ and also a divisor of integer $b$, then $d$ is called a **common divisor** of $a$ and $b$. For example, 3 is a common divisor of 36 and 48. These two numbers, however, have other common divisors. Here is a list of all the common divisors of 36 and 48:

$$\pm 1, \quad \pm 2, \quad \pm 3, \quad \pm 4, \quad \pm 6, \quad \pm 12$$

The greatest number among all these common divisors is 12. It is called the **greatest common divisor** (abbreviated GCD).

In general, if $a$ and $b$ are integers, not both zero, there will be only a finite number of common divisors of $a$ and $b$. Among these common divisors there will therefore be a greatest. There can be at most one such greatest number among the common divisors because if, say, $g_1$ and $g_2$ were each greatest, then $g_1 \leq g_2$ and $g_2 \leq g_1$ from which it follows that $g_1 = g_2$.

**Definition.** Let $a$ and $b$ be integers, not both zero. The greatest of all the common divisors of $a$ and $b$ is called the **greatest common divisor (GCD)** of $a$ and $b$.

Determining the GCD of two given integers would be a tedious task indeed if it were necessary to list all their common divisors. Fortunately, the GCD of two integers can be computed by a far more efficient, systematic procedure which was known to Euclid over 2000 years ago. It is called the *Euclidean algorithm* and is closely related to the Division Theorem (Theorem 3.1). To illustrate the Euclidean algorithm consider the following example.

*Example*  Determine the GCD of 2328 and 1080.

*Solution:*  Divide 2328 by 1080 to obtain the quotient 2 and the remainder 168.
$$2328 = 1080 \cdot 2 + 168$$

Next divide 1080 by 168 to obtain the quotient 6 and the remainder 72.
$$1080 = 168 \cdot 6 + 72$$

Continue this procedure (using the Division Theorem) until a zero remainder is obtained.
$$168 = 72 \cdot 2 + 24$$
$$72 = 24 \cdot 3$$

The last *nonzero* remainder, 24, turns out to be the desired GCD of 2328 and 1080.

To prove this we need merely observe that, according to the last equation, 24 is a divisor of 72. Since 24 is a divisor of 72, it then follows from the next to the last equation that 24 is also a divisor of 168. Then going back to the preceding equation it follows that 24 is a divisor of 1080, and finally, from the first equation, that 24 is also a divisor of 2328. Thus we have proved that 24 is certainly a common divisor of 1080 and 2328.

To prove it is the GCD we observe that, according to the first equation, if $d$ is any common divisor of 2328 and 1080, then $d$ must be a divisor of 168. From the second equation it then follows that $d$ is a divisor of 72 and, proceeding to the third equation, we find that $d$ is also a divisor of 24. Since every common divisor of 2328 and 1080 is a divisor of 24, it follows that each such divisor must be $\leq 24$. Hence 24 is indeed the *greatest* common divisor.

*Note:* in this proof we have used the following lemma.

LEMMA.  If $m$ and $n$ are positive integers and $m$ is a divisor of $n$, then $m \leq n$.

This lemma is easily shown to be true by the following argument. Since $m$ is a divisor of $n$, then $n = km$, where $k$ is a positive integer, that is, $k \geq 1$. Since $k \geq 1$, $km \geq m$, that is, $n \geq m$.

This example illustrates the method by which we prove the following theorem.

**THEOREM 3.2** Any two integers $a_1$ and $a_2$ (with $a_2 \neq 0$) have a (unique) GCD which can be obtained by the Euclidean algorithm.

*Proof:* Observe first that the divisors of any integer $m$ are the same as the divisors of $-m$. Hence we may assume, without loss of generality, that both $a_1$ and $a_2$ are nonnegative (that is, $a_1 \geq 0$ and $a_2 > 0$).

In accordance with the Euclidean algorithm let us start by dividing $a_1$ by $a_2$. We obtain

$$a_1 = a_2 q_1 + a_3$$

where the quotient $q_1$ and the remainder $a_3$ are whole numbers and $0 \leq a_3 < a_2$ (see Theorem 3.1). Now, if $a_3 = 0$, we stop. If $a_3 > 0$, we divide $a_2$ by $a_3$ to obtain

$$a_2 = a_3 q_2 + a_4$$

where once again $q_2$ and $a_4$ are whole numbers and $0 \leq a_4 < a_3$. This process cannot be repeated indefinitely because the remainders $a_3, a_4, \ldots$, etc., form a decreasing sequence of whole numbers. A zero remainder must therefore be obtained after a finite number of steps, for otherwise the remainders would form a set of whole numbers without a least member and that would violate the Well-Ordering Axiom. Suppose that a zero remainder is obtained after $n$ steps:

$$
\begin{aligned}
a_1 &= a_2 q_1 + a_3 &&\text{where } 0 \leq a_3 < a_2 \\
a_2 &= a_3 q_2 + a_4 &&\phantom{\text{where }} 0 \leq a_4 < a_3 \\
&\cdots &&\phantom{\text{where }} \cdots \\
a_k &= a_{k+1} q_k + a_{k+2} &&\phantom{\text{where }} 0 \leq a_{k+2} < a_{k+1} \\
&\cdots &&\phantom{\text{where }} \cdots \\
a_{n-1} &= a_n q_{n-1} + a_{n+1} &&\phantom{\text{where }} 0 \leq a_{n+1} < a_n \\
a_n &= a_{n+1} q_n
\end{aligned}
$$

We now show that the last nonzero remainder $a_{n+1}$ is the GCD of $a_1$ and $a_2$.

From the first equation we see that if $d$ is any common divisor of $a_1$ and $a_2$, then $d$ must also be a divisor of $a_3$. From the second equation it then follows that $d$ is a divisor of $a_4$. By referring to each successive equation in turn, we see that $d$ is a divisor of each of the remainders

$$a_3, a_4, \ldots, a_n, a_{n+1}$$

Thus, in particular, $d$ is a divisor of $a_{n+1}$. Since $d$ is a divisor of $a_{n+1}$ and $a_{n+1} > 0$, it follows by the lemma proved above that $d \leq a_{n+1}$.

Conversely, from the last equation, we see that $a_{n+1}$ is a divisor of $a_n$. From the next to the last equation, it then follows that $a_{n+1}$ is also a divisor of $a_{n-1}$, and by referring in turn to each equation (proceeding upward), we deduce that $a_{n+1}$ is a divisor of each of the numbers

$$a_n, a_{n-1}, \ldots, a_4, a_3, a_2, \text{ and } a_1$$

Thus, in particular, $a_{n+1}$ is a common divisor of $a_1$ and $a_2$. Therefore $a_{n+1}$ satisfies the requirements for a GCD of $a_1$ and $a_2$.

It is worth observing that we have proved the following additional result.

COROLLARY (TO THEOREM 3.2). If $g$ is the GCD of integers $a$ and $b$, then every common divisor of $a$ and $b$ is also a divisor of $g$.

In Section 3.4 we study some other very fundamental consequences of Theorem 3.2.

## EXERCISES 3.3

1. Find the GCD for each of the following pairs.
   (a) (100, 144)      (b) (1000, 144)
   (c) (300, 144)      (d) (1024, 288)
   (e) (3456, 123)
2. How would you define GCD for a triple of integers? Using your definition, compute the GCD for each of the following:
   (a) (100, 144, 200)
   (b) (234, 288, 400)
   (c) (1024, 288, 316)
3. If $p$ and $q$ are distinct primes, what is the GCD of $p$ and $q$?

4. If $a$ is any integer, what is the GCD of 0 and $a$? (Consider first the case $a \neq 0$ and then the case $a = 0$.)
5. Prove that if $g$ is the GCD of $a$ and $b$, then $g$ need not be the GCD of $a$ and $a + kb$, where $k$ is an integer.

## 3.4 THE FUNDAMENTAL THEOREM OF ARITHMETIC

We can gain additional valuable information concerning the greatest common divisor of two integers by looking more closely at certain details of the proof of Theorem 3.2 where we encountered the equations

$$a_1 = a_2 q_1 + a_3$$
$$a_2 = a_3 q_2 + a_4$$
$$\cdots$$
$$a_k = a_{k+1} q_k + a_{k+2}$$
$$\cdots$$
$$a_{n-1} = a_n q_{n-1} + a_{n+1}$$
$$a_n = a_{n+1} q_n$$

Using the first of these equations, we can express the first remainder $a_3$ in terms of the original integers $a_1$ and $a_2$ as follows:

$$(1) \qquad a_3 = 1 \cdot a_1 - q_1 a_2$$

Using the second equation, we can express the next remainder $a_4$ in terms of $a_2$ and $a_3$:

$$a_4 = 1 \cdot a_2 - q_2 a_3$$

In this last equation let us replace "$a_3$" by its expression in terms of $a_1$ and $a_2$ obtained above. The equation for $a_4$ then becomes

$$a_4 = 1 \cdot a_2 - q_2(a_1 - q_1 a_2)$$

The terms on the right can be rearranged as follows:

$$(2) \qquad a_4 = (-q_2) a_1 + (1 + q_1 q_2) a_2$$

which expresses the remainder $a_4$ in terms of the *original* integers $a_1$ and $a_2$. In each of the Equations (1) and (2) the coefficients of $a_1$ and $a_2$ are integers. In other words, we have succeeded in expressing each of the remainders $a_3$ and $a_4$ in the form

$$x a_1 + y a_2$$

where $x$ and $y$ are integers.

We now show that this is possible in turn with each of the succeeding remainders. For example, from the equation

$$a_3 = a_4 q_3 + a_5$$

we obtain
$$a_5 = 1 \cdot a_3 - q_3 a_4$$
and since we already know that $a_3$ and $a_4$ are expressible as
$$a_3 = x_3 a_1 + y_3 a_2 \qquad a_4 = x_4 a_1 + y_4 a_2$$
where $x_3$, $y_3$, $x_4$, $y_4$ are integers, it follows that
$$a_5 = 1 \cdot (x_3 a_1 + y_3 a_2) - q_3(x_4 a_1 + y_4 a_2)$$
This last equation is readily rewritten as
$$a_5 = (x_3 - q_3 x_4)a_1 + (y_3 - q_3 y_4)a_2$$
showing that the remainder $a_5$ is expressible as $x_5 a_1 + y_5 a_2$ where $x_5$ and $y_5$ are integers defined by
$$x_5 = (x_3 - q_3 x_4), \qquad y_5 = (y_3 - q_3 y_4)$$
Now observe that the $k$th remainder, $a_{k+2}$, is expressible in terms of the two preceding remainders $a_k$ and $a_{k+1}$ as follows:
$$a_{k+2} = 1 \cdot a_k - q_k a_{k+1}$$
Hence, if we have already somehow managed to express these two previous remainders in terms of $a_1$ and $a_2$, let us say
$$a_k = x_k a_1 + y_k a_2, \qquad a_{k+1} = x_{k+1} a_1 + y_{k+1} a_2$$
where $x_k$, $y_k$, $x_{k+1}$, and $y_{k+1}$ are all integers, then it will follow that
$$a_{k+2} = 1(x_k a_1 + y_k a_2) - q_k(x_{k+1} a_1 + y_{k+1} a_2)$$
or equivalently,
$$a_{k+2} = (x_k - q_k x_{k+1})a_1 + (y_k - q_k y_{k+1})a_2$$
In other words, we have proved that the $k$th remainder $a_{k+2}$ can be expressed in terms of $a_1$ and $a_2$ whenever the two preceding remainders $a_k$ and $a_{k+1}$ are so expressible. The expression for $a_{k+2}$ takes the form
$$a_{k+2} = x_{k+2} a_1 + y_{k+2} a_2$$
where $x_{k+2}$ and $y_{k+2}$ are integers defined by
$$x_{k+2} = x_k - q_k x_{k+1}, \qquad y_{k+2} = y_k - q_k y_{k+1}$$
Since this repetitive (inductive) argument applies to each of the remainders that appear in the Euclidean algorithm, in particular it will apply to the last nonzero remainder, $a_{n+1}$. In short, this proves that
$$a_{n+1} = xa_1 + ya_2$$
where $x$ and $y$ are integers. We summarize this important result as follows.

## 3.4 The Fundamental Theorem of Arithmetic

**THEOREM 3.3** If $a_{n+1}$ is the GCD of integers $a_1$ and $a_2$, then there exist integers $x$ and $y$ such that

$$a_{n+1} = xa_1 + ya_2$$

In Theorem 3.3 we merely assert the existence of the integers $x$ and $y$. However, our proof of Theorem 3.3 was "constructive" in the sense that we gave a definite procedure for computing the integers $x$ and $y$. This does not mean that there cannot be other pairs of integers $x$ and $y$ that fulfill the requirements of Theorem 3.3. (See Exercise 7.)

As an illustration let us recall that in Section 3.3 we used the Euclidean algorithm to compute the GCD of 2328 and 1080 to be 24. We should therefore be able to express 24 in terms of 2328 and 1080, that is, we should be able to determine $x$ and $y$ so that

$$24 = 2328x + 1080y$$

To do this, we need only reverse the steps of the Euclidean algorithm. Recall that these steps were as follows:

$$2328 = 1080 \cdot 2 + 168$$
$$1080 = 168 \cdot 6 + 72$$
$$168 = 72 \cdot 2 + 24$$

From the last equation we obtain

$$24 = 168 - 72 \cdot 2$$

and from the second equation

$$72 = 1080 - 168 \cdot 6$$

Hence $\qquad 24 = 168 - (1080 - 168 \cdot 6) \cdot 2$

that is, $\qquad 24 = 168 \cdot 13 - 1080 \cdot 2$

But from the first equation we have

$$168 = 2328 - 1080 \cdot 2$$

Hence $\qquad 24 = (2328 - 1080 \cdot 2) \cdot 13 - 1080 \cdot 2$

Therefore $\qquad 24 = 2328 \cdot 13 - 1080 \cdot 28$

We can express this last equality as follows:

$$24 = 2328 \cdot x + 1080 \cdot y$$

where
$$x = 13 \quad \text{and} \quad y = -28$$

Observe that this method is constructive. It does more than merely show the existence of the integers $x$ and $y$—it actually produces them.

**Definition.** Two integers $a$ and $b$ are called **relatively prime** whenever their GCD is 1. (We also say "$a$ is relatively prime to $b$" or "$b$ is relatively prime to $a$.")

For example, the integers 63 and 20 are relatively prime because (using the Euclidean algorithm)
$$63 = 20 \cdot 3 + 3$$
$$20 = 3 \cdot 6 + 2$$
$$3 = 2 \cdot 1 + 1$$
from which we see that the GCD of 63 and 20 (the last nonzero remainder) is 1.

Clearly, if two integers are relatively prime, then they have no common factors other than the trivial ones $\pm 1$. For example, the divisors of 63 are
$$\pm 1, \quad \pm 3, \quad \pm 7, \quad \pm 9, \quad \pm 21, \quad \pm 63$$
whereas the divisors of 20 are
$$\pm 1, \quad \pm 2, \quad \pm 4, \quad \pm 5, \quad \pm 10, \quad \pm 20$$
The only common divisors are $\pm 1$.

Any two distinct primes are relatively prime. More generally, if an integer $n$ is not divisible by a prime $p$, then $n$ and $p$ are relatively prime. (See Exercise 2.)

If $a$ and $b$ are integers, Theorem 3.3 asserts that there exist integers $x$ and $y$ such that $xa + yb = g$ (the GCD of $a$ and $b$). It is interesting to observe that the integers $x$ and $y$ must always be relatively prime. For example, we showed previously that the GCD of 2328 and 1080 was 24, and we expressed this GCD in terms of 2328 and 1080 as follows:
$$24 = 2328x + 1080y$$
where $x = 13$ and $y = -28$. Notice that 13 and $-28$ are indeed relatively prime. This is no accident. (See Exercise 7d.)

**THEOREM 3.4** Two integers $a$ and $b$ are relatively prime if and only if there exist integers $x$ and $y$ such that
$$ax + by = 1$$

*Proof:* If $a$ and $b$ are relatively prime, then their GCD is 1. Hence, by Theorem 3.3, there exist integers $x$ and $y$ such that $ax + by = 1$. Conversely, if there exist integers $x$ and $y$ such that $ax + by = 1$, then any common divisor of $a$ and $b$ must be a divisor of 1. But the only divisors of 1 are $\pm 1$ and hence the GCD of $a$ and $b$ is 1. This completes the proof.

As an example, we saw that the GCD of 63 and 20 is 1. If we "work upward" from the last equation of the Euclidean algorithm, we obtain

$$1 = 3 - (2) \cdot 1$$
$$\therefore 1 = 3 - (20 - 3 \cdot 6) \cdot 1 = 3 - 20 \cdot 1 + 3 \cdot 6$$
$$\therefore 1 = (3) \cdot 7 - (20) \cdot 1$$
$$\therefore 1 = (63 - 20 \cdot 3) \cdot 7 - (20) \cdot 1$$
$$\therefore 1 = (63) \cdot 7 - (20) \cdot 22$$

that is,

$$1 = (63) \cdot x + (20)y$$

where $x = 7$ and $y = -22$ (integers).

**THEOREM 3.5** If a given integer is relatively prime to each of several others, then it is relatively prime to their product.

*Proof:* Let $a$ be relatively prime to $b$ and also to $c$. Then by Theorem 3.4, there exist integers $x_1, y_1, x_2, y_2$ such that

$$ax_1 + by_1 = 1 \quad \text{and} \quad ax_2 + cy_2 = 1$$

Hence

$$ax_1 + by_1(ax_2 + cy_2) = 1$$

and therefore

$$a(x_1 + by_1 x_2) + bc(y_1 y_2) = 1$$

so by Theorem 3.4 it follows that $a$ and $bc$ are relatively prime.

It now follows that if $a$ and $d$ are also relatively prime, then (since $a$ and $bc$ are relatively prime) $a$ and $(bc)d$ must also be relatively prime. Clearly, this argument can be repeated so that it applies to a product of any number of factors.

**THEOREM 3.6** If a product of several integers is divisible by a prime $p$, then at least one of the integers is divisible by $p$.

*Proof:* If none of the integers were divisible by $p$, then each integer would be relatively prime to $p$. Then, by Theorem 3.5, their product would be relatively prime to $p$, that is, their product would not be divisible by $p$.

A somewhat more general version of Theorem 3.6 is the following.

**THEOREM 3.7** If an integer $a$ divides the product $bc$ of integers $b$ and $c$ and if $a$ is relatively prime to $b$, then $a$ must be a divisor of $c$.

*Proof:* Since $a$ is relatively prime to $b$, it follows (by Theorem 3.4) that there exist integers $x$ and $y$ such that $ax + by = 1$. Multiply both members of this equation by $c$.

$$acx + bcy = c$$

Now $a$ is certainly a divisor of $acx$. Hence, if $a$ is a divisor of $bc$, then $a$ will be a divisor of $acx + bcy$, that is, $a$ will be a divisor of $c$.

We now proceed to prove a very basic theorem of number theory often called **The Fundamental Theorem of Arithmetic**.

**THEOREM 3.8** (*The Unique Factorization Theorem*) Every composite positive integer $N$ can be expressed as a product of (positive) primes in one and only one way, except for the order of the factors.

*Proof:* Since $N$ is not a prime, let $p_1$ be the least positive divisor of $N$ other than 1. $p_1$ must itself be a prime, for otherwise it would not be the least such divisor. We now know that

$$N = p_1 N_1$$

If $N_1$ is a prime, we can stop at this point. If $N_1$ is not a prime, let $p_2$ be the least positive divisor of $N_1$ other than 1. Then $p_2$ must also be a prime. (Note that $p_2$ may or may not be distinct from $p_1$. We do not exclude the possibility that the same prime factor might appear more than once with $N$ expressed as a product of primes.)

## 3.4 The Fundamental Theorem of Arithmetic

We now have
$$N_1 = p_2 N_2$$
and hence
$$N = p_1 p_2 N_2$$

If $N_2$ is a prime, we have finished. If $N_2$ is not a prime, we continue the argument to obtain $N_2 = p_3 N_3$ and hence
$$N = p_1 p_2 p_3 N_3, \text{ etc.}$$

Since $N_1 > N_2 > N_3 > \cdots$ etc., this procedure must terminate after a finite number, say $n$ steps, for otherwise the Well-Ordering Axiom would be violated. We obtain in this manner a factorization of $N$ into a finite number of primes
$$N = p_1 p_2 \cdots p_n$$

To show that this factorization of $N$ into primes is unique, suppose that
$$N = q_1 q_2 \cdots q_m$$
is another factorization of $N$ into primes. By Theorem 3.6 the prime $p_1$ must be a divisor of at least one of the integers $q_1, q_2, \ldots, q_m$; let us say $q_1$. But the only positive divisors of $q_1$ are 1 and $q_1$ (because $q_1$ is a prime). Hence $p_1 = q_1$ and we can now deduce that
$$p_2 p_3 \cdots p_n = q_2 q_3 \cdots q_m$$

If we repeat the argument using in turn $p_2, p_3$, etc., we see that $p_2 = q_2$ (say), $p_3 = q_3$ (say), ..., etc., showing that the factors
$$q_1, q_2, \ldots, q_m$$
are identical in some order with the prime factors
$$p_1, p_2, \ldots, p_n$$

This completes the proof of the Unique Factorization Theorem.

As an example, suppose $N = 252$. Let us express $N$ as a product of primes. The least positive prime divisor of $N$ is 2. Hence $p_1 = 2$ (note that this is a prime) and $N = p_1 N_1$ becomes
$$252 = 2 \cdot 126$$

When we repeat the process with $N_1 = 126$, the least positive prime divisor of 126 is (again) 2. Hence $p_2 = 2$ and $N_1 = p_2 N_2$ becomes

$$126 = 2 \cdot 63$$

So far we have $252 = 2 \cdot 2 \cdot 63$, that is, $N = p_1 p_2 N_2$. Continuing this procedure, we find

$$252 = 2 \cdot 2 \cdot 63 = 2 \cdot 2 \cdot 3 \cdot 21 = 2 \cdot 2 \cdot 3 \cdot 3 \cdot 7$$

which expresses 252 as a product of positive primes. This expression is unique except for the order of the factors. It is usual to abbreviate the expression by writing

$$252 = 2^2 \cdot 3^2 \cdot 7^1$$

In general, any integer $N$ is expressible in the form

$$N = p_1^{e_1} p_2^{e_2} \cdots p_k^{e_k}$$

where $p_1, p_2, \ldots p_k$ are distinct primes and the exponents $e_1, e_2, \ldots, e_k$ indicate how often each prime appears as a factor.

The Unique Factorization Theorem can be used to find the *greatest common divisor* of a pair of integers. For example, to find the GCD of 2328 and 1080 we express each of these integers in the form

$$N = p_1^{e_1} p_2^{e_2} \cdots p_k^{e_k}$$

Thus
$$2328 = 2^3 \cdot 3^1 \cdot 97^1 \quad \text{and} \quad 1080 = 2^3 \cdot 3^3 \cdot 5^1$$

The GCD of these two integers is therefore

$$2^3 \cdot 3^1 = 24$$

The Unique Factorization Theorem can also be used to find the *least common multiple* of a pair of integers. To see what this means let us first consider a simple illustration. The equality

$$24 = 6 \times 4$$

tells us that 6 and 4 are divisors of 24. We also say

24 is a *multiple* of 6

and

24 is a *multiple* of 4

Since 24 is a multiple of both 6 and 4, we call 24 a **common multiple** of 6 and 4. However, 24 is not the only common multiple of 6 and 4. There are many numbers, for example,

$$0, \pm 12, \pm 24, \pm 36, \ldots$$

that are common multiples of 6 and 4. Among all these common multiples it is convenient to single out the least positive common multiple, 12. We call this particular common multiple, 12, the **least common multiple** (LCM) of 6 and 4.

More generally, if $a$ and $b$ are nonzero integers, the least positive common multiple of $a$ and $b$ is called the least common multiple (LCM) of $a$ and $b$. If $a = 0$ or $b = 0$, we define the LCM of $a$ and $b$ to be 0.

To see how the Unique Factorization Theorem can be used to find a LCM let us consider again the illustration used previously

$$2328 = 2^3 \cdot 3^1 \cdot 97 \quad \text{and} \quad 1080 = 2^3 \cdot 3^3 \cdot 5^1$$

The LCM of 2328 and 1080 must contain each prime factor at least as often as it appears in either expression. Hence the LCM of 2328 and 1080 is

$$2^3 \cdot 3^3 \cdot 5^1 \cdot 97^1$$

which has the value 104,760.

## EXERCISES 3.4

1. Express as a product of primes:
   (a) 144      (b) 1000      (c) 576
   (d) 234      (e) 324      (f) 423
2. (a) Prove that any two distinct primes are relatively prime.
   (b) Prove that if an integer $n$ is not divisible by a prime $p$, then $n$ and $p$ are relatively prime.
3. Prove that if the GCD of $b$ and $c$ is 1 and if both $b$ and $c$ are divisors of $a$, then $bc$ is also a divisor of $a$.
4. Show that if $b$ and $c$ are relatively prime to $a$, then $b + c$ need not be relatively prime to $a$.
5. (a) Prove that if $b$ and $c$ are relatively prime, then $b^2$ and $c^3$ are relatively prime.
   (b) Discuss whether $b^r$ and $c^s$ are relatively prime for any positive integers $r$ and $s$, given that $b$ and $c$ are relatively prime.
6. (a) Prove that if $b^2$ and $c^3$ are relatively prime, then $b$ and $c$ are also relatively prime.
   (b) If $r$ and $s$ are positive integers, argue that if $b^r$ and $c^s$ are relatively prime, then $b$ and $c$ are relatively prime.
7. Let us define $a \wriggle b$ ($a$ wriggle $b$) to mean that $a$ is relatively prime to $b$.
   (a) Prove: If $x \wriggle y$, then $y \wriggle x$.
   (b) Express Theorems 3.4 and 3.5 using the wriggle symbol.

(c) Prove: If $ax + by = 1$, then $a \gtrless b$, $a \gtrless y$, $x \gtrless b$, and $x \gtrless y$.
(d) Prove: If $ax + by = g$ where $g$ is the GCD of $a$ and $b$, then $x \gtrless y$.
(e) Prove: If $(ab) \gtrless c$, then $a \gtrless c$ and $b \gtrless c$.

8. Prove that if $ax + by = 1$, then there are infinitely many values corresponding to $x$ and $y$ for which the equality holds. For example,
$$4(-2) + 3(+3) = 1$$
$$4(+1) + 3(-1) = 1$$

9. Prove Exercise 7c if $ax + by = -1$.
10. How can prime factorization be used to obtain
    (a) GCD of 2 integers?
    (b) LCM of 2 integers?
11. Using 10, find the GCD and LCM of the following pairs.
    (a) (48, 63)           (b) (98, 144)
    (c) (144, 50)          (d) (69, 92)
12. Prove that the product of the GCD and LCM for the positive integers $a$ and $b$ is $ab$.
13. If $d$ is the GCD of integers $a$ and $b$, prove that $\dfrac{a}{d}$ and $\dfrac{b}{d}$ are relatively prime.

## 3.5 CONGRUENCE MODULO AN INTEGER

The binary operations of modular arithmetic which we studied in Chapter 1 are closely related to the Division Theorem and notions of divisibility for integers. For example, in Section 1.4 (page 14) we stated

There exists in $W_4$ the number $y = 3$ such that
$$11 = y + (\text{an integer} \times 4), \text{ namely}$$
$$11 = 3 + (2 \times 4)$$

Hence in modulo 4 arithmetic adding 3 is equivalent to adding 11.

The equality $11 = (4 \times 2) + 3$ used here is clearly a specific illustration of the Division Theorem. Again, in Section 1.5 (page 17) we stated

"7" and "10" both represent the same number, namely 1, in the modulo 3 system of arithmetic.

The integers 7 and 10 are not really the same number—they are two distinct integers. Neither of these integers is actually a member of the modulo

3 number system (0, 1, and 2 are the only numbers in that system). The above statements therefore must be regarded merely as a way of saying that 7 and 10 can each be reduced to 1 in the modulo 3 system and that 11 can be reduced to 3 in the modulo 4 system. In each instance the reduction is performed by using the Division Theorem. Thus, if $m$ is a positive integer and $a$ is any integer and if

$$a = mq_1 + r \quad \text{where} \quad 0 \le r < m$$

in accordance with the Division Theorem, then we say that the integer $a$ "reduces to $r$" in the modulo $m$ system. Now suppose that the integer $b$ also reduces to $r$ (modulo $m$), that is,

$$b = mq_2 + r \quad \text{where} \quad 0 \le r < m$$

Then it follows that

$$a - b = m(q_1 - q_2)$$

that is, in this case $m$ is a *divisor* of $a - b$. Whenever this is so, we say that *a is congruent to b modulo m*. In general, we have the following definition.

**Definition.** Integer $a$ is said to be **congruent** to integer $b$ modulo integer $m$ if and only if $a - b$ is divisible by $m$, that is, if and only if

$$a - b = qm$$

where $q$ is some integer.

It is convenient to abbreviate the sentence "$a$ is congruent to $b$ modulo $m$" by writing

$$a \equiv b \pmod{m}$$

The notation (first introduced by Gauss), $a \equiv b \pmod{m}$, is therefore equivalent to each of the following sentences:

$a - b$ is divisible by $m$
$a - b = qm$ where $q$ is an integer
$a = b + qm$ where $q$ is an integer

**Examples.**
(a) $12 \equiv 2 \pmod{5}$
(b) $13 \equiv -1 \pmod{7}$
(c) $6 \equiv 0 \pmod{3}$
(d) $-1 \equiv 3 \pmod{4}$
(e) $3 \not\equiv 1 \pmod{4}$

Observe that if $m$ is any fixed positive integer and if

$$W_m = \{0, 1, 2, \ldots, m-1\}$$

is the set of whole numbers less than $m$, then every integer $a$ must be congruent modulo $m$ to exactly one member of $W_m$. This assertion is a direct consequence of the Division Theorem, for if

$$a = qm + r \quad \text{where} \quad 0 \leq r < m$$

then

$$a - r = qm$$

It then follows that

$$a \equiv r \pmod{m}$$

where $r$ is the unique remainder or *residue* when $a$ is divided by $m$. For this reason the set $W_m$ is often called the set of **least residues** modulo $m$.

In so-called modular arithmetic we usually confine our attention to a set of least residues modulo some positive integer $m$. In number theory, however, we study all integers, but we are frequently interested in whether the integers of a given pair are, or are not congruent modulo $m$. Clearly, integers $a$ and $b$ will be congruent modulo $m$, or not congruent modulo $m$, according to whether they have, or do not have the same least residue modulo $m$.

## EXERCISES 3.5

1. Label true or false, whichever applies, to the following:
   (a) $7 \equiv 7 \pmod{5}$
   (b) $7 \equiv 2 \pmod{5}$
   (c) $7 \equiv 5 \pmod{2}$
   (d) $7 \equiv 0 \pmod{5}$
   (e) $7 \equiv -2 \pmod{5}$
   (f) $7 \equiv -5 \pmod{2}$
   (g) $-7 \equiv -2 \pmod{5}$
   (h) $-7 \equiv -5 \pmod{2}$
   (i) $-7 \equiv 2 \pmod{5}$
   (j) $-7 \equiv 5 \pmod{2}$
   (k) $7 + 11 \equiv 2 + 11 \pmod{5}$
   (l) $7 - 11 \equiv 2 - 11 \pmod{5}$
   (m) $7 \times 11 \equiv 2 \times 11 \pmod{5}$
   (n) $11 - 7 \equiv 11 - 2 \pmod{5}$
   (o) $7^2 \equiv 2^2 \pmod{5}$
   (p) $7^3 \equiv 2^3 \pmod{5}$
   (q) $7^4 \equiv 2^4 \pmod{5}$
   (r) $7^2 \equiv 5^2 \pmod{2}$
   (s) $7^3 \equiv 5^3 \pmod{2}$
   (t) $2^4 \equiv 1 \pmod{5}$
   (u) $3^4 \equiv 1 \pmod{5}$
   (v) $4^4 \equiv 1 \pmod{5}$
   (w) $2^7 \equiv 2^2 \pmod{5}$

2. Find the smallest nonnegative integer value of $x$ that makes each of the following sentences true:
   (a) $x \equiv 47 \pmod{13}$
   (b) $47 \equiv x \pmod{13}$
   (c) $2x \equiv 47 \pmod{13}$
   (d) $x \equiv -47 \pmod{13}$
   (e) $2x \equiv -47 \pmod{13}$
   (f) $3x \equiv 47 \pmod{13}$
   (g) $3x \equiv -47 \pmod{13}$
   (h) $4x \equiv 47 \pmod{13}$
   (i) $4x \equiv -47 \pmod{13}$
   (j) $5x \equiv 47 \pmod{13}$

3. Find the set of least residues satisfying the condition.
   (a) $x^2 \equiv 2 \pmod{7}$
   (b) $x^2 \equiv 4 \pmod{7}$
   (c) $x^2 \equiv 6 \pmod{7}$
   (d) $x^2 \equiv -1 \pmod{7}$
   (e) $x^2 \equiv -3 \pmod{7}$
   (f) $x^2 \equiv 1 \pmod{7}$
   (g) $x^2 \equiv -5 \pmod{7}$

4. Compute the least residue for each of the following:
   (a) $2 \pmod{7}$
   (b) $2^2 \pmod{7}$
   (c) $2^3 \pmod{7}$
   (d) $2^4 \pmod{7}$
   (e) $2^5 \pmod{7}$
   (f) $2^6 \pmod{7}$
   (g) $2^7 \pmod{7}$
   (h) $2^8 \pmod{7}$
   What seems to be a pattern?

5. On the basis of a pattern compute the least residue for
   (a) $2^{10} \pmod{7}$
   (b) $2^{12} \pmod{7}$
   (c) $2^{15} \pmod{7}$
   (d) $2^{99} \pmod{7}$
   (e) $2^{100} \pmod{7}$
   (f) $2^{200} \pmod{7}$

## 3.6 BASIC THEOREMS ABOUT CONGRUENCES

It is important to realize that although congruences resemble equations in many ways, the relation of congruence (modulo $m$) is quite different from the relation of equality. When we write

$$a = b$$

we assert that *a and b are actually the same number*. On the other hand, when we write

$$a \equiv b \pmod{m}$$

we are merely asserting that the difference $a - b$ is exactly divisible by $m$, or equivalently, that $a = b + qm$ (where $q$ is an integer). The integers $a$ and $b$ can therefore be quite different, yet still be congruent modulo $m$. The following theorems stipulate some of the ways in which congruences resemble equations and also some of the ways in which congruences differ from equations. To avoid excessive repetition, let it be understood that all letters (such as $a$, $b$, $c$, $d$, $m$, etc.) used below represent integers.

**THEOREM 3.9** If $a$, $b$, $c$, and $m \neq 0$ are integers, then
  (a) (Reflexivity): $a \equiv a \pmod{m}$
  (b) (Symmetry): If $a \equiv b \pmod{m}$, then $b \equiv a \pmod{m}$
  (c) (Transitivity): If $a \equiv b \pmod{m}$ and $b \equiv c \pmod{m}$, then $a \equiv c \pmod{m}$

*Proof:* (a) Since $a = a + 0m$, it follows that $a \equiv a \pmod{m}$.
  (b) If $a \equiv b \pmod{m}$, then $a = b + qm$ where $q$ is an integer. It then follows that

$$b = a + (-q)m$$

where $-q$ is, of course, also an integer. Hence $b \equiv a \pmod{m}$ whenever $a \equiv b \pmod{m}$.
  (c) If $a \equiv b \pmod{m}$ and $b \equiv c \pmod{m}$, then

$$a = b + q_1 m \quad \text{and} \quad b = c + q_2 m$$

where $q_1$ and $q_2$ are integers. It then follows that

$$a = c + q_2 m + q_1 m$$

that is, $a = c + (q_1 + q_2)m$ and since $q_1 + q_2$ is also an integer, this means that

$$a \equiv c \pmod{m}$$

**THEOREM 3.10** If $a \equiv b \pmod{m}$ and $c \equiv d \pmod{m}$, then
  (a) $a + c \equiv b + d \pmod{m}$
  (b) $a - c \equiv b - d \pmod{m}$
  (c) $ac \equiv bd \pmod{m}$

*Proof:* If $a \equiv b \pmod{m}$ and $c \equiv d \pmod{m}$, then $a = b + q_1 m$, $c = d + q_2 m$ where $q_1$ and $q_2$ are integers. It then follows that
  (a) $a + c = (b + q_1 m) + (d + q_2 m)$
  $\therefore a + c = (b + d) + (q_1 + q_2)m$
  $\therefore a + c \equiv (b + d) \pmod{m}$
  (b) $a - c = (b + q_1 m) - (d + q_2 m)$
  $\therefore a - c = (b - d) + (q_1 - q_2)m$
  $\therefore a - c \equiv (b - d) \pmod{m}$

## 3.6 Basic Theorems About Congruences 115

(c) $\quad ac = (b + q_1 m)(d + q_2 m)$
$\therefore ac = bd + bq_2 m + dq_1 m + q_1 q_2 m^2$
$\therefore ac = bd + (bq_2 + dq_1 + q_1 q_2 m) \cdot m$
$\therefore ac \equiv bd \pmod{m}$

COROLLARY. If $a \equiv b \pmod{m}$ and if $k$ is any integer, then $ka \equiv kb \pmod{m}$.

*Proof:* This is a special case of Theorem 3.10c where $c = d = k$. As an illustration of Theorem 3.10 suppose that we have already verified

$$30 \equiv 24 \pmod{3}$$
$$15 \equiv 6 \pmod{3}$$

We may now conclude that

$$30 + 15 \equiv 24 + 6 \pmod{3}$$

that is, that

$$45 \equiv 30 \pmod{3}$$

Similarly, we may conclude that

$$30 \cdot 15 \equiv 24 \cdot 6 \pmod{3}$$

that is, that

$$450 \equiv 144 \pmod{3}$$

Observe, however, that we may not conclude that

$$\frac{30}{15} \equiv \frac{24}{6} \pmod{3}$$

In fact, this asserts that $2 \equiv 4 \pmod{3}$, which is clearly false. Thus, although it is perfectly legitimate to add, subtract, or multiply congruences as we often do with equations, it is not correct (in general) to divide congruences. In fact, it is not in general correct to divide both members of a congruence by the same integer even if this integer is a divisor of both members. For example, if we try to divide both members of the congruence

$$30 \equiv 24 \pmod{3}$$

by the common divisor 6, we obtain

$$\frac{30}{6} \equiv \frac{24}{6} \pmod{3}$$

that is,

$$5 \equiv 4 \pmod{3}$$

which is false. On the other hand, if we divide both members by the common divisor 2, we obtain

$$\frac{30}{2} \equiv \frac{24}{2} \pmod{3}$$

that is,

$$15 \equiv 12 \pmod{3}$$

which is true. This seemingly erratic behavior of congruences in regard to division is explained by the following theorem.

**THEOREM 3.11** If $d$ is the GCD of $k$ and $m$, then $ka \equiv kb \pmod{m}$ if and only if $a \equiv b \left( \bmod \frac{m}{d} \right)$.

In other words, whenever both members of a congruence are divided by a common factor $k$, an equivalent congruence results when the modulus $m$ is divided by the GCD of $k$ and $m$.

*Proof:* If $\quad ka \equiv kb \pmod{m}$
then $\quad ka - kb = qm \quad$ where $q$ is an integer

$$k(a - b) = qm$$

Hence, if $d$ is the GCD of $k$ and $m$,

$$\left(\frac{k}{d}\right)(a - b) = q\left(\frac{m}{d}\right)$$

So the integer $\frac{m}{d}$ is a divisor of $\left(\frac{k}{d}\right)(a - b)$. But the integer $\frac{m}{d}$ is relatively prime to the integer $\frac{k}{d}$ (because otherwise $d$ would not be the GCD; also see Exercise 13 in Exercises 3.4). Hence by Theorem 3.7 the integer $\frac{m}{d}$ must be a divisor of $a - b$, which means that

$$a \equiv b \left( \bmod \frac{m}{d} \right)$$

## 3.6 Basic Theorems About Congruences

To complete the proof, we must show that, conversely,

$$\text{if } a \equiv b \left(\bmod \frac{m}{d}\right), \quad \text{then} \quad ka \equiv kb \;(\bmod\; m)$$

This is readily proved by reversing the above steps. (See Exercise 12.)

In the above illustration, namely

$$30 \equiv 24 \;(\bmod\; 3) \quad \text{where} \quad m = 3$$

if we take $k = 2$, the GCD of $k$ and $m$ is 1, so we obtain

$$\frac{30}{2} \equiv \frac{24}{2} \left(\bmod \frac{3}{1}\right)$$

$$15 \equiv 12 \;(\bmod\; 3)$$

On the other hand, if we take $k = 6$, then the GCD of $k$ and $m$ is 3. Hence for this case we get

$$\frac{30}{6} \equiv \frac{24}{6} \left(\bmod \frac{3}{3}\right)$$

that is,
$$5 \equiv 4 \;(\bmod\; 1)$$

which is of course (trivially) true.

Although congruences do not behave exactly like equations, it is easy to keep track of the points of resemblance as well as the points of difference. Theorems 3.9 and 3.10 show that congruences may be manipulated essentially like equations as far as addition, subtraction, and multiplication are concerned. Theorem 3.11, however, indicates that care must be taken when we want to divide both sides of a congruence by the same number $k$. To be certain of an equivalent congruence, we must also divide the modulus $m$ by the GCD of $k$ and $m$. In the special cases where $m$ is a prime or where $k$ and $m$ are relatively prime, this GCD is simply 1, and so these cases also resemble the situation with equations.

As an example, let us solve the congruence

$$2x \equiv 6 \;(\bmod\; 7)$$

Because 2 is relatively prime to 7, we may divide both sides by 2 and obtain immediately

$$x \equiv 3 \;(\bmod\; 7)$$

This result signifies that each solution is congruent to 3 (modulo 7). Conversely, if $x \equiv 3 \;(\bmod\; 7)$, it follows from Theorem 3.10 (corollary) that

$$2x \equiv 6 \;(\bmod\; 7)$$

Hence every integer congruent to 3 (mod 7) is actually a solution. These integers can all be obtained by adding to 3 any integer multiple of 7, that is,

$$x = 3 + 7q \quad \text{where} \quad q = 0, \pm 1, \pm 2, \pm 3, \ldots$$

The solution set therefore contains

$$3 + 7(0) = 3$$
$$3 + 7(1) = 10$$
$$3 + 7(2) = 17$$
$$3 + 7(3) = 24$$
$$\cdots$$
$$3 + 7(-1) = -4$$
$$3 + 7(-2) = -11$$
$$3 + 7(-3) = -18$$

This infinite solution set is most conveniently expressed by writing simply

$$x \equiv 3 \pmod{7}$$

As a second illustration, let us solve

$$3x + 5 \equiv 7 \pmod{8}$$

By Theorem 3.10b we may subtract 5 from each side and obtain

$$3x \equiv 2 \pmod{8}$$

It is very tempting to write

$$x \equiv \frac{2}{3} \pmod{8}$$

but at the moment this has no meaning. In Exercise 6 we show that we can justify and give meaning to such expressions, but here we use a different approach. Let us multiply both members of $3x \equiv 2 \pmod{8}$ by some integer $k$ (to be determined later)

$$(3k)x \equiv 2k \pmod{8}$$

We now try to find $k$ so that

$$3k \equiv 1 \pmod{8}$$

When

$$k = 1: \quad 3k = 3 \equiv 3 \pmod{8}$$

When

$$k = 2: \quad 3k = 6 \equiv 6 \pmod{8}$$

When
$$k = 3: \quad 3k = 9 \equiv 1 \pmod{8}$$

The value we sought is $k = 3$. Using $k = 3$, we go back to the congruence
$$3x \equiv 2 \pmod{8}$$
and multiply both sides by 3. We get
$$9x \equiv 6 \pmod{8}$$
But
$$1 \equiv 9 \pmod{8} \qquad \text{(Why?)}$$
Therefore
$$1x \equiv 9x \pmod{8} \qquad \text{(Why?)}$$
and since
$$9x \equiv 6 \pmod{8}$$
it follows (by Theorem 3.9) that
$$x \equiv 6 \pmod{8}$$

This is the solution we sought.

The following exercises provide further practice in dealing with congruences.

## EXERCISES 3.6

1. Label each sentence true or false, whichever is appropriate. Whenever there is even one counterexample, label the sentence false and give the counterexample.
   (a) Since $9 \equiv 5 \pmod{2}$, $9a \equiv 5a \pmod{2}$ for all integer values $a$.
   (b) If $9a \equiv 5a \pmod{3}$, then $9 \equiv 5 \pmod{3}$.
   (c) Whenever $m$ is such that $2^3 \equiv 1 \pmod{m}$, then
   $$2 \equiv 1 \pmod{m}.$$
   (d) Whenever $a \equiv 1 \pmod{m}$, then $a^3 \equiv 1 \pmod{m}$.
   (e) Whenever $a \equiv -1 \pmod{m}$, then $a^3 \equiv 1 \pmod{m}$.
   (f) Whenever $a \equiv -1 \pmod{m}$, then $a^3 \equiv -1 \pmod{m}$.
   (g) Whenever $a^2 \equiv 1 \pmod{m}$, then $a \equiv 1 \pmod{m}$.
   ★(h) Whenever $a^2 \equiv 1 \pmod{m}$, then either $a \equiv 1 \pmod{m}$ or $a \equiv -1 \pmod{m}$.
   (i) Whenever $a^3 \equiv 1 \pmod{m}$, then $a \equiv 1 \pmod{m}$.

(j) Whenever $a \equiv b \pmod{m}$, then $a^3 \equiv b^3 \pmod{m}$.
(k) Whenever $a^2 \equiv b^2 \pmod{m}$, then $a \equiv b \pmod{m}$.
(l) $2^{100} \equiv 1 \pmod{7}$.
(m) $3^{100} \equiv 1 \pmod{7}$.

2. (a) Prove that if $a \equiv b \pmod{m}$, then
   (1) $a^2 \equiv b^2 \pmod{m}$
   (2) $a^3 \equiv b^3 \pmod{m}$
   (3) $a^4 \equiv b^4 \pmod{m}$
   (b) What generalization is suggested by part a? Try to prove this generalization.

3. Let $k$ be any integer other than 0. Prove the following:
   (a) If $a \equiv b \pmod{m}$, then $ka \equiv kb \pmod{km}$.
   (b) If $a \equiv b \pmod{km}$, then $a \equiv b \pmod{m}$.
   (c) If $ka \equiv kb \pmod{km}$ then $a \equiv b \pmod{m}$.

4. (a) For each of the following find the smallest whole number $n > 0$ that will make it a true statement.
   (1) $2^n \equiv 1 \pmod{5}$         (2) $2^n \equiv 1 \pmod{9}$
   (3) $2^n \equiv 1 \pmod{11}$        (4) $3^n \equiv 1 \pmod{4}$
   (5) $3^n \equiv 1 \pmod{5}$         (6) $3^n \equiv 1 \pmod{10}$
   (7) $4^n \equiv 1 \pmod{5}$         (8) $4^n \equiv 1 \pmod{9}$
   (9) $6^n \equiv 1 \pmod{5}$         (10) $6^n \equiv 1 \pmod{25}$

   ★(b) Prove that if the GCD of $a$ and $b$ is 1, then there is some whole number $n > 0$ such that $a^n \equiv 1 \pmod{b}$.

5. Solve each of the following:
   (a) $2x \equiv 1 \pmod{7}$          (b) $2x \equiv -1 \pmod{7}$
   (c) $2x \equiv 9 \pmod{7}$          (d) $3x \equiv 1 \pmod{7}$
   (e) $3x \equiv -1 \pmod{7}$         (f) $3x \equiv 2 \pmod{7}$
   (g) $3x \equiv -2 \pmod{7}$         (h) $3x \equiv 6 \pmod{7}$
   (i) $4x \equiv -1 \pmod{7}$         (j) $4x \equiv 2 \pmod{7}$
   (k) $2x \equiv 2 \pmod{7}$          (l) $2x \equiv 4 \pmod{7}$
   (m) $5x \equiv 1 \pmod{7}$          (n) $5x \equiv -1 \pmod{7}$
   (o) $5x \equiv 2 \pmod{7}$          (p) $5x \equiv -2 \pmod{7}$
   (q) $6x \equiv 3 \pmod{9}$

6. Interpret "$\frac{a}{b}$" as "$a \times \frac{1}{b}$," that is, as "$a$ times the multiplicative inverse of $b$." With this interpretation, determine the smallest whole number that satisfies each of the following conditions:

   (a) $x \equiv \frac{1}{3} \pmod{7}$          (b) $x \equiv \frac{2}{3} \pmod{7}$
   (c) $x \equiv \frac{1}{5} \pmod{7}$          (d) $x \equiv \frac{4}{5} \pmod{7}$

(e) $x \equiv \dfrac{1}{6} \pmod{7}$  (f) $x \equiv \dfrac{5}{6} \pmod{7}$

(g) $x \equiv \left(\dfrac{1}{3} \times \dfrac{1}{6}\right) \pmod{7}$  (h) $x \equiv \left(\dfrac{2}{3} \times \dfrac{5}{6}\right) \pmod{7}$

(i) $x \equiv \left(\dfrac{2}{3} + \dfrac{5}{6}\right) \pmod{7}$  (j) $x \equiv \left(\dfrac{5}{6} - \dfrac{2}{3}\right) \pmod{7}$

(k) $x \equiv \left(\dfrac{5}{6} \div \dfrac{2}{3}\right) \pmod{7}$

**7.** Using the interpretation in Exercise 6, prove the following:

(a) $\left(\dfrac{2}{3} \times \dfrac{4}{7}\right) \equiv \left(\dfrac{8}{21}\right) \pmod{11}$

(b) $\left(\dfrac{2}{3} \div \dfrac{5}{7}\right) \equiv \left(\dfrac{14}{15}\right) \pmod{11}$

**8.** Solve each of the following systems of simultaneous congruences:

(a) $\begin{cases} x + y \equiv 1 \pmod{7} \\ x - y \equiv 2 \pmod{7} \end{cases}$  (b) $\begin{cases} x + 2y \equiv 1 \pmod{7} \\ x - y \equiv 2 \pmod{7} \end{cases}$

(c) $\begin{cases} 2x + y \equiv 1 \pmod{7} \\ x - 2y \equiv 2 \pmod{7} \end{cases}$  (d) $\begin{cases} 2x + 3y \equiv 1 \pmod{7} \\ 3x - 2y \equiv 2 \pmod{7} \end{cases}$

**9.** Find an example to show that if $r \equiv s \pmod{m}$, then

$$a^r \equiv a^s \pmod{m}$$

may be false.

**10.** Try to find three pairs of integers that satisfy each of the following congruences:

(a) $2x + 3y \equiv 1 \pmod{7}$
(b) $2x - 3y \equiv 1 \pmod{7}$
(c) $2x + 3y \equiv -1 \pmod{7}$

**11.** Try to obtain a general solution for $2x + 3y \equiv 1 \pmod{7}$.

**12.** Complete the proof of Theorem 3.11 by showing that if $d$ is the GCD of $k$ and $m$ and if $a \equiv b \left(\bmod \dfrac{m}{d}\right)$, then $ka \equiv kb \pmod{m}$.

# REVIEW EXERCISES

1. Show that the set of rational numbers from 1 to 3 inclusive is not well-ordered (with respect to the "less than" order relation).
2. Find the GCD and LCM for each of the following pairs of numbers:
   (a) (243, 732)    (b) (423, 732)
   (c) (234, 372)    (d) (227, 367)

3. Express each GCD in Exercise 2 in the form $ax + by$, where $a$ and $b$ are the original numbers.
4. Solve the following systems of congruences:
   (a) $\begin{cases} 2x + y \equiv 5 \pmod{7} \\ 3x - y \equiv 1 \pmod{7} \end{cases}$
   (b) $\begin{cases} 2x + 3y \equiv 4 \pmod{7} \\ 4x + 5y \equiv -1 \pmod{7} \end{cases}$
   (c) $\begin{cases} 2x - 3y \equiv 1 \pmod{7} \\ 2x + 4y \equiv 2 \pmod{7} \end{cases}$
   (d) $\begin{cases} 3x - 6y \equiv 2 \pmod{7} \\ -2x + 4y \equiv 3 \pmod{7} \end{cases}$
   (e) $\begin{cases} 3x - 6y \equiv 2 \pmod{7} \\ -2x + 4y \equiv 1 \pmod{7} \end{cases}$

5. (a) Determine the smallest whole number for each of the following conditions. (See Exercise 6 in Exercises 3.6.)
$$x \equiv \frac{2}{5} \pmod{7} \qquad y \equiv \frac{3}{5} \pmod{7} \qquad z \equiv \frac{6}{25} \pmod{7}$$
   (b) Show that $xy \equiv z \pmod 7$ and $x + y \equiv 1 \pmod 7$
6. Try to formulate a generalization suggested by Exercise 5 and then prove your generalization.
   *Hint:* Let $x \equiv \dfrac{a}{b} \pmod m$, $y \equiv \dfrac{c}{d} \pmod m$.

★7. Let $N = p^2 q^5$, where $p$ and $q$ are distinct positive primes.
   (a) Find all the positive divisors of $N$.
   (b) Observe that the number of positive divisors of $N$ is
$$(1 + 2)(1 + 5) \text{ or } 18.$$
   What is a generalization suggested here?
   (*Hint:* Show that every partial product for the expansion of $(1 + p + p^2)(1 + q + q^2 + q^3 + q^4 + q^5)$ is a divisor of $N = p^2 q^5$.)
   (c) Try to formulate a more inclusive generalization than than given in (b).

★8. Let $r, s, t, u$ be integers, not necessarily distinct. Prove that at least one of the following statements is true:
   (a) One of the integers $r, s, t, u$ is divisible by 4.
   (b) Two of these integers have a sum that is divisible by 4.
   (c) Three of these integers have a sum that is divisible by 4.
   (d) $r + s + t + u$ is divisible by 4.
   *Hint:* Consider the numbers
$$r, r + s, r + s + t, r + s + t + u \pmod 4.$$

*J. J. Sylvester (1814-1897)*

# 4 | Elementary Rings

> The mathematician's patterns, like the painter's or poet's, must be beautiful... There is no permanent place in the world for ugly mathematics.
>
> G. H. HARDY

## 4.1 EXAMPLES AND DEFINITION OF A RING

In the study of groups, we fix our attention on a single binary operation $\circ$ in a given set $S$. To qualify as a group, the system $\{S, \circ\}$ must have the four group properties: the closure property, the associative property, the identity property, and the inverse property. (See Chapter 2, page 58.)

Many groups, but not all, possess a fifth property, commutativity. When a group possesses the commutative property, we call the group a commutative group or an **abelian** group after the brilliant Norwegian mathematician Neils Henrik Abel (1802–1829). Like his brilliant contemporary, Galois, young Abel made profound contributions to algebra.

The system $\{Z, +\}$, consisting of the integers together with the binary operation addition, is a classic example of an abelian group with infinitely many members. Other familiar examples of infinite abelian groups are $\{Q, +\}$, $\{R, +\}$, and $\{C, +\}$, where $Q$, $R$, and $C$ are, respectively, the set of rational numbers, the set of real numbers, and the set of complex numbers. Since the set of integers $Z$ may be regarded as a subset of the rational numbers

$Q$, $Q$ as a subset of $R$, and $R$ as a subset of $C$, we have a hierarchy in which each group mentioned is a *subgroup* of the following one. The system $\{Z, +\}$ is not only a subgroup of the other groups listed here but $\{Z, +\}$ has also subgroups of its own. For example, the system $\{E, +\}$, where $E$ is the set of even integers, is precisely such a subgroup. It is an infinite abelian subgroup of $\{Z, +\}$. In Exercise 1, Exercises 4.1, you will discover other infinite abelian subgroups of $\{Z, +\}$.

The usefulness of number systems such as $\{E, +\}$, $\{Z, +\}$, and $\{Q, +\}$ is enhanced greatly by adjoining to each of these systems the multiplication operation. We then obtain the more versatile mathematical systems

$$\{E, +, \times\}, \quad \{Z, +, \times\}, \quad \{Q, +, \times\}, \quad \ldots, \text{etc.}$$

Each of these systems now consists of a set of numbers together with *two* binary operations on the given set of numbers. In all of these systems, the multiplication operation exhibits some of the group properties. Specifically, the multiplication operation exhibits closure and associativity, but it does not possess the identity property in the system $\{E, +, \times\}$ because 1 is not an even number and no even number can serve as an identity for multiplication. Moreover, multiplication does not possess the inverse property in any of the systems $\{E, +, \times\}$, $\{Z, +, \times\}$, and $\{Q, +, \times\}$. In fact, the number 0 does not have a multiplicative inverse in any of these systems and the even integers have no multiplicative inverses in the first two systems. In Section 4.2 there is another example of a system in which multiplication is not commutative.

In view of these facts we shall not demand in this section that a multiplication operation possess all the group properties. We shall require only that it have both the closure property and the associative property but not necessarily the identity nor the inverse properties. Neither shall we require that multiplication be commutative.

However, in order to establish some connection between addition and multiplication in the various systems under study, we shall stipulate that multiplication *distributes* over addition in both directions, that is, that

$$x(y + z) = xy + xz \quad \text{and} \quad (y + z)x = yx + zx$$

both hold for all $x, y, z$ under consideration. (We shall frequently use the traditional convention of indicating multiplication either by juxtaposition or by the elevated dot. For example, $x(y + z)$ and $x \cdot (y + z)$ both mean $x \times (y + z)$.)

We can now say that each of the more versatile mathematical systems mentioned consists of a set $S$ together with *two* binary operations in $S$. These operations, addition (+) and multiplication (×), are such that

    (1) $\{S, +\}$ is a commutative group.
    (2) × is a binary operation on $S$.
    (3) × is an associative operation on $S$.
    (4) × distributes over + (in both directions).

All such mathematical systems are known as **rings**.

Although the terms "addition" and "multiplication" have been borrowed from ordinary arithmetic, it is essential to understand that the elements of a ring need not be the usual numbers of arithmetic. Hence in a ring, addition and multiplication need not be the usual operations of arithmetic. For convenience, however, we shall continue to use the symbols "+" and " × " with the corresponding terms "addition" and "multiplication" to designate the two ring operations regardless of whether they do or do not denote ordinary addition and multiplication. With this broader interpretation of + and × clearly understood, we now formulate a precise definition of a ring as follows.

**Definition of a Ring.** Let $R = \{S, +, \times\}$ be a mathematical system consisting of a set $S$ together with two binary operations, + and ×, in $S$. The system $R$ is called a **ring** if and only if $R$ has the following properties (axioms for a ring):

1. $\{S, +\}$ is an abelian group.
2. × is a binary operation on $S$.
3. × is an associative operation on $S$.
4. × distributes over addition in both directions, that is, for all $x, y, z$ in $S$,

$$x \cdot (y + z) = (x \cdot y) + (x \cdot z)$$

and

$$(y + z) \cdot x = (y \cdot x) + (z \cdot x)$$

We have already mentioned three systems that are rings: $\{E, +, \times\}$, $\{Z, +, \times\}$, and $\{Q, +, \times\}$. Let us verify in detail that $\{E, +, \times\}$ is indeed a ring.

1. $\{E, +\}$ is an abelian group because
    (a) (Closure) + is a binary operation on $E$. This follows from the fact that a sum of even numbers is even.

(b) (Associativity) + is an associative operation on $E$ because addition of integers (and hence certainly addition of even integers) is associative.
(c) (Identity property) The (even) number 0 is an identity element for + in $E$.
(d) (Inverse property) $-x$ is the inverse of each number $x$ in $E$.
(e) (Commutativity) Addition is a commutative operation on $E$ because addition of integers (and hence of even integers) is commutative.

2. $\times$ is a binary operation on $E$ because a product of even integers is even.
3. $\times$ is an associative operation on $E$ because multiplication of integers is associative.
4. $\times$ distributes over addition in both directions because multiplication of integers has this property.

Notice that to prove that $\{E, +, \times\}$ is a ring we had to accept certain familiar properties of the system of integers. In a more formal approach it would be necessary to derive these properties from other basic assumptions. Such a development, however, is beyond our present scope. Here we shall use freely the usual properties of integers, rational numbers, real numbers, and, to some extent, the complex numbers as well.

Because there are infinitely many elements in each of the sets $E$, $Z$, and $Q$, the systems $\{E, +, \times\}$, $\{Z, +, \times\}$, and $\{Q, +, \times\}$ are infinite rings. It is also possible to have *finite rings*. The simplest finite ring is the system

$$\{\{0\}, +, \times\}$$

In this system the number 0 is the only element, but under the operations of ordinary addition and ordinary multiplication all the ring properties are satisfied.

A less trivial example of a finite ring is the modulo 4 system

$$\{W_4, + \text{ (mod 4)}, \times \text{ (mod 4)}\}$$

Recall that this system contains exactly four elements 0, 1, 2, 3 (the members of $W_4$). Recall also that the addition and multiplication tables for this system are

| +(mod 4) | 0 | 1 | 2 | 3 |
|---|---|---|---|---|
| 0 | 0 | 1 | 2 | 3 |
| 1 | 1 | 2 | 3 | 0 |
| 2 | 2 | 3 | 0 | 1 |
| 3 | 3 | 0 | 1 | 2 |

| ×(mod 4) | 0 | 1 | 2 | 3 |
|---|---|---|---|---|
| 0 | 0 | 0 | 0 | 0 |
| 1 | 0 | 1 | 2 | 3 |
| 2 | 0 | 2 | 0 | 2 |
| 3 | 0 | 3 | 2 | 1 |

Let us verify that this system is a ring.

1. The addition table clearly indicates that $\{W_4, +\,(\text{mod } 4)\}$ is an abelian group.
2. The multiplication table clearly indicates that $\times\,(\text{mod } 4)$ is a binary operation on $W_4$.
3. To check associativity of multiplication for every triple is, of course, lengthy. We therefore consider only two fairly representative cases.

$$(1 \times 2) \times 3 = 2 \times 3 \qquad\qquad 1 \times (2 \times 3) = 1 \times 2$$
$$= 2 \qquad\qquad\qquad\qquad\qquad = 2$$

Hence $\qquad (1 \times 2) \times 3 = 1 \times (2 \times 3)$.

$$(3 \times 2) \times 2 = 2 \times 2 \qquad\qquad 3 \times (2 \times 2) = 3 \times 0$$
$$= 0 \qquad\qquad\qquad\qquad\qquad = 0$$

Hence $\qquad (3 \times 2) \times 2 = 3 \times (2 \times 2)$

4. To check distributivity we also present only two cases.

$$3 \times (2+3) = 3 \times (1) \qquad\qquad (3 \times 2) + (3 \times 3) = 2+1$$
$$= 3 \qquad\qquad\qquad\qquad\qquad\qquad = 3$$

Hence $\qquad 3 \times (2+3) = (3 \times 2) + (3 \times 3)$

$$2 \times (3+1) = 2 \times (0) \qquad\qquad (2 \times 3) + (2 \times 1) = 2+2$$
$$= 0 \qquad\qquad\qquad\qquad\qquad\qquad = 0$$

Hence $\qquad 2 \times (3+1) = (2 \times 3) + (2 \times 1)$

It is possible to give general proofs of the associativity and distributivity requirements for systems such as this one.*

## EXERCISES 4.1

1. (a) If $T = \{\text{All integers that are divisible by 3}\}$, show that $\{T, +\}$ is an abelian group and is therefore an abelian subgroup of $\{Z, +\}$.
   (b) If $F = \{\text{All integers that are divisible by 4}\}$, show that $\{F, +\}$ is also an abelian subgroup of $\{Z, +\}$.
   (c) Generalize (a) and (b) and show that $\{Z, +\}$ has infinitely many abelian subgroups.
2. Show that each of the following mathematical systems is a ring by verifying that all the requirements for a ring are satisfied. Use

---
* See N. McCoy, *Rings and Ideals;* Carus Mathematical Monograph No. 8 published by the Mathematical Association of America, 1948, Second Printing 1956.

the following notation:

$3Z = \{0, \pm 3, \pm 6, \ldots\} = \{$All the integer multiples of 3$\}$.
$5Z = \{0, \pm 5, \pm 10, \ldots\} = \{$All the integer multiples of 5$\}$.
$nZ = \{0, \pm n, \pm 2n, \ldots\} = \{$All the integer multiples of $n\}$.

In general, if $S$ is any set of numbers and $r$ is any number, we define
$$rS = \{rx \mid x \in S\}$$

(a) $\{3Z, +, \times\}$
(b) $\{5Z, +, \times\}$
(c) $\{10Z, +, \times\}$
(d) $\{(-2)Z, +, \times\}$
(e) $\{W_3, +(\text{mod } 3), \times(\text{mod } 3)\}$
(f) $\{W_5, +(\text{mod } 5), \times(\text{mod } 5)\}$
(g) $\{W_{10}, +(\text{mod } 10), \times(\text{mod } 10)\}$
(h) $\{W_n, +(\text{mod } n), \times(\text{mod } n)\}$
(i) $\{Q, +, \times\}$

Note: $Z(\sqrt{2}) = \{a + b\sqrt{2} \mid a, b \in Z\}$. In general, if $S$ is any set of numbers and $r$ is any number, we define
$$S(r) = \{a + br \mid a, b \in S\}$$

(j) $\{Z(\sqrt{2}), +, \times\}$
(k) $\{Z(\sqrt{3}), +, \times\}$
(l) $\{Q(\sqrt{2}), +, \times\}$
(m) $\{Q(\sqrt{3}), +, \times\}$
★(n) $\{(3Z)(\sqrt{2}), +, \times\}$

3. State all the ring requirements that are not met by each of the following. (*Note:* See Exercise 2 for the meaning of expressions such as $rS$ and $S(r)$ where $r$ is any number. $S^+$ is the set of positive elements of $S$. $S^-$ is the set of negative elements of $S$. $S^*$ is the set of nonzero elements of $S$.)

(a) $\{Z^+, +, \times\}$
(b) $\{Q^+, +, \times\}$
(c) $\{Z^*, +, \times\}$
(d) $\{Q^*, +, \times\}$
(e) $\{W_4, +, \times\}$
(f) $\{Z^-, +, \times\}$
(g) $\{Z(\tfrac{1}{2}), +, \times\}$
(h) $\{Z(\sqrt[3]{2}), +, \times\}$
(i) $\{Q(\sqrt[3]{2}), +, \times\}$
(j) $\{\sqrt{2}Z, +, \times\}$
(k) $\{\sqrt{3}Z, +, \times\}$
(l) $\{\sqrt{2}Q, +, \times\}$
(m) $\{Q(\sqrt[3]{5}), +, \times\}$
★(n) $\{\pi Q, +, \times\}$

4. Suppose that $\{S, +, \times\}$ is a ring.
   (a) What requirements must $\{S, +\}$ meet?
   (b) What requirements must $\{S, \times\}$ meet?
   (c) Which requirement involves both operations?

5. Let $S = \{(0, 0), (0, 1), (1, 0), (1, 1)\}$.

   Prove that $\{S, +, \cdot\}$ is a ring where addition is defined as follows:
   $$(a, b) + (c, d) = (a + c \,(\text{mod } 2), \, b + d \,(\text{mod } 2))$$
   Multiplication is defined as follows:
   $$(a, b) \cdot (c, d) = (ac, ad)$$
   Thus $\quad (1, 0) + (1, 1) = (0, 1)$
   and $\quad (1, 0) \cdot (1, 1) = (1, 1)$

★6. Let $T$ be the set $\{0, 1\}$ and let $S$ be the set of all subsets of $T$, $S = \{\varnothing, \{0\}, \{1\}, \{0, 1\}\}$. Let addition and multiplication be defined on $S$ as follows. For any elements $A$ and $B$ of $S$

   $A + B = \{$All the elements in either $A$ or $B$ but not in both$\}$

   This is often written as $A + B = (A \cup B) - (A \cap B)$.

   $A \times B = A \cap B$

   (a) Show that $\{S, +\}$ is an abelian group.
   (b) Is $\times$ a binary operation on $S$?
   (c) Is $\times$ an associative operation on $S$?
   (d) Check whether $\times$ distributes over $+$ in both directions.
   (e) Is $\{S, +, \times\}$ a ring?

7. The Gaussian integers are defined as the set $G$ of all complex numbers $a + bi$, where $a$ and $b$ are ordinary integers and where $i^2 = -1$. Addition $(+)$ and multiplication $(\cdot)$ of elements of $G$ are defined as follows:
   $$(a + bi) + (c + di) = (a + c) + (b + d)i$$
   $$(a + bi) \cdot (c + di) = (ac - bd) + (ad + bc)i$$
   Show that the system $\{G, +, \cdot\}$ is a commutative ring with a unit element.

## 4.2 RINGS OF TWO BY TWO MATRICES

The notion of a matrix was first introduced in the middle of the nineteenth century by the brilliant British mathematician Arthur Cayley (1821–1895). As often happens with abstract mathematical tools, matrices turned out more than half a century later to be precisely the technique needed by modern physicists in quantum mechanics. Today matrices are used in many diverse fields of scientific endeavor, ranging from engineering and atomic physics to sociology, economics, and genetics.

In this section we restrict ourselves to a very simple type, the *two by two matrix*. Much of our discussion, however, applies with equal validity to matrices of higher order. The following are examples of two by two matrices.

$$\begin{bmatrix} 1 & -2 \\ 3 & 0 \end{bmatrix} \quad \begin{bmatrix} 0 & 0 \\ 0 & 0 \end{bmatrix} \quad \begin{bmatrix} \tfrac{1}{2} & \sqrt{2} \\ 3.14 & -\tfrac{3}{4} \end{bmatrix}$$

Roughly speaking, a matrix is a rectangular array of elements, usually numbers. In a two by two matrix the numbers are arranged in two rows and two columns. Thus, if $a, b, c, d$ are numbers, then

$$\begin{bmatrix} a & b \\ c & d \end{bmatrix}$$

is a two by two matrix whose elements are $a, b, c,$ and $d$. Here $a, b$ are in the first row and $c, d$ are in the second row. The numbers $a, c$ are in the first column and $b, d$ in the second column. *Unless otherwise specified, we use the term matrix to mean two by two matrix throughout this section.*

We consider here only matrices whose elements belong to some specified ring such as the ring of integers, the ring of rational numbers, and one of the modular rings $\{W_n, +\;(\bmod\; n), \times\;(\bmod\; n)\}$. Matrices may be regarded as a new kind of number because they can be added and multiplied. Before discussing these operations with matrices, we pause to remark that an equality of matrices such as

$$\begin{bmatrix} a & b \\ c & d \end{bmatrix} = \begin{bmatrix} e & f \\ g & h \end{bmatrix}$$

simply means that

$$\begin{aligned} a &= e & b &= f \\ c &= g & d &= h \end{aligned}$$

> Matrices are equal if and only if corresponding entries are equal.

Suppose $A$ and $B$ are matrices, say

(1) $$A = \begin{bmatrix} a & b \\ c & d \end{bmatrix} \quad B = \begin{bmatrix} p & q \\ r & s \end{bmatrix}$$

where $a, b, c, d, p, q, r,$ and $s$ are all elements of some specified ring $\{R, +, \times\}$. We then define the sum of the matrices $A$ and $B$ as

(2) $$A + B = \begin{bmatrix} a & b \\ c & d \end{bmatrix} + \begin{bmatrix} p & q \\ r & s \end{bmatrix} = \begin{bmatrix} a+p & b+q \\ c+r & d+s \end{bmatrix}$$

## 4.2 Rings of Two by Two Matrices

*Thus to add matrices we simply add elements that occupy corresponding positions.*

Multiplication of matrices is defined in a somewhat more elaborate fashion. Merely multiplying corresponding elements does not produce a particularly useful definition. It turns out to be more interesting and far more useful to adopt a row by column multiplication rule. In Chapter 10 we gain a deeper insight into the reasons for adopting such a multiplication rule in a far more general setting than considered here. For the present, if $A$ and $B$ are the matrices defined in (1), then we define the product of the matrices $A$ and $B$ as

(3) $$A \cdot B = \begin{bmatrix} a & b \\ c & d \end{bmatrix} \cdot \begin{bmatrix} p & q \\ r & s \end{bmatrix} = \begin{bmatrix} ap+br & aq+bs \\ cp+dr & cq+ds \end{bmatrix}$$

Thus to find a particular element of the product matrix $A \cdot B$ we multiply the elements of a row of matrix $A$ by the corresponding elements of a column of matrix $B$ and then add the products obtained. Examples will help make these operations clear.

**Example 1.** Let the elements of $A$ and $B$ be chosen from the ring $\{Z, +, \times\}$ and suppose that

$$A = \begin{bmatrix} 2 & 1 \\ -3 & 4 \end{bmatrix} \quad \text{and} \quad B = \begin{bmatrix} 5 & 0 \\ 3 & -1 \end{bmatrix}$$

Then for the sum $A + B$ we have

$$A + B = \begin{bmatrix} 2+5 & 1+0 \\ -3+3 & 4+(-1) \end{bmatrix} = \begin{bmatrix} 7 & 1 \\ 0 & 3 \end{bmatrix}$$

and for the product $AB$ we obtain

$$AB = \begin{bmatrix} 2 & 1 \\ -3 & 4 \end{bmatrix} \cdot \begin{bmatrix} 5 & 0 \\ 3 & -1 \end{bmatrix} = \begin{bmatrix} 2(5)+1(3) & 2(0)+1(-1) \\ -3(5)+4(3) & -3(0)+4(-1) \end{bmatrix}$$

$$= \begin{bmatrix} 13 & -1 \\ -3 & -4 \end{bmatrix}$$

**Example 2.** Let the elements of $A$ and $B$ be chosen from the ring $\{W_6, +\pmod 6, \times \pmod 6\}$. We must remember to compute all sums and products modulo 6. As a specific illustration suppose that

$$A = \begin{bmatrix} 2 & 1 \\ 3 & 4 \end{bmatrix} \quad \text{and} \quad B = \begin{bmatrix} 1 & 0 \\ 3 & 5 \end{bmatrix}$$

Then using mod 6 addition and multiplication, we have

$$A + B = \begin{bmatrix} 2+1 & 1+0 \\ 3+3 & 4+5 \end{bmatrix} = \begin{bmatrix} 3 & 1 \\ 0 & 3 \end{bmatrix}$$

and

$$A \times B = \begin{bmatrix} 2(1)+1(3) & 2(0)+1(5) \\ 3(1)+4(3) & 3(0)+4(5) \end{bmatrix}$$

$$= \begin{bmatrix} 5 & 5 \\ 3 & 2 \end{bmatrix}$$

With addition and multiplication of matrices defined as in Equations (2) and (3), it is now possible to prove the following important theorem.

**THEOREM 4.1** Let $M$ be the set of all (two by two) matrices whose elements belong to a given ring and let $+$ and $\times$ be addition and multiplication of these matrices. Then the system $\{M, +, \times\}$ is also a ring.

*Proof:* 1. We must first show that $\{M, +\}$ is an abelian group. Let $A, B, C$ be any matrices in $M$, say

$$A = \begin{bmatrix} a & b \\ c & d \end{bmatrix} \quad B = \begin{bmatrix} p & q \\ r & s \end{bmatrix} \quad C = \begin{bmatrix} x & y \\ u & v \end{bmatrix}$$

where $a, b, c, d, p, q, r, s, x, y, u, v$ are all members of the specified ring.

(a) (Closure) $+$ is a binary operation on $M$ because the sum of any two matrices in $M$

$$A + B = \begin{bmatrix} a+p & b+q \\ c+r & d+s \end{bmatrix}$$

is again a matrix in $M$ because the elements $a+p$, $b+q$, etc., are again members of the given ring.

(b) (Associativity) $+$ is an associative operation on $M$ because for any matrices $A, B, C$ in $M$ (as defined

4.2 Rings of Two by Two Matrices     135

above) we have

$$A+(B+C) = \begin{bmatrix} a & b \\ c & d \end{bmatrix} + \begin{bmatrix} p+x & q+y \\ r+u & s+v \end{bmatrix}$$

$$= \begin{bmatrix} a+(p+x) & b+(q+y) \\ c+(r+u) & d+(s+v) \end{bmatrix}$$

$$= \begin{bmatrix} (a+p)+x & (b+q)+y \\ (c+r)+u & (d+s)+v \end{bmatrix}$$

$$= \begin{bmatrix} a+p & b+q \\ c+r & d+s \end{bmatrix} + \begin{bmatrix} x & y \\ u & v \end{bmatrix}$$

$$= (A+B)+C$$

(c) (Identity property) The identity element for $+$ in $M$ is the matrix

$$O = \begin{bmatrix} 0 & 0 \\ 0 & 0 \end{bmatrix}$$

To show that $O$ is indeed an identity element for $+$, observe that for any matrix $A$ in $M$

$$A+O = \begin{bmatrix} a & b \\ c & d \end{bmatrix} + \begin{bmatrix} 0 & 0 \\ 0 & 0 \end{bmatrix} = \begin{bmatrix} a+0 & b+0 \\ c+0 & d+0 \end{bmatrix} = A$$

and similarly $O+A = A$.

(d) (Inverse property) Each matrix $A$ in $M$ has an additive inverse matrix $-A$ in $M$, which is defined as follows

$$\text{If} \quad A = \begin{bmatrix} a & b \\ c & d \end{bmatrix}, \quad \text{then} \quad -A = \begin{bmatrix} -a & -b \\ -c & -d \end{bmatrix}$$

The matrix $-A$ is a matrix in $M$ because $-a, -b, -c, -d$ are elements of the given ring. Moreover,

$$A+(-A) = \begin{bmatrix} 0 & 0 \\ 0 & 0 \end{bmatrix} = O$$

and similarly $(-A)+A = O$.

(e) (Commutativity) Addition is a commutative operation on $M$ because if $A$ and $B$ are any matrices in $M$,

## 4 || Elementary Rings

then

$$A + B = \begin{bmatrix} a & b \\ c & d \end{bmatrix} + \begin{bmatrix} p & q \\ r & s \end{bmatrix}$$

$$= \begin{bmatrix} a+p & b+q \\ c+r & d+s \end{bmatrix}$$

$$= \begin{bmatrix} p+a & q+b \\ r+c & s+d \end{bmatrix}$$

$$= \begin{bmatrix} p & q \\ r & s \end{bmatrix} + \begin{bmatrix} a & b \\ c & d \end{bmatrix} = B + A$$

We have shown thus far that $\{M, +\}$ is an abelian group.

2. $\times$ is a binary operation on $M$ because a product of matrices in $M$ is again a matrix in $M$:

$$A \times B = \begin{bmatrix} a & b \\ c & d \end{bmatrix} \cdot \begin{bmatrix} p & q \\ r & s \end{bmatrix}$$

$$= \begin{bmatrix} ap+br & aq+bs \\ cp+dr & cq+ds \end{bmatrix}$$

Each of the elements $ap + br$, $aq + bs$, and so on, is an element of the given ring.

3. $\times$ is an associative operation on $M$, because if $A, B, C$ are any matrices in $M$, then

$(A \times B) \times C$

$$= \begin{bmatrix} ap+br & aq+bs \\ cp+dr & cq+ds \end{bmatrix} \cdot \begin{bmatrix} x & y \\ u & v \end{bmatrix}$$

$$= \begin{bmatrix} (ap+br)x+(aq+bs)u & (ap+br)y+(aq+bs)v \\ (cp+dr)x+(cq+ds)u & (cp+dr)y+(cq+ds)v \end{bmatrix}$$

Using the distributive and associative properties of the underlying ring to which the elements $a, b, p, r, x$, and so on, belong, this may be written as

$(A \times B) \times C$

$$= \begin{bmatrix} apx+brx+aqu+bsu & apy+bry+aqv+bsv \\ cpx+drx+cqu+dsu & cpy+dry+cqv+dsv \end{bmatrix}$$

## 4.2 Rings of Two by Two Matrices

By a similar use of properties of the underlying ring we obtain

$A \times (B \times C)$

$$= \begin{bmatrix} apx + brx + aqu + bsu & apy + bry + aqv + bsv \\ cpx + drx + cqu + dsu & cpy + dry + cqv + dsv \end{bmatrix}$$

Hence

$(A \times B) \times C = A \times (B \times C)$ for all $A, B, C$ in $M$.

4. Finally, to show that $\times$ distributes over addition, we observe that

$A \times (B + C)$

$$= \begin{bmatrix} a & b \\ c & d \end{bmatrix} \cdot \begin{bmatrix} p+x & q+y \\ r+u & s+v \end{bmatrix}$$

$$= \begin{bmatrix} a(p+x) + b(r+u) & a(q+y) + b(s+v) \\ c(p+x) + d(r+u) & c(q+y) + d(s+v) \end{bmatrix}$$

$$= \begin{bmatrix} (ap+br) + (ax+bu) & (aq+bs) + (ay+bv) \\ (cp+dr) + (cx+du) & (cq+ds) + (cy+dv) \end{bmatrix}$$

$$= \begin{bmatrix} ap+br & aq+bs \\ cp+dr & cq+ds \end{bmatrix} + \begin{bmatrix} ax+bu & ay+bv \\ cx+du & cy+dv \end{bmatrix}$$

$$= \begin{bmatrix} a & b \\ c & d \end{bmatrix} \cdot \begin{bmatrix} p & q \\ r & s \end{bmatrix} + \begin{bmatrix} a & b \\ c & d \end{bmatrix} \cdot \begin{bmatrix} x & y \\ u & v \end{bmatrix}$$

$\therefore A \times (B + C) = (A \times B) + (A \times C)$

This clearly holds for all $A, B, C$ in $M$. Similarly, we can prove that

$(B + C) \times A = (B \times A) + (C \times A)$

Our proof of Theorem 4.1 is now complete.

The significance of Theorem 4.1 is simply this: *No matter what ring we start with, we can always build a new one by forming the set of all (two by two) matrices whose elements are members of the given ring.* This process of constructing new rings from former ones is of considerable importance in mathematics.

It is worth noting that although addition of matrices is commutative, multiplication of matrices is not commutative. For example, consider the matrices

$$A = \begin{bmatrix} 1 & 1 \\ 2 & 0 \end{bmatrix} \quad \text{and} \quad B = \begin{bmatrix} 0 & 1 \\ 0 & 1 \end{bmatrix}$$

Then

$$A \times B = \begin{bmatrix} 1 & 1 \\ 2 & 0 \end{bmatrix} \cdot \begin{bmatrix} 0 & 1 \\ 1 & 1 \end{bmatrix} = \begin{bmatrix} 1 & 2 \\ 0 & 2 \end{bmatrix}$$

But

$$B \times A = \begin{bmatrix} 0 & 1 \\ 1 & 1 \end{bmatrix} \cdot \begin{bmatrix} 1 & 1 \\ 2 & 0 \end{bmatrix} = \begin{bmatrix} 2 & 0 \\ 3 & 1 \end{bmatrix}$$

Hence $A \times B \neq B \times A$ in this particular case.

Thus although in some rings, for example, $\{Z, +, \times\}$, $\{W_4, +, \times\}$, etc., the multiplication operation is commutative, there are also rings such as $\{M, +, \times\}$ in which the multiplication operation is not commutative. Rings for which the multiplication operation is commutative are called **commutative rings**. Rings for which the multiplication operation is not commutative are called **noncommutative rings.**

Although most finite rings are commutative rings, it is possible to build a finite noncommutative ring using matrices. The smallest such ring is described here.

### A Finite Noncommutative Ring

Consider the following matrices:

$$e = \begin{bmatrix} 0 & 0 \\ 0 & 0 \end{bmatrix} \quad a = \begin{bmatrix} 0 & 1 \\ 0 & 0 \end{bmatrix} \quad b = \begin{bmatrix} 1 & 0 \\ 0 & 0 \end{bmatrix} \quad c = \begin{bmatrix} 1 & 1 \\ 0 & 0 \end{bmatrix}$$

If we use addition modulo 2 as the + operation, we obtain the following addition table:

| + | e | a | b | c |
|---|---|---|---|---|
| e | e | a | b | c |
| a | a | e | c | b |
| b | b | c | e | a |
| c | c | b | a | e |

## 4.2 Rings of Two by Two Matrices

We see immediately that the system $\{\{e, a, b, c\}, +\}$ is an abelian group. (In fact, it is a "Klein Four Group.")

The multiplication table (actually it makes no difference whether we use ordinary multiplication or modulo 2 multiplication) becomes

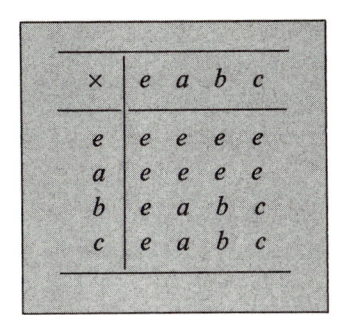

| × | e | a | b | c |
|---|---|---|---|---|
| e | e | e | e | e |
| a | e | e | e | e |
| b | e | a | b | c |
| c | e | a | b | c |

Clearly, × is a binary operation on $\{e, a, b, c\}$ (the closure property holds). Notice, however, that × is not commutative on $\{e, a, b, c\}$ because

$$b \times c = c \quad \text{and} \quad c \times b = b$$

and since $c \neq b$, it follows that

$$b \times c \neq c \times b$$

On the other hand, × is an associative operation on $\{e, a, b, c\}$. To prove this directly from the table would require examining 64 cases such as, for example,

$$(a \times b) \times c = e \times c = e$$
$$a \times (b \times c) = a \times c = e$$
$$\therefore (a \times b) \times c = a \times (b \times c)$$

Fortunately, we can avoid checking all these cases by a general argument. We have already proved (see pages 136–137) that multiplication of two by two matrices with elements in a given ring is associative. (Indeed, so is addition, but we have already observed that $\{\{e, a, b, c\}, +\}$ is an abelian group.) In the present case we are dealing with matrices whose elements belong to the ring

$$\{W_2, +(\text{mod } 2), \times(\text{mod } 2)\}$$

Hence every triple is an associative triple with respect to the multiplication operation. Similarly, the distributivity of × with respect to + follows from the corresponding property (also proved) for matrices.

This completes the proof that the system

$$\{\{e, a, b, c\}, +, \times\}$$

## 4 || Elementary Rings

is a ring. Since there are only four elements in the system and since multiplication is noncommutative in this system, we have an example of a finite, noncommutative ring.

## EXERCISES 4.2

1. State the value of each variable that makes each of the following true statements:

   (a) $\begin{bmatrix} x+3 & 2y-1 \\ 3z & w-4 \end{bmatrix} = \begin{bmatrix} -1 & 9 \\ z+4 & 2w+3 \end{bmatrix}$

   (b) $\begin{bmatrix} 2x & y+x \\ 3z & z+2w \end{bmatrix} = \begin{bmatrix} 6 & 5 \\ 12 & 0 \end{bmatrix}$

   (c) $\begin{bmatrix} 2x+y & y-1 \\ z-2u & z+u \end{bmatrix} = \begin{bmatrix} x-y & x+2 \\ -z-u & 9 \end{bmatrix}$

2. Let

   $A = \begin{bmatrix} 3 & 2 \\ -4 & 0 \end{bmatrix} \quad B = \begin{bmatrix} -5 & 4 \\ 3 & -2 \end{bmatrix} \quad C = \begin{bmatrix} -1 & -3 \\ 4 & 5 \end{bmatrix}$

   Compute:
   (a) $(A + B) + C$         (b) $A + (B + C)$
   (c) $AB$                  (d) $AC$
   (e) $A(B + C)$            (f) $BA$
   (g) $CA$                  (h) $(B + C)A$

3. Using the matrices $A$, $B$, $C$ of Exercise 2, show that
   (a) $AB \ne BA$           (b) $(AB)C = A(BC)$
   (c) $A(B + C) = AB + AC$  (d) $(B + C)A = BA + CA$
   (e) $BC \ne CB$           (f) $(BC)A = B(CA)$
   (g) $C(B + A) = CB + CA$  (h) $(B + A)C = BC + AC$

4. Let $a$, $b$, $c$, $d$, and $n$ be elements of set $S$ in the ring $\{S, +, \times\}$. We define the product of $n$ and matrix $\begin{bmatrix} a & b \\ c & d \end{bmatrix}$ as follows:

   $n\begin{bmatrix} a & b \\ c & d \end{bmatrix} = \begin{bmatrix} na & nb \\ nc & nd \end{bmatrix}$

   For example, if the matrix $A$ is defined as in Exercise 2, then

   $3A = 3\begin{bmatrix} 3 & 2 \\ -4 & 0 \end{bmatrix} = \begin{bmatrix} 9 & 6 \\ -12 & 0 \end{bmatrix}$

   Using the matrices $A$, $B$, and $C$ of Exercise 2 compute:
   (a) $2A + 3B$             (b) $2A - 3B$
   (c) $(2A)(3B)$            (d) $7(A + B)$
   (e) $7(2C - B)$           (f) $3(2C - 3B)$
   (g) $-3A + 2B$            (h) $2A(3B - 4C)$

## 4.2 Rings of Two by Two Matrices

5. By referring to Exercises 2 and 4 show that
   (a) $(2A)(3B) = 6(AB)$
   (b) $2A(3B + 4C) = 6(AB) + 8(AC)$
   (c) $2A + 3A = 5A$
   (d) $(2A + B) + 3A = 5A + B$

6. Complete the proof (page 135) that $\begin{bmatrix} 0 & 0 \\ 0 & 0 \end{bmatrix}$ is the additive identity.

7. Complete the proof (page 135) that $-A$ is the additive inverse of $A$.

8. Complete the proof (page 137) that multiplication distributes over addition in both directions.

9. Let $\{M, +, \times\}$ be the ring of all two by two matrices whose elements belong to a given ring. Show that each of the following subsets of $M$ also forms a ring under matrix addition and multiplication.

   (a) The subset of all matrices of the form $\begin{bmatrix} x & 0 \\ 0 & 0 \end{bmatrix}$

   (b) The subset of all matrices of the form $\begin{bmatrix} x & 0 \\ 0 & x \end{bmatrix}$

   (*Note:* Matrices of this form are often called *scalar* matrices.)

   (c) The subset of all matrices of the form $\begin{bmatrix} x & x \\ 0 & 0 \end{bmatrix}$

   (d) The subset of all matrices of the form $\begin{bmatrix} x & 0 \\ 0 & y \end{bmatrix}$

   (e) The subset of all matrices of the form $\begin{bmatrix} x & y \\ 0 & 0 \end{bmatrix}$

   (f) The subset of all matrices of the form $\begin{bmatrix} 0 & 0 \\ 0 & x \end{bmatrix}$

10. Consider the ring of all two by two matrices whose elements are rational numbers.

    (a) Prove $\begin{bmatrix} 1 & 0 \\ 0 & 1 \end{bmatrix}$ is a multiplicative identity.

    (b) Show that a multiplicative inverse of
    $$\begin{bmatrix} 5 & 2 \\ 2 & 1 \end{bmatrix} \text{ is } \begin{bmatrix} 1 & -2 \\ -2 & 5 \end{bmatrix}$$

(c) Show that a multiplicative inverse of

$$\begin{bmatrix} 3 & 5 \\ 1 & 2 \end{bmatrix} \text{ is } \begin{bmatrix} 2 & -5 \\ -1 & 3 \end{bmatrix}$$

(d) Guess a multiplicative inverse of

(1) $\begin{bmatrix} 7 & 2 \\ 3 & 1 \end{bmatrix}$  (2) $\begin{bmatrix} 3 & 4 \\ 5 & 7 \end{bmatrix}$  (3) $\begin{bmatrix} 4 & 2 \\ 3 & 2 \end{bmatrix}$

★(e) Let $I = \begin{bmatrix} 1 & 0 \\ 0 & 1 \end{bmatrix}$ be the identity matrix and let $X^{-1}$ denote a multiplicative inverse of matrix $X$ (if $X$ has an inverse), that is, let

$$XX^{-1} = I = X^{-1}X$$

Show that if matrices $A$ and $B$ have, respectively, the inverses $A^{-1}$ and $B^{-1}$, then the product $AB$ also has an inverse, namely $B^{-1}A^{-1}$.

(f) If $O = \begin{bmatrix} 0 & 0 \\ 0 & 0 \end{bmatrix}$, find matrices $A$ and $B$ such that $A \neq O$, $B \neq O$, but $AB = O$.

(g) Show that there is an infinite number of matrix solutions to the matrix equation $X \cdot X = I$ where $I = \begin{bmatrix} 1 & 0 \\ 0 & 1 \end{bmatrix}$ and $X$ is a matrix.

$\left(\text{Hint:} \text{ One such matrix is } \begin{bmatrix} 0 & 3 \\ \frac{1}{3} & 0 \end{bmatrix}.\right)$

★(h) An element of a ring $x$ is called an *idempotent element* if $x \cdot x = x$. Find some matrices that are indempotent.

★(i) An element of a ring $x$ is a *nilpotent element* if for some positive integer $n$, $x^n = 0$. Find some matrices that are nilpotent.

## 4.3  BASIC THEOREMS FOR RINGS

We have seen in Chapter 2 that although there is an endless variety of groups, they nevertheless all share certain common properties, namely, those properties which are described by the theorems that we derived in Section 2.4 from the four group axioms (see pages 68–71). Similarly, although there is an infinite variety of rings, all of them must necessarily share all properties that are embodied in any theorems that can be derived from the

## 4.3 Basic Theorems for Rings

ring axioms (see page 127). In this section we derive some of the simpler, basic theorems about rings.

First, it should be very clear that any ring $\{S, +, \times\}$ incorporates within itself a group—the so-called additive group $\{S, +\}$. Certainly, all the previously proved group theorems apply to this group. This observation already supplies us with quite a few important theorems that apply to the ring as well. For example, Theorem 2.1 implies that in any ring there is precisely one (unique) additive identity. It is customary to denote the additive identity of a ring by "0," and to call this element "zero."

Theorem 2.3 implies that every element $x$ in a ring has a unique additive inverse that we usually denote by $-x$. This means that

$$x + (-x) = (-x) + x = 0 \quad \text{for every element } x \text{ in the ring.}$$

Theorem 2.4 (*Unique Solutions*) implies that if $a$ and $b$ are any elements of a ring, then the equation $x + a = b$ has the unique solution $x = b + (-a)$. By definition this is also written $x = b - a$ so that

$$x = b - a \quad \text{if and only if} \quad x + a = b$$

In this manner subtraction is defined in any ring.

> Subtraction is, in fact, a binary operation on the elements of any ring.

Theorem 2.5 (*Cancellation Theorem*) implies that for any elements $x$, $y$, $z$ in a ring

$$\text{if } x + z = y + z, \text{ then } x = y$$
$$\text{and if } z + x = z + y, \text{ then } x = y$$

Thus Theorem 2.5 permits us to cancel (that is, to omit) a common (additive) term appearing on each side of an equation.

Theorem 2.6 (*Inverse of an Inverse*) implies that $-(-x) = x$ and Theorem 2.7 (*Inverse of a Product*) implies that $-(a+b) = (-b) + (-a)$. In view of the fact that $\{S, +\}$ is an abelian (commutative) group, this further implies that

$$-(a+b) = (-a) + (-b)$$

Other properties of rings, which are merely translations of group properties, will no doubt suggest themselves (see Exercise 2, Exercises 4.3).

To obtain somewhat less trivial theorems about rings, we must of course turn to those ring axioms that deal with the multiplication operation of the ring. Of particular significance is the Distributivity Axiom.

**Distributivity Axiom.** *For all $x, y, z$ in the system*
$$x(y+z) = xy+xz \quad \text{and} \quad (y+z)x = yx+zx$$

Using this axiom we can derive some very important properties.

**THEOREM 4.2** In any ring $\{S, +, \times\}$, for every $x$ in $S$
$$x \cdot 0 = 0 \quad \text{and} \quad 0 \cdot x = 0$$

*Proof:* Since $0 + x = x$, it follows that
$$x \cdot (0+x) = x \cdot x$$
Hence, by "left" distributivity,
$$x \cdot 0 + x \cdot x = x \cdot x$$
that is,
$$x \cdot 0 + x \cdot x = 0 + x \cdot x$$
We may now cancel the term $x \cdot x$ from each side by using Theorem 2.5 (Cancellation Theorem). We obtain
$$x \cdot 0 = 0$$
Similarly, we prove that
$$0 \cdot x = 0$$

Theorem 4.2 may be conveniently expressed as follows:

> A product $ab$ is zero if either $a = 0$ or $b = 0$ (or both).

It is important to observe that the converse of this result need not be true. For example, in the ring
$$\{W_6, + \text{ (mod 6)}, \times \text{ (mod 6)}\}$$
the product $2 \times 3 = 0$, but neither of the factors 2 or 3 is zero. In this case the numbers 2 and 3 are called **divisors of zero**. In general, we define a divisor of zero as follows.

**Definition.** An element $b$ of a ring $\{S, +, \times\}$ is called a **divisor of zero** if and only if there exists in $S$ an element $c \neq 0$ such that $b \times c = 0$ or and element $d \neq 0$ such that $d \times b = 0$.

In view of Theorem 4.2, 0 is always a divisor of zero, except in the trivial ring which consists solely of the element 0.

> Any element of a ring, other than 0, which is a divisor of zero is called a **proper divisor** of zero.

For example, in the ring

$$\{W_6, +(\text{mod } 6), \times(\text{mod } 6)\}$$

the numbers 2 and 3 are proper divisors of zero because in this system $2 \times 3 = 0$ even though $2 \neq 0$ and $3 \neq 0$. An even more startling example is the finite noncommutative ring mentioned in Section 4.2. In this ring each of the four elements $e, a, b, c$ is a divisor of zero. This is so because in this system $e = 0$ and $a \neq 0$, but $ax = 0$ for each $x$, namely, for $x = e, a, b, c$. On the other hand, the ring of integers, $\{Z, +, \times\}$, has no proper divisors of zero because if $b$ and $c$ are integers and if $b \neq 0$ and $c \neq 0$, then $b \times c \neq 0$.

The following theorems, like Theorem 4.2, express important rules of ordinary algebra.

**THEOREM 4.3** In any ring $\{S, +, \times\}$, for every $x, y$ in $S$

$$x(-y) = -(xy) \quad \text{and} \quad (-x)y = -(xy)$$

*Proof:* Since $y + (-y) = 0$, it follows that

$$x(y + (-y)) = x \cdot 0$$

and hence $\quad xy + x(-y) = 0$
But $\quad xy + (-(xy)) = 0$
Therefore $\quad xy + x(-y) = xy + (-(xy))$

and hence using Theorem 2.5 (the Cancellation Theorem), it now follows that

$$x(-y) = -(xy)$$

In a similar fashion we prove that

$$(-x)y = -(xy)$$

**THEOREM 4.4** In any ring $\{S, +, \times\}$, for every $x, y$ in $S$

$$(-x)(-y) = xy$$

*Proof:* By Theorem 4.3 $\qquad (-x)(-y) = -(x(-y))$
which (again by Theorem 4.3) $\qquad\qquad\ = -(-(xy))$
which by Theorem 2.6 (Inverse of an Inverse) $= xy$

If we recall that in any ring we define $x - y = x + (-y)$, then we can easily prove the following corollary.

COROLLARY. For all $x, y, z$ in $S$ (see Exercise 5, Exercises 4.3)

$$x(y - z) = xy - xz$$
$$(y - z)x = yx - zx$$

It is important to observe that if $\{S, +, \times\}$ is a ring, then although $\{S, +\}$ is always a group, the system $\{S, \times\}$ is not. There are several reasons for this. First, although $S$ always contains the additive identity 0, $S$ need not contain a multiplicative identity. For example, in the ring of even integers $\{E, +, \times\}$ the set $E$ contains only even integers and hence does not contain the number 1.

Second, Theorem 4.2 shows that in any nontrivial ring, 0 does not have a multiplicative inverse because

$$0 \cdot x = x \cdot 0 = 0 \quad \text{(not 1)}$$

Third, even if the ring does possess a multiplicative identity, it may fail to contain a multiplicative inverse even for elements other than 0. An example of such a ring is the system

$$\{W_4, +(\text{mod } 4), \times(\text{mod } 4)\}$$

Here $W_4 = \{0, 1, 2, 3\}$ certainly contains 1 (the multiplicative identity), but $W_4$ fails to contain a multiplicative inverse for the number 2. There is no number $n$ in $W_4$ such that $2 \times n = 1$. (Of course, as in any ring there is no multiplicative inverse for 0 either.)

If a ring does contain a multiplicative identity, the multiplicative identity is also called a **unit element**. The unit element is often designated by the familiar symbol " 1," as in the rings $\{Z, +, \times\}$, $\{W_n, +(\text{mod } n), \times(\text{mod } n)\}$, etc. On the other hand, in the ring of two by two matrices, $\{M, +, \times\}$, the unit element is the matrix $\begin{bmatrix} 1 & 0 \\ 0 & 1 \end{bmatrix}$ and this *identity matrix* is usually designated by "$I$" rather than by " 1."

## 4.3 Basic Theorems for Rings

A ring with a unit element may or may not contain a multiplicative inverse for various elements. If an element $c$ in such a ring does have a multiplicative inverse, then the inverse element must be unique. The reason for this uniqueness is simple. In any ring $\{S, +, \times\}$ the operation $\times$ is associative on $S$; thus if $S$ also contains a multiplicative identity, then the two requirements of Theorem 1.2 (Uniqueness of Inverses) are satisfied.

It is customary to represent the multiplicative inverse of an element $c$ by either of the symbols

$$\frac{1}{c} \quad \text{or} \quad c^{-1}$$

so that

$$c \times \frac{1}{c} = \frac{1}{c} \times c = 1$$

or

$$c \times c^{-1} = c^{-1} \times c = 1$$

$\frac{1}{c}$ (or $c^{-1}$) is also often called the **reciprocal** of $c$.

In any ring that contains more than one element, the additive identity 0 has no reciprocal. In any ring that contains a unit 1, this unit is its own reciprocal, that is,

$$1^{-1} = \frac{1}{1} = 1$$

In the ring $\{Z, +, \times\}$ the numbers 1 and $-1$ each has a reciprocal,

$$\frac{1}{1} = 1 \quad \text{and} \quad \frac{1}{-1} = -1$$

but all other integers have no reciprocal.

In the ring $\{W_5, +, \times\}$ each element other than 0 has a multiplicative inverse. The reciprocals are

$$\frac{1}{1} = 1, \quad \frac{1}{2} = 3, \quad \frac{1}{3} = 2, \quad \frac{1}{4} = 4$$

On the other hand, in the ring $\{W_4, +\text{(mod 4)}, \times \text{(mod 4)}\}$ only the elements 1 and 3 have reciprocals

$$\frac{1}{1} = 1 \quad \text{and} \quad \frac{1}{3} = 3$$

but 0 and 2 have no multiplicative inverse (no reciprocal).

## 4 ‖ Elementary Rings

Because multiplicative inverses (reciprocals) may not exist for nonzero elements of a ring $\{S, +, \times\}$, the Cancellation Theorem (Theorem 2.5) may or may not apply to the system $\{S, \times\}$ although this theorem is always valid for the system $\{S, +\}$. In a ring it may happen that

$$c \times a = c \times b$$

with $c \neq 0$ and yet $a \neq b$. For example, in the ring

$$\{W_6, +(\text{mod } 6), \times(\text{mod } 6)\}$$

we know that

$$2 \times 2 = 4 \quad \text{and} \quad 2 \times 5 = 4$$

Hence $2 \times 2 = 2 \times 5$; yet it would certainly be incorrect to cancel the left factor 2 on each side of this equation and conclude that $2 = 5$. Even more obvious is the case $0 \times 3 = 0 \times 4$. We certainly may not cancel the 0 on each side of this equality. On the other hand, in the modulo 6 ring, if we know that $5x = 5y$ we may validly cancel the 5's and conclude that $x = y$. To prove this we need merely multiply both sides of the equation $5x = 5y$ by 5

$$5(5x) = 5(5y) \quad (\text{mod } 6)$$

and (using the associativity of multiplication on $W_6$) we get

$$(5 \times 5)x = (5 \times 5)y \quad (\text{mod } 6).$$

But, in this system, $5 \times 5 = 1 \pmod{6}$. Hence

$$1x = 1y$$

and

$$x = y$$

The failure of cancellation in some cases and its success in others may appear puzzling at first. The following theorem shows that there is actually a very simple explanation.

**THEOREM 4.5** *(Cancellation Theorem)* In any ring $\{S, +, \times\}$ if $b$ is not a divisor of zero, then either of the equations

$$bx = by \quad \text{or} \quad xb = yb$$

implies $x = y$.

*Proof:* If $\qquad bx = by$

then $\qquad bx - by = 0$

Hence $\qquad b(x - y) = 0$

## 4.3 Basic Theorems for Rings

But $b$ is not a divisor of zero. Hence (by definition of a divisor of zero) it follows that $x - y = 0$, that is, that $x = y$. Similarly, we prove if $xb = yb$, then $x = y$.

Notice, however, that the equation $bx = yb$ need not imply $x = y$ even when $b$ is not a divisor of zero. For example, in the ring of two by two matrices let

$$b = \begin{bmatrix} 1 & 1 \\ 0 & 1 \end{bmatrix} \quad x = \begin{bmatrix} 1 & 1 \\ 1 & 1 \end{bmatrix} \quad y = \begin{bmatrix} 2 & 0 \\ 1 & 0 \end{bmatrix}$$

Here, although $b$ is not a divisor of zero (see Exercise 11), nevertheless we find

$$bx = \begin{bmatrix} 1 & 1 \\ 0 & 1 \end{bmatrix} \begin{bmatrix} 1 & 1 \\ 1 & 1 \end{bmatrix} = \begin{bmatrix} 2 & 2 \\ 1 & 1 \end{bmatrix}$$

and

$$yb = \begin{bmatrix} 2 & 0 \\ 1 & 0 \end{bmatrix} \begin{bmatrix} 1 & 1 \\ 0 & 1 \end{bmatrix} = \begin{bmatrix} 2 & 2 \\ 1 & 1 \end{bmatrix}$$

Therefore $$bx = yb$$

but $$x \neq y$$

Of course, in a commutative ring this situation cannot occur because $yb = by$ and hence Theorem 4.5 would also apply to this case. Clearly, therefore a commutative ring in which there are no proper divisors of zero will "behave" very well as far as the Cancellation Theorem (Theorem 4.5) is concerned. The ring of integers $\{Z, +, \times\}$ is an example of such a well-behaved ring. Another example is the ring of even integers $\{E, +, \times\}$. Such rings are called **integral domains**.

**Definition.** An **integral domain** is a commutative ring in which there are no proper divisors of zero.

We remark here that some mathematicians also require that an integral domain have a unit element. There appears, however, to be no universal agreement on this point.*

---

* For example, Van der Waerden, *Modern Algebra*, McCoy, *Rings and Ideals*, and Herstein, *Topics in Algebra*, do not require that an integral domain have a unit element. On the other hand, Birkhoff and Maclane, *A Survey of Modern Algebra*, Dubisch, *Introduction to Abstract Algebra*, Dean, *Elements of Abstract Algebra*, and Mostow, Sampson, and Meyer, *Fundamental Structures of Algebra*, do impose this additional requirement.

## 4 || Elementary Rings

In this book, we have chosen the more general, less restrictive definition of an integral domain in order to emphasize the other two essential requirements for a ring to be an integral domain. These are that (1) multiplication be commutative and (2) there be no proper divisors of zero.

## EXERCISES 4.3

1. In any ring $\{S, +, \times\}$ we have defined $b - a = b + (-a)$ for for all $a, b \in S$. Calculate each of the following:
   (a) $2 - 5$ in the ring $\{Z, +, \times\}$.
   (b) $2 - 5$ in the ring $\{W_6, + \pmod 6), \times \pmod 6)\}$.
   (c) $0 - 3$ in the ring $\{Z, +, \times\}$.
   (d) $0 - 3$ in the ring $\{W_4, +\pmod 4), \times \pmod 4)\}$.
   (e) $0 - 3$ in the ring $\{W_5, +\pmod 5), \times \pmod 5)\}$.
   (f) $0 - 3$ in the ring $\{W_n, +\pmod n), \times \pmod n)\}$, where $n$ is a positive integer $\geq 3$.
   (g) $b - c$ in the finite noncommutative ring of Section 4.2.

2. Establish that each of the following is true in any ring $\{S, +, \times\}$ by first determining an appropriate group property and then translating this group property into notation that applies to the ring.
   (a) $-0 = 0$
   (b) $0 + 0 = 0$
   (c) $x - 0 = x$ for all $x$ in the ring
   (d) $0 - x = -x$ for all $x$ in the ring
   (e) $y - x = -(x - y)$ for all $x, y$ in the ring

3. Provide the missing reasons for the following alternative proof of Theorem 4.3:
   (a) $x(-y) + [xy + (-(xy))] = [x(-y) + xy] + (-(xy))$
   (b) $x(-y) + 0 \qquad\qquad\qquad = [x(-y+y)] + (-(xy))$
   (c) $x(-y) \qquad\qquad\qquad\qquad = (x \cdot 0) \qquad + (-(xy))$
   (d) $x(-y) \qquad\qquad\qquad\qquad = 0 \qquad\qquad + (-(xy))$
   (e) $x(-y) \qquad\qquad\qquad\qquad = \qquad\qquad\quad -(xy)$

4. Prove the second part of Theorem 4.3, $(-x)y = -(xy)$, by the method used
   (a) in the text.
   (b) in Exercise 3.

5. Another proof of Theorem 4.3 uses the fact that each element of a group has exactly one inverse. Set up such a proof by showing that $x(-y)$, $(-x)y$, and $-(xy)$ are each additive inverses of $xy$.

## †4.4 Isomorphic Rings

6. Try to find an alternate proof of Theorem 4.4 using a method similar to
   (a) Exercise 3.
   (b) Exercise 5.
7. Prove the following corollaries to Theorem 4.4:
   (a) $x(y - z) = (xy) - (xz)$
   (b) $(y - z)x = (yx) - (zx)$
★8. In Exercise 10h of Exercises 4.2 we defined an idempotent element of a ring to be an element $x$ such that $x \cdot x = x$. Find all idempotent elements of each of these rings.
   (a) $\{W_{12}, +(\text{mod } 12), \times(\text{mod } 12)\}$
   (b) $\{W_{18}, +(\text{mod } 18), \times(\text{mod } 18)\}$
   (c) $\{W_{36}, +(\text{mod } 36), \times(\text{mod } 36)\}$
9. Find all the proper divisors of zero in the rings of Exercise 8.
10. For each of the rings of Exercise 8 find the solution set of
    (a) $x^2 = 0$
    ★(b) $x^2 - 4 = 0$
    ★(c) $x^2 - 4x = 0$
11. In Exercise 10i of Exercises 4.2 we defined a nilpotent element of a ring to be an element $x$ for which there exists a positive integer $n$ such that $x^n = 0$. Find all nilpotent elements for the rings of Exercise 8.
12. Show that if $bx = by$ with $b \neq 0$, we need not have $x = y$ by using examples from the rings in Exercise 8.
13. Show that the matrix $\begin{bmatrix} 1 & 1 \\ 0 & 1 \end{bmatrix}$ is not a divisor of zero by proving that

    if $\begin{bmatrix} 1 & 1 \\ 0 & 1 \end{bmatrix} \cdot \begin{bmatrix} a & b \\ c & d \end{bmatrix} = \begin{bmatrix} 0 & 0 \\ 0 & 0 \end{bmatrix}$

    then $a = 0$, $b = 0$, $c = 0$, and $d = 0$.

    And if $\begin{bmatrix} p & q \\ r & s \end{bmatrix} \cdot \begin{bmatrix} 1 & 1 \\ 0 & 1 \end{bmatrix} = \begin{bmatrix} 0 & 0 \\ 0 & 0 \end{bmatrix}$

    then $p = 0$, $q = 0$, $r = 0$, and $s = 0$.
★14. Show that commutativity for addition might be deduced for a system $\{S, +, \times\}$ where
    $\{S, +\}$ is a group.
    $S$ contains a multiplicative identity 1.
    $\times$ distributes over addition in both directions.
    (*Hint:* Compute $(1 + 1)(x + y)$ in two different ways and use the Cancellation Theorem for $\{S, +\}$.)

15. For each of the following rings determine which elements have multiplicative inverses (reciprocals) and which elements do not.
 (a) $\{W_2, +(\text{mod } 2), \times(\text{mod } 2)\}$
 (b) $\{W_3, +(\text{mod } 3), \times(\text{mod } 3)\}$
 (c) $\{W_6, +(\text{mod } 6), \times(\text{mod } 6)\}$
 (d) $\{W_{10}, +(\text{mod } 10), \times(\text{mod } 10)\}$
 (e) $\{Z, +, \times\}$
 (f) $\{E, +, \times\}$
 ★(g) $\{M_s, +, \times\}$ where $M_s$ is the set of all matrices of the form $\begin{bmatrix} x & 0 \\ 0 & x \end{bmatrix}$ where (1) $x \in Q$ (2) $x \in Z$

16. Prove that if a ring has a unit element 1, then $(-1)x = -x$ for all $x$ in the ring.
17. Complete the proof of Theorem 4.5 on pages 148–149.

## †4.4 ISOMORPHIC RINGS

In Chapter 2, Section 2.4, we studied isomorphic systems, particularly isomorphic groups. Each of the systems considered there involved just one operation on a given set. At present we are considering rings and these are systems involving two operations on a given set. Nevertheless the notion of isomorphism can also be applied to rings even though these systems involve two operations on a set. Let us see how this is done.

Suppose that $\{S, +, \times\}$ is a ring and that $\{\dot{S}, \dot{+}, \dot{\times}\}$ is also a ring. We call these two rings isomorphic if

 (1) the systems $\{S, +\}$ and $\{\dot{S}, \dot{+}\}$ are isomorphic and
 (2) the systems $\{S, \times\}$ and $\{\dot{S}, \dot{\times}\}$ are also isomorphic under the same correspondence.

Referring back to our previous definition for isomorphic systems with just one operation (see page 78) we see that this simply means the following:

 (1) There is a one-to-one correspondence between the sets $S$ and $\dot{S}$ (that is, for each $x \in S$ there is exactly one $\dot{x} \in \dot{S}$ and conversely).
 (2) Whenever $x \in S$ and $y \in S$, then $x + y$ corresponds to $\dot{x} \dot{+} \dot{y}$ and $x \times y$ corresponds to $\dot{x} \dot{\times} \dot{y}$.

Since our notation here is a bit cumbersome, we shall adapt a notational convention that, although not really accurate, is simpler. We shall always use the same two symbols $+$ and $\times$, to denote the addition and multiplication operations regardless of which ring may be under consideration. Thus although multiplication of numbers is certainly not the same operation as multiplication of say matrices, nevertheless we shall use "$a \times b$" (or also simply "$ab$") to denote the product of $a$ and $b$, regardless of whether $a$ and $b$ happen to be numbers or matrices. This will not cause any confusion since it will always be clear from context which type of product is involved. Let us now look at some examples of isomorphic rings.

**Example 1.** Let $R_1$ denote the ring $\{Z, +, \times\}$ consisting of the integers together with ordinary addition and multiplication. Let $R_2$ denote the ring $\{M_s, +, \times\}$ where $M_s$ is the set of all (two by two) **scalar matrices**. These are matrices of the form $\begin{bmatrix} n & 0 \\ 0 & n \end{bmatrix}$ where $n$ is an integer (that is, where $n \in Z$), and where "$+$" and "$\times$" now refer, respectively, to addition and multiplication of these scalar matrices. (That $R_2$ is indeed a ring was shown in Exercise 9b of Exercises 4.2.) The two rings $R_1$ and $R_2$ are isomorphic under the following one-to-one correspondence.

$$x \leftrightarrow \begin{bmatrix} x & 0 \\ 0 & x \end{bmatrix} \quad \text{for each } x \in Z$$

This is so because if $x \in Z$ and $y \in Z$, then

$$x + y \leftrightarrow \begin{bmatrix} x+y & 0 \\ 0 & x+y \end{bmatrix} = \begin{bmatrix} x & 0 \\ 0 & x \end{bmatrix} + \begin{bmatrix} y & 0 \\ 0 & y \end{bmatrix}$$

and

$$x \cdot y \leftrightarrow \begin{bmatrix} xy & 0 \\ 0 & xy \end{bmatrix} = \begin{bmatrix} x & 0 \\ 0 & x \end{bmatrix} \cdot \begin{bmatrix} y & 0 \\ 0 & y \end{bmatrix}$$

Stated verbally, we may say that the image of the sum is the sum of the images and the image of a product is the product of the images. Hence the correspondence is indeed an isomorphism.

In view of this isomorphism we may state that the scalar matrices, that is, the members of the set $M_s$, behave exactly like the ordinary integers, the members of $Z$. The two rings $R_1$ and $R_2$ are only notationally different. Structurally they are identical.

***Example 2.*** Let $R_1$ denote the noncommutative ring whose elements are the matrices

$$e = \begin{bmatrix} 0 & 0 \\ 0 & 0 \end{bmatrix} \qquad a = \begin{bmatrix} 0 & 1 \\ 0 & 0 \end{bmatrix}$$

$$b = \begin{bmatrix} 1 & 0 \\ 0 & 0 \end{bmatrix} \quad \text{and} \quad c = \begin{bmatrix} 1 & 1 \\ 0 & 0 \end{bmatrix}$$

(See Section 4.2.) Let $R_2$ denote the ring $\{S, +, \times\}$ described in Exercise 5 of Exercises 4.1,

$$S = \{(0, 0), (0, 1), (1, 0), (1, 1)\}$$

where addition was defined by

$$(a, b) + (c, d) = (a + c \ (\text{mod } 2), b + d \ (\text{mod } 2))$$

and multiplication was defined by

$$(a, b) \times (c, d) = (ac, ad)$$

There is an obvious one-to-one correspondence between the elements of these two systems. Under this one-to-one correspondence the rings $R_1$ and $R_2$ are isomorphic.

As with isomorphic groups much can be said about isomorphic rings. At this point, however, we shall merely mention the following items which are obviously closely related to the observations made in Section 2.4 about isomorphic groups. (See Exercise 5 in Exercises 2.4.)

(1) If two rings are isomorphic, the additive identity of one ring corresponds to the additive identity of the other (that is, the "image" of the zero element of one ring is the zero element of the other).
(2) If $x$ and $\bar{x}$ are corresponding elements of two isomorphic rings, then $-x$ and $-(\bar{x})$ are also corresponding elements (that is, the image of the additive inverse of an element is the additive inverse of its image, or that $\overline{(-x)} = -(\bar{x})$).
(3) If either of two isomorphic rings has a multiplicative identity element (that is, a unit element), then the image of this unit element is the multiplicative identity element of the other ring.
(4) If the element $x$ of either of two isomorphic rings has a multiplicative inverse, that is, a reciprocal $\dfrac{1}{x}$, then its image $\bar{x}$ also has a multiplicative

inverse $\frac{1}{\bar{x}}$, and this reciprocal is the image of $\frac{1}{x}$. (The image of a reciprocal of an element is the reciprocal of the image.)

In Exercise 3 you are asked to prove these theorems about isomorphic rings.

## †EXERCISES 4.4

1. Show that the following pairs of rings are isomorphic:
   (a) $\{Z, +, \cdot\}$ and $\{M, +, \cdot\}$ where $M$ is the set of matrices $\begin{bmatrix} x & 0 \\ 0 & 0 \end{bmatrix}$ with $x$ an integer.
   (b) $\{Z, +, \cdot\}$ and $\{M, +, \cdot\}$ where $M$ is the set of matrices $\begin{bmatrix} x & x \\ 0 & 0 \end{bmatrix}$ with $x$ an integer.
   (c) $\{M, +, \cdot\}$ and $\{N, +, \cdot\}$ where $M$ is the set of matrices $\begin{bmatrix} x & 0 \\ 0 & y \end{bmatrix}$ with $x$ and $y$ integers and where $N$ is the set of matrices $\begin{bmatrix} x & y \\ 0 & 0 \end{bmatrix}$ with $x$ and $y$ integers.
   (d) $\{Q, +, \cdot\}$ and $\{M, +, \cdot\}$ where $M$ is the set of matrices $\begin{bmatrix} 0 & 0 \\ x & x \end{bmatrix}$ with $x$ a rational number.
   (e) $\{Z(\sqrt{2}), +, \cdot\}$ and itself under the correspondence $a + b\sqrt{2} \leftrightarrow a - b\sqrt{2}$.

2. Let $R_1$ and $R_2$ be isomorphic rings. Prove each of the following theorems:
   (a) If 0 is the additive identity of $R_1$, then the image of zero is the additive identity of $R_2$.
   (b) If $x$ is an element of $R_1$ and $\bar{x}$ is the image of $x$ in $R_2$, then the image of $-x$ is $-\bar{x}$.
   (c) If $R_1$ has a unit (call it 1) and if $\bar{1}$ is the image of 1 in $R_2$, then $\bar{1}$ is the unit element of $R_2$.
   (d) If $x$ is an element of $R_1$ which has a reciprocal $\frac{1}{x}$ and if $\bar{x}$ and $\overline{\left(\frac{1}{x}\right)}$ are, respectively, images of $x$ and $\frac{1}{x}$ in $R_2$, then $\bar{x}$ has a recipocal $\frac{1}{\bar{x}}$ in $R_2$ and furthermore $\frac{1}{\bar{x}} = \overline{\left(\frac{1}{x}\right)}$.

(e) The image of $(xy)z$ is $(\bar{x}\bar{y})\bar{z}$.
(f) The image of $x(y+z)$ is $\bar{x}(\bar{y}+\bar{z})$.

★3. Show that the following pairs of rings are not isomorphic:
  (a) $\{Z, +, \cdot\}$ and $\{Q, +, \cdot\}$
  (b) $\{Z, +, \cdot\}$ and $\{2Z, +, \cdot\}$ (See Section 4.1, Exercise 2.)
  (c) $\{2Z, +, \cdot\}$ and $\{3Z, +, \cdot\}$ under the correspondence $2x \leftrightarrow 3x$.
  (d) $\{Z(\sqrt{2}), +, \cdot\}$ and $\{Z(\sqrt{3}), +, \cdot\}$ under the correspondence $x + y\sqrt{2} \leftrightarrow x + y\sqrt{3}$.

# REVIEW EXERCISES

1. Show that each of the following mathematical systems is a ring.
   (a) $\{W_6, +\pmod 6, \cdot \pmod 6\}$
   (b) $\{Z(\sqrt{5}), +, \cdot\}$
   (c) $\{Q(\sqrt{5}), +, \cdot\}$
   (d) $\{M, +, \cdot\}$ where $M$ is the set of all matrices $\begin{bmatrix} x & y \\ 0 & z \end{bmatrix}$ with $x, y, z$ elements of some fixed ring. (Of course, the ring operations of addition and multiplication are to be used when adding and multiplying elements of the matrices.)

2. For each of the following rings determine which elements have multiplicative inverses and which elements do not.
   (a) $\{W_7, +\pmod 7, \cdot \pmod 7\}$
   (b) $\{W_{10}, +\pmod{10}, \cdot \pmod{10}\}$
   (c) $\{Q, +, \cdot\}$
   (d) $\{M, +, \cdot\}$ where $M$ is the set of all matrices $\begin{bmatrix} x & 0 \\ 0 & y \end{bmatrix}$ with $x, y$ rational numbers.

3. Prove that the ring $\{Z(\sqrt{3}), +, \cdot\}$ is isomorphic to itself under the correspondence $a + b\sqrt{3} \leftrightarrow a - b\sqrt{3}$.

4. Prove that the rings $\{Q, +, \cdot\}$ and $\{2Q, +, \cdot\}$ are not isomorphic under the correspondence $x \leftrightarrow 2x$.

5. Prove: If a ring has a unit element 1, every element $x \neq 1$ for which $xx = x$ must be a divisor of zero.

6. Prove: If a ring has a unit element 1, then the solution set for $x^2 = 1$ need not be $\{1, -1\}$. (*Hint:* Find a ring in which $x^2 = 1$ has another solution in addition to 1 and $-1$.)

7. Find the solution set for $x^2 \equiv -1 \pmod{10}$.

8. Find the multiplicative inverse of each of the following matrices.

(*Hint:* See Exercise 10, Exercises 4.2.)

(a) $\begin{bmatrix} 7 & 8 \\ 6 & 7 \end{bmatrix}$ (b) $\begin{bmatrix} a & a+1 \\ a-1 & a \end{bmatrix}$

(c) $\begin{bmatrix} 7 & 9 \\ 5 & 7 \end{bmatrix}$ (d) $\begin{bmatrix} a & a+2 \\ a-2 & a \end{bmatrix}$

9. Let $\{S, +, \cdot\}$ be a ring such that for every $x \in S$, $x \cdot x = x$. A ring such as this is called a *Boolean ring*.
   (a) Verify that $\{S, +, \cdot\}$ is a Boolean ring if $S = \{(0, 0), (0, 1), (1, 0), (1, 1)\}$, and

   $$(a, b) + (c, d) = (a + c \pmod 2, b + d \pmod 2),$$
   $$(a, b) \cdot (c, d) = (ac, bd).$$

   (b) Prove: $x = -x$ for every $x$ in a Boolean ring.
   (c) Prove: $xy = yx$ for every pair $x$, $y$ in a Boolean ring.
   (d) Find another example of a Boolean ring.

10. Let $\{S, +, \cdot\}$ be a ring. Let $T$ be a nonempty subset of $S$ such that
    (a) $\{T, +\}$ is a subgroup of $\{S, +\}$.
    (b) For every $s \in S$ and $t \in T$, both $st$ and $ts$ are in $T$.
    We then call $\{T, +, \cdot\}$ an *ideal* of the ring $\{S, +, \cdot\}$.
    For $\{W_{24}, +\pmod{24}, \cdot \pmod{24}\}$, find
    (a) all the ideals
    (b) all the idempotents
    (c) all the nilpotents
    (For (b) and (c) see Exercises 10h and i, Exercises 4.2.)

★11. The *parabolic integers* are defined as the set of $P$ of parabolic numbers $a + bj$ where $a$ and $b$ are ordinary integers and where $j^2 = 0$. Addition ($+$) and multiplication ($\cdot$) of elements of $P$ are defined as follows:

$$(a + bj) + (c + dj) = (a + c) + (b + d)j$$
$$(a + bj) \cdot (c + dj) = ac + (ad + bc)j$$

Show that the system $\{G, +, \cdot\}$ is a commutative ring with a unit element.

★12. The *hyperbolic integers* are defined as the set $H$ of hyperbolic numbers $a + br$ where $a$ and $b$ are ordinary integers and where $r^2 = 1$ (but $r$ is not an ordinary integer, that is, $r \neq \pm 1$). Addition ($+$) and multiplication ($\cdot$) of elements of $H$ are defined as follows:

$$(a + br) + (c + dr) = (a + c) + (b + d)r$$
$$(a + br) \cdot (c + dr) = (ac + bd) + (ad + bc)r$$

Show that the system $\{G, +, \cdot\}$ is a commutative ring with a unit element.

13. Let $\{S, +, \cdot\}$ be a ring for which $xy = -yx$ for all $x, y$ in $S$. We call such a ring *anticommutative*.

    For example, every Boolean ring is anticommutative (see Exercise 9 above). For an anticommutative ring prove each of the following:
    (a) $x^2 + x^2 = 0$ for all $x$ in $S$.
    (b) $xyz = -yzx = yzx$ for all $x, y, z$ in $S$.
    (c) $ab^2 = b^2 a$ for all $a, b$ in $S$.
    (d) If $S$ has a multiplicative identity 1, then
        (1) $x = -x$ for all $x$ in $S$.
        (2) $xy = yx$ for all $x, y$ in $S$ so that the ring is also commutative.
    (e) If this ring has no proper divisors of 0, then the ring is also commutative.

14. Problem for investigation: Must an anticommutative ring be commutative?

*Emmy Noether (1882-1935)*

# 5 Fields

> From the axiomatic point of view, mathematics appears ... as a storehouse of abstract forms—the mathematical structures; and it so happens—without our knowing why—that certain aspects of empirical reality fit themselves into these forms, as if through a kind of preadaption.
>
> N. BOURBAKI

## 5.1 EXAMPLES AND DEFINITION OF A FIELD

Every ring $\{S, +, \times\}$ may be viewed as a system built on the two simpler mathematical systems

$$\{S, +\} \quad \text{and} \quad \{S, \times\}$$

According to the definition of a ring, the first of these simpler structures, $\{S, +\}$, is always an abelian group. The structure $\{S, \times\}$, however, is not a group (except in the trivial case where the set $S$ consists solely of the element 0).

Although $\{S, \times\}$ is not a group, there are nevertheless many rings where the system $\{S, \times\}$ "comes very close" to being a group. For example, consider the ring

$$\{W_5, +(\text{mod } 5), \times(\text{mod } 5)\}$$

Recall the addition and multiplication tables for this system:

$\{W_5, +(\text{mod } 5)\}$ $\qquad\qquad$ $\{W_5, \times(\text{mod } 5)\}$

| +(mod 5) | 0 | 1 | 2 | 3 | 4 |
|---|---|---|---|---|---|
| 0 | 0 | 1 | 2 | 3 | 4 |
| 1 | 1 | 2 | 3 | 4 | 0 |
| 2 | 2 | 3 | 4 | 0 | 1 |
| 3 | 3 | 4 | 0 | 1 | 2 |
| 4 | 4 | 0 | 1 | 2 | 3 |

| ×(mod 5) | 0 | 1 | 2 | 3 | 4 |
|---|---|---|---|---|---|
| 0 | 0 | 0 | 0 | 0 | 0 |
| 1 | 0 | 1 | 2 | 3 | 4 |
| 2 | 0 | 2 | 4 | 1 | 3 |
| 3 | 0 | 3 | 1 | 4 | 2 |
| 4 | 0 | 4 | 3 | 2 | 1 |

The addition table, of course, represents a group, as it must for any ring. On the other hand, the multiplication table shows clearly that $\{W_5, \times(\text{mod } 5)\}$ is not a group because 0 has no multiplicative inverse in the set $W_5$. However, suppose that we remove 0 temporarily from the set $W_5$. Let us call the resulting set $W_5^*$, that is,

$$W_5^* = \{1, 2, 3, 4\}$$

and let us consider the "reduced" system $\{W_5^*, \times(\text{mod } 5)\}$ which has the following reduced table.

$\{W_5^*, \times(\text{mod } 5)\}$

| ×(mod 5) | 1 | 2 | 3 | 4 |
|---|---|---|---|---|
| 1 | 1 | 2 | 3 | 4 |
| 2 | 2 | 4 | 1 | 3 |
| 3 | 3 | 1 | 4 | 2 |
| 4 | 4 | 3 | 2 | 1 |

This table reveals that $\{W_5^*, \times(\text{mod } 5)\}$ is a group (indeed, it is an abelian group). Apparently, it is solely the presence of the number 0 in $W_5$ that prevents the system $\{W_5, \times(\text{mod } 5)\}$ from being a group.

By way of contrast let us consider another simple ring

$$\{W_4, +(\text{mod } 4), \times(\text{mod } 4)\}$$

## 5.1 Examples and Definition of a Field

Its addition and multiplication tables are:

$\{W_4, +(\text{mod } 4)\}$ 	$\{W_4, \times(\text{mod } 4)\}$

| +(mod 4) | 0 | 1 | 2 | 3 |
|---|---|---|---|---|
| 0 | 0 | 1 | 2 | 3 |
| 1 | 1 | 2 | 3 | 0 |
| 2 | 2 | 3 | 0 | 1 |
| 3 | 3 | 0 | 1 | 2 |

| ×(mod 4) | 0 | 1 | 2 | 3 |
|---|---|---|---|---|
| 0 | 0 | 0 | 0 | 0 |
| 1 | 0 | 1 | 2 | 3 |
| 2 | 0 | 2 | 0 | 2 |
| 3 | 0 | 3 | 2 | 1 |

This time we see that not only does $\{W_4, \times(\text{mod } 4)\}$ fail to be a group but even if we remove the element 0 from the set $W_4$, the resulting reduced system $\{W_4^*, \times(\text{mod } 4)\}$ will still fail to be a group. In fact, the reduced table shows this very clearly.

$\{W_4^*, \times(\text{mod } 4)\}$

| ×(mod 4) | 1 | 2 | 3 |
|---|---|---|---|
| 1 | 1 | 2 | 3 |
| 2 | 2 | 0 | 2 |
| 3 | 3 | 2 | 1 |

Observe that $\times(\text{mod } 4)$ is not even a binary operation on $W_4^*$, hence the system is not closed. Moreover, 2 fails to have a multiplicative inverse in this system. The removal of 0 from this system has hardly improved its behavior under multiplication, as in the previous system.

We see therefore that there is a considerable difference in the way these two rings behave, particularly with regard to multiplication. Rings that exhibit the type of behavior shown in our first illustration play such a prominent role in mathematics that they have a special name, **fields**.

**Definition.** A ring $\{S, +, \times\}$ is called a **field** if the set consisting of all elements of $S$, excluding 0, forms an abelian group under the operation $\times$.

The ring
$$\{W_5, +(\bmod 5), \times(\bmod 5)\}$$
is therefore a field, but the ring
$$\{W_4, +(\bmod 4), \times(\bmod 4)\}$$
is not a field.

Let us examine some other rings from this point of view.

***Example 1.*** Is the ring of integers $\{Z, +, \times\}$ a field? To answer this question we ask another: Does the set consisting of all the integers, excluding 0, form an abelian group under multiplication? Clearly, the answer is No because there are integers such as 2, 3, etc., that have no multiplicative inverse in $Z$. Since every element of a group must have an inverse, the failure of this particular group requirement shows that $\{Z, +, \times\}$ is not a field.

***Example 2.*** How about the ring of rational numbers $\{Q, +, \times\}$? Is this ring a field? The answer is Yes because the set of all rational numbers, excluding zero, does indeed form an abelian group under multiplication.

***Example 3.*** Let us consider the ring $\{M, +, \times\}$ of all two by two matrices whose elements are *rational* numbers, that is, members of $Q$. Is this system a field? The answer is No because

(a) the multiplication operation is not commutative in this system and
(b) some nonzero matrices do not have reciprocals (multiplicative inverses.)

(For example, $\begin{bmatrix} 1 & 1 \\ 0 & 0 \end{bmatrix}$ is such a matrix (see Exercise 1b).) Thus the set of all nonzero elements of $M$ fails to form an abelian group under $\times$ because two of the requirements for an abelian group are lacking.

***Example 4.*** Let $M_S$ be the set of all matrices of the form $\begin{bmatrix} x & 0 \\ 0 & x \end{bmatrix}$ where $x \in Q$ (that is, where $x$ is a rational number). We call matrices of this form **scalar matrices**. From Exercise 9b of Exercises 4.2 it follows that $\{M_S, +, \times\}$ is a ring. This is a "subring" of the ring in Example 3. Let us verify in detail that this subring is a field. The zero element is the matrix $\begin{bmatrix} 0 & 0 \\ 0 & 0 \end{bmatrix}$. Let us remove this zero matrix from the set $M_S$ and call the resulting set $M_S^*$. We must show that $\{M_S^*, \times\}$ is an abelian group.

(a) (Closure Property) Let $\begin{bmatrix} x & 0 \\ 0 & x \end{bmatrix}$ and $\begin{bmatrix} y & 0 \\ 0 & y \end{bmatrix}$ be matrices in $M_S^*$. Their product is

$$\begin{bmatrix} x & 0 \\ 0 & x \end{bmatrix} \times \begin{bmatrix} y & 0 \\ 0 & y \end{bmatrix} = \begin{bmatrix} xy & 0 \\ 0 & xy \end{bmatrix}$$

Since $x \neq 0$ and $y \neq 0$, it follows that $xy \neq 0$. Hence the product matrix is also a member of $M_S^*$. The operation $\times$ is therefore a binary operation on $M_S^*$.

(b) (Associative Property) This property follows immediately from the closure established in (a) and from the fact that multiplication of matrices in general is associative.

(c) (Identity Property) The unit matrix $\begin{bmatrix} 1 & 0 \\ 0 & 1 \end{bmatrix}$ is a member of $M_S^*$ and is the identity element for $\{M_S^*, \times\}$.

(d) (Inverse Property) if $\begin{bmatrix} x & 0 \\ 0 & x \end{bmatrix}$ $(x \neq 0)$ is any member of $M_S^*$, then the matrix $\begin{bmatrix} \frac{1}{x} & 0 \\ 0 & \frac{1}{x} \end{bmatrix}$ is clearly also a member of $M_S^*$ and furthermore

$$\begin{bmatrix} x & 0 \\ 0 & x \end{bmatrix} \begin{bmatrix} \frac{1}{x} & 0 \\ 0 & \frac{1}{x} \end{bmatrix} = \begin{bmatrix} 1 & 0 \\ 0 & 1 \end{bmatrix} = \begin{bmatrix} \frac{1}{x} & 0 \\ 0 & \frac{1}{x} \end{bmatrix} \begin{bmatrix} x & 0 \\ 0 & x \end{bmatrix}$$

showing that $\begin{bmatrix} \frac{1}{x} & 0 \\ 0 & \frac{1}{x} \end{bmatrix}$ is the inverse matrix for $\begin{bmatrix} x & 0 \\ 0 & x \end{bmatrix}$.

(e) (Commutative Property) If $\begin{bmatrix} x & 0 \\ 0 & x \end{bmatrix}$ and $\begin{bmatrix} y & 0 \\ 0 & y \end{bmatrix}$ are any elements of $M_S^*$, then (noting that $xy = yx$ since $x$ and $y$ are rational numbers)

$$\begin{bmatrix} x & 0 \\ 0 & x \end{bmatrix}\begin{bmatrix} y & 0 \\ 0 & y \end{bmatrix} = \begin{bmatrix} xy & 0 \\ 0 & xy \end{bmatrix} = \begin{bmatrix} yx & 0 \\ 0 & yx \end{bmatrix} = \begin{bmatrix} y & 0 \\ 0 & y \end{bmatrix}\begin{bmatrix} x & 0 \\ 0 & x \end{bmatrix}$$

This shows that multiplication is indeed commutative in this particular ring (even though multiplication of matrices in general is not commutative).

Example 4 indicates that it may be convenient to have available a more detailed list of the defining properties for a field by breaking down the definition as follows.

**Alternate Definition for a Field.** A field is a system $\{S, +, \times\}$ consisting of a set of elements $S$ together with two binary operations $+$ and $\times$ in $S$ such that the following properties hold.

### FIELD AXIOMS

1. $+$ is a binary operation on $S$.
2. $+$ is associative on $S$.
3. There exists an element $0$ in $S$ such that $0 + x = x + 0 = x$ for every $x \in S$.
4. For each $x$ in $S$, there exists an element $-x$ in $S$ such that $x + (-x) = (-x) + x = 0$.
5. $+$ is commutative on $S$.

6. $\times$ is a binary operation on $S$.
7. $\times$ is associative on $S$.
8. There exists an element $1 (\neq 0)$ in $S$ such that $1 \times x = x \times 1 = x$ for every $x \in S$.
9. For each $x (\neq 0)$ in $S$, there exists an element $\frac{1}{x}$ in $S$ such that $x \times \frac{1}{x} = \frac{1}{x} \times x = 1$.
10. $\times$ is commutative on $S$.

11. For all $x, y, z$ in $S$    $x(y + z) = xy + xz$.

Viewed in this manner, a field $\{S, +, \times\}$ comprises both an additive group $\{S, +\}$ and a multiplicative group $\{S^*, \times\}$, where $S^*$ is the set consisting of all elements of $S$ excluding 0. Both of these groups are abelian and they are related to each other by the Distributivity Axiom

$$x \cdot (y + z) = (x \cdot y) + (x \cdot z)$$

## EXERCISES 5.1

1. Prove that the following matrices, whose elements are rational numbers, have no multiplicative inverses. Recall that matrices $A$ and $B$ are inverses if and only if $A \times B = \begin{bmatrix} 1 & 0 \\ 0 & 1 \end{bmatrix}$.

   (a) $\begin{bmatrix} 1 & 0 \\ 0 & 0 \end{bmatrix}$  (b) $\begin{bmatrix} 1 & 1 \\ 0 & 0 \end{bmatrix}$  (c) $\begin{bmatrix} 1 & 1 \\ 1 & 1 \end{bmatrix}$  (d) $\begin{bmatrix} 6 & 2 \\ 3 & 1 \end{bmatrix}$

2. Show that each of the following is not a field by listing in each case all the field requirements that are not satisfied.
   (a) $\{W_6, +\pmod{6}, \times\pmod{6}\}$
   (b) $\{2Z, +, \times\}$
   (c) $\{Z(\sqrt{2}), +, \times\}$
   (d) $\{Q(\sqrt[3]{2}), +, \times\}$
   (e) $\left\{\left[\begin{smallmatrix}0&0\\0&0\end{smallmatrix}\right], \left[\begin{smallmatrix}0&1\\0&0\end{smallmatrix}\right], \left[\begin{smallmatrix}1&0\\0&0\end{smallmatrix}\right], \left[\begin{smallmatrix}1&1\\0&0\end{smallmatrix}\right]\right\}, +\pmod{2}, \times\right\}$
   (f) $\{S, +, \times\}$ where the addition and multiplication tables are

   | + | 0 | a | b | c |
   |---|---|---|---|---|
   | 0 | 0 | a | b | c |
   | a | a | 0 | c | b |
   | b | b | c | 0 | a |
   | c | c | b | a | 0 |

   | × | 0 | a | b | c |
   |---|---|---|---|---|
   | 0 | 0 | 0 | 0 | 0 |
   | a | 0 | a | 0 | a |
   | b | 0 | 0 | b | b |
   | c | 0 | a | b | c |

3. Let $M_1$ be the set of all matrices $\begin{bmatrix} a & b \\ c & d \end{bmatrix}$ where $a, b, c, d$ are rational numbers such that $ad - bc = 1$.
   (a) Show that $M_1$ is closed under matrix multiplication.
   (b) Show that $M_1$ is associative under multiplication.
   (c) What is the identity matrix for multiplication?
   (d) Show that the multiplicative inverse of $\begin{bmatrix} a & b \\ c & d \end{bmatrix}$ is $\begin{bmatrix} d & -b \\ -c & a \end{bmatrix}$
   (e) Is $\{M_1, \times\}$ a group?
   (f) Is $\{M_1, \times\}$ an abelian group?
   (g) Is $\{M_1, +\}$ a group? Why?

(h) Is $\{M_1, +, \times\}$
   (1) a ring?   (2) a field?

4. Show that the following mathematical systems are fields:
   (a) $\{W_2, +(\bmod 2), \times(\bmod 2)\}$
   (b) $\{W_3, +(\bmod 3), \times(\bmod 3)\}$
   (c) $\{Q(\sqrt{2}), +, \times\}$

   *Hint:* Note that $\dfrac{1}{a+b\sqrt{2}} = \dfrac{1}{(a+b\sqrt{2})} \cdot \dfrac{(a-b\sqrt{2})}{(a-b\sqrt{2})}$.

   (d) $\{Q(\sqrt{3}), +, \times\}$
   (e) $\{2Q, +, \times\}$
   (f) $\{(-2)Q, +, \times\}$
   (g) $\{\{0, 2, 4, 6, 8\}, +(\bmod 10), \times(\bmod 10)\}$
   (h) $\{\{0, 1, a, b\}, +, \times\}$ where $+$ and $\times$ are defined by the tables:

   | + | 0 | 1 | a | b |
   |---|---|---|---|---|
   | 0 | 0 | 1 | a | b |
   | 1 | 1 | 0 | b | a |
   | a | a | b | 0 | 1 |
   | b | b | a | 1 | 0 |

   | × | 0 | 1 | a | b |
   |---|---|---|---|---|
   | 0 | 0 | 0 | 0 | 0 |
   | 1 | 0 | 1 | a | b |
   | a | 0 | a | b | 1 |
   | b | 0 | b | 1 | a |

   (i) $\{\{\text{Complex numbers}\}, +, \times\}$

## 5.2 BASIC FIELD THEOREMS

We have already defined subtraction in any ring $\{S, +, \times\}$ as follows:

$$x - y = x + (-y) \quad \text{for all } x, y \text{ in } S$$

As defined here subtraction ($-$) is a binary operation on $S$ because every element $y$ in $S$ has a unique additive inverse in $S$, so that $x - y$ is well defined for every pair of elements $x, y$ in $S$.

Although every element of a ring has a unique additive inverse, it is not true that every element of a ring has a unique multiplicative inverse. In fact, we have seen that in many rings there may even be nonzero elements, each of which fails to have a multiplicative inverse.

On the other hand, if a ring happens to be a field, then the nonzero elements form an abelian group under multiplication. Consequently, in a field every nonzero element $y$ has a unique multiplicative inverse $y^{-1}$ in the system. This enables us to define division ($\div$) in a field.

## 5.2 Basic Field Theorems

> If $x$ is any element of a field $\{S, +, \times\}$ and if $y$ is any nonzero element in this field, we define
> $$x \div y = x \times y^{-1} \qquad \text{for all } x, y \text{ in } S \text{ such that } y \neq 0$$

$x \div y$ is called a **quotient** and is often designated by $\dfrac{x}{y}$. Division assigns a quotient $\dfrac{x}{y}$ to every ordered pair $(x, y)$ of elements in $S$ for which $y \neq 0$. Division is therefore a binary operation *in* $S$ although not *on* $S$. (Division is, however, a binary operation on the set $S^*$ consisting of all elements of $S$ except 0.) Because multiplication is commutative in a field,

$$x \times y^{-1} = y^{-1} \times x$$

Hence it is also true that

$$\frac{x}{y} = y^{-1} \times x$$

The definition of division has an interesting consequence. Suppose that we let $x = 1$ in the definition of $\dfrac{x}{y}$; we obtain

$$\frac{1}{y} = 1 \times y^{-1} = y^{-1} \qquad \text{for all } y \neq 0$$

This means that the element obtained by dividing 1 by $y$ is the same as the multiplicative inverse of $y$. It is customary to call $\dfrac{1}{y}$ the *reciprocal* of $y$. We, have shown here that

> The **reciprocal** of each nonzero element of a field is the same as its multiplicative inverse.

We shall include this fact in the list of field theorems to be discussed (see Theorem 5.2h).

It should be clear that all theorems that apply to rings also apply to fields because every field is a ring. Moreover, since the nonzero elements of a field form an abelian group under multiplication, there are additional theorems valid for fields but not necessarily valid for rings that are not fields.

It is convenient at this point to assemble some of the most useful field theorems. Many of them have already been proved as ring theorems. Others follow directly from appropriate group theorems or previous theorems about binary operations. It is therefore unnecessary to prove each of these theorems in detail. However, the key ideas of each proof will be supplied and details left for the reader. In stating these theorems it is convenient to represent the multiplication operation by · instead of ×.

**THEOREM 5.1** (*Uniqueness of Identity and Inverses*) If $\{S, +, \cdot\}$ is a field, then (a) 0 is the only identity element for addition.
(b) 1 is the only identity element for multiplication.
(c) for each $x$ in $S$, $-x$ is the only additive inverse of $x$ in $S$.
(d) for each $x \neq 0$ in $S$, $\dfrac{1}{x}$ is the only multiplicative inverse of $x$ in $S$.

Theorem 5.1 is essentially a restatement of earlier group theorems; for example, part d follows from Theorem 2.3 (Uniqueness of Inverses in a Group).

The reader will recognize many of the items in the following theorems. They express the usual rules of arithmetic and algebra that he has studied in more elementary mathematics courses.

**THEOREM 5.2** (*Special Properties of* 0 *and* 1) In any field $\{S, +, \cdot\}$

(a) $-0 = 0$
(b) $\dfrac{1}{1} = 1$
(c) $x - 0 = x$
(d) $0 - x = -x$       for all $x$ in $S$
(e) $\dfrac{x}{1} = x$
(f) $x \cdot 0 = 0 \cdot x = 0$

(g) $\dfrac{0}{x} = 0$
(h) $\dfrac{1}{x} = x^{-1}$       for all nonzero $x$ in $S$
(i) $\dfrac{x}{x} = 1$

## 5.2 Basic Field Theorems

Note that Theorems 5.2a and b follow immediately from the fact that in any group $e' = e$ (the identity element is its own inverse). To prove (c) and (d) recall the definition of subtraction. Prove (e) by recalling the definition of division. Part f has already been proved as a ring theorem and (g) follows immediately from (f) using the definition of division. Part h has already been proved in our discussion of division; (i) is proved by using the definition of division.

**THEOREM 5.3** (*Preservation of Equality under Addition, Subtraction, Multiplication, and Division*) Let $\{S, +, \cdot\}$ be a field. If $x = y$ and $u = v$, then

(a) $x + u = y + v$
(b) $x \cdot u = y \cdot v$   for all $x, y, u, v$ in $S$
(c) $x - u = y - v$

(d) $\dfrac{x}{u} = \dfrac{y}{v}$   for all $x, y$ in $S$ and all nonzero $u, v$ in $S$

Each part of Theorem 5.3 is actually a direct consequence of Theorem 1.3 (see page 47). For example, to prove Theorem 5.3c, observe that subtraction is a binary operation on $S$, so if $x = y$, then by Theorem 1.3,

$$x - u = y - u$$

and hence if $u = v$, we may apply the Axiom of Replacement to obtain

$$x - u = y - v$$

To prove (d), we observe that division is a binary operation on $S^*$ (the set of all elements of $S$ excluding 0). Applying Theorem 1.3 and the Axiom of Replacement, we obtain

$$x \div u = y \div v \quad \text{for all } x, y, u, v \text{ in } S^*$$

However, by Theorem 5.2g, the last equality holds even when $x = 0$ and $y = 0$, that is, (d) holds for all $x, y$ in $S$ and all $u, v$ in $S^*$.

**THEOREM 5.4** (*Inverses*) If $\{S, +, \cdot\}$ is a field, then

(a) $-(-x) = x$   for all $x$ in $S$

(b) $\dfrac{1}{\frac{1}{x}} = x$   for all nonzero $x$ in $S$

(c) $-(x + y) = (-x) + (-y)$   for all $x, y$ in $S$

(d) $\dfrac{1}{x \cdot y} = \dfrac{1}{x} \cdot \dfrac{1}{y}$

(e) $\dfrac{1}{\frac{x}{y}} = \dfrac{y}{x}$

$\left.\begin{array}{c}\\ \\ \\ \\ \end{array}\right\}$ for all nonzero $x, y$ in $S$

Observe that Theorems 5.4a and b are direct consequences of Theorem 2.6 (Inverse of an Inverse), whereas (c) and (d) follow immediately from Theorem 2.7 (Inverse of a Product) and the commutativity of $+$ and $\cdot$ on $S$. Theorem 5.4e follows from the fact that

$$\dfrac{x}{y} \cdot \dfrac{y}{x} = \dfrac{xy}{xy} = 1$$

and that inverses are unique.

**THEOREM 5.5** (*Zero Products and Zero Quotients*) Let $\{S, +, \cdot\}$ be a field. For all $x, y$ in $S$

(a) If $x \cdot y = 0$, then $x = 0$ or $y = 0$

(b) If $x \neq 0$ and $\dfrac{x}{y} = 0$, then $x = 0$

To prove Theorem 5.5a, we use the fact that the set of all nonzero elements of $S$ forms a group under multiplication. This implies that whenever $x \neq 0$ and $y \neq 0$, then $x \cdot y \neq 0$ (closure); so if $x \cdot y = 0$, then either $x = 0$ or $y = 0$ (or both). Theorem 5.5b follows from (a) by noting that

$$\dfrac{x}{y} = x \cdot \dfrac{1}{y}$$

**THEOREM 5.6** (*Operations and Fractions*) If $\{S, +, \cdot\}$ is a field, then

(a) $\dfrac{x}{y} = \dfrac{x \cdot z}{y \cdot z}$ for all $x$ in $S$ and all nonzero $y, z$ in $S$

(b) $\dfrac{x}{y} \cdot \dfrac{u}{v} = \dfrac{x \cdot u}{y \cdot v}$ for all $x, u$ in $S$ and all nonzero $y, v$ in $S$

(c) $\dfrac{x}{y} \div \dfrac{u}{v} = \dfrac{x}{y} \cdot \dfrac{v}{u} = \dfrac{x \cdot v}{y \cdot u}$ for all $x$ in $S$ and all nonzero $y, u, v$ in $S$

(d) $\dfrac{x}{z} + \dfrac{y}{z} = \dfrac{x+y}{z}$ for all $x, y$ in $S$ and all nonzero $z$ in $S$

## 5.2 Basic Field Theorems

(e) $\dfrac{x}{u} + \dfrac{y}{v} = \dfrac{x \cdot v + y \cdot u}{u \cdot v}$ for all $x$, $y$ in $S$ and all nonzero $u$, $v$ in $S$

Theorem 5.6a is proved as follows. By definition

$$\frac{x \cdot z}{y \cdot z} = (x \cdot z) \cdot \frac{1}{y \cdot z} = (x \cdot z) \cdot \left(\frac{1}{y} \cdot \frac{1}{z}\right)$$

Using the associativity and commutativity of multiplication in $S$, this implies

$$\frac{x \cdot z}{y \cdot z} = \left(x \cdot \frac{1}{y}\right) \cdot \left(z \cdot \frac{1}{z}\right) = \left(x \cdot \frac{1}{y}\right) \cdot 1 = \frac{x}{y}$$

Theorem 5.6b is proved in a similar fashion; Theorem 5.6c is proved as follows by definition of division.

$$\frac{x}{y} \div \frac{u}{v} = \frac{x}{y} \cdot \frac{1}{\dfrac{u}{v}}$$

Using Theorem 5.4e and 5.6b, this becomes

$$\frac{x}{y} \div \frac{u}{v} = \frac{x}{y} \cdot \frac{v}{u} = \frac{x \cdot v}{y \cdot u}$$

Theorem 5.6d is proved as follows:

$$\frac{x}{z} + \frac{y}{z} = x \cdot \frac{1}{z} + y \cdot \frac{1}{z} = (x+y) \cdot \frac{1}{z} = \frac{x+y}{z}$$

(The reader should supply a reason for each step.) Theorem 5.6e follows from (a) and (d).

$$\frac{x}{u} + \frac{y}{v} = \frac{x \cdot v}{u \cdot v} + \frac{u \cdot y}{u \cdot v} = \frac{x \cdot v}{u \cdot v} + \frac{y \cdot u}{u \cdot v} = \frac{x \cdot v + y \cdot u}{u \cdot v}$$

**THEOREM 5.7** (*Signs and Fractions*) If $\{S, +, \cdot\}$ is a field, then

(a) $\dfrac{-x}{y} = -\dfrac{x}{y}$

(b) $\dfrac{x}{-y} = -\dfrac{x}{y}$  for all $x$ in $S$ and all $y$ in $S$ except 0

(c) $\dfrac{-x}{y} = \dfrac{x}{-y}$

(d) $\dfrac{-x}{-y} = \dfrac{x}{y}$

We prove Theorem 5.7a as follows:

$$\frac{-x}{y} = (-x) \cdot \frac{1}{y} = -\left(x \cdot \frac{1}{y}\right) = -\frac{x}{y} \quad \text{(The reader should supply reasons.)}$$

To prove (c), we start by applying Theorem 5.6a to the fraction $\frac{-x}{y}$ using $z = -1$. (Supply reasons.)

$$\frac{-x}{y} = \frac{(-x) \cdot (-1)}{y \cdot (-1)} = \frac{x \cdot 1}{-(y \cdot 1)} = \frac{x}{-y}$$

Theorems 5.7b and d may be proved in a similar manner. (Note also that (b) follows immediately from (a) and (c).)

## EXERCISES 5.2

1. Let $\{S, +, \cdot\}$ be a field. Prove that
   (a) $(x - y) + y = x$   for all $x, y$ in $S$
   (b) $(x + y) - y = x$   for all $x, y$ in $S$
   (c) $(x - y) + (y - x) = 0$   for all $x, y$ in $S$
   (d) $x + (y - z) = (x + y) - z = (x - z) + y$   for all $x, y, z$ in $S$
   (e) $-(x - y) = y - x$   for all $x, y$ in $S$
   (f) $(x - y) - z = x - (y + z)$   for all $x, y, z$ in $S$
   (g) $x - (y - z) = (x + z) - y$   for all $x, y, z$ in $S$
   (h) $(x + z) - (y + z) = x - y$   for all $x, y, z$ in $S$
   (i) $(x - z) - (y - z) = x - y$   for all $x, y, z$ in $S$
   (j) $(x - y) - (u - v) = (x + v) - (y + u)$   for all $x, y, u, v$ in $S$
   (k) $(x - y) + (u - v) = (x + u) - (y + v)$   for all $x, y, u, v$ in $S$
   (l) $(x - y) - (u - v) = (x - u) - (y - v)$   for all $x, y, u, v$ in $S$

2. Let $\{S, +, \cdot\}$ be a field. Prove that
   (a) $\dfrac{x}{y} \cdot y = x$   for all $x$ in $S$ and all nonzero $y$ in $S$
   (b) $\dfrac{x \cdot y}{y} = x$   for all $x$ in $S$ and all nonzero $y$ in $S$
   (c) $\dfrac{x}{y} \cdot \dfrac{y}{x} = 1$   for all nonzero $x, y$ in $S$
   (d) $x \cdot \dfrac{y}{z} = \dfrac{x \cdot y}{z} = \dfrac{x}{z} \cdot y$   for all $x, y$ in $S$ and all nonzero $z$ in $S$
   (e) $\dfrac{1}{\frac{x}{y}} = \dfrac{y}{x}$   for all nonzero $x, y$ in $S$

(f) $\dfrac{\dfrac{x}{y}}{z} = \dfrac{x}{y \cdot z}$

(g) $\dfrac{x}{\dfrac{y}{z}} = \dfrac{x \cdot z}{y}$

$\quad$ for all $x$ in $S$ and all nonzero $y, z$ in $S$

(h) $\dfrac{x}{z} \div \dfrac{y}{z} = \dfrac{x}{y}$ for all $x$ in $S$ and all nonzero $y, z$ in $S$

(i) $\dfrac{x}{y} \div \dfrac{u}{v} = \dfrac{x \cdot v}{y \cdot u}$

(j) $\dfrac{x}{y} \div \dfrac{u}{v} = \dfrac{x \div u}{y \div v}$

$\quad$ for all $x$ in $S$ and all nonzero $y, u, v$ in $S$

(k) $\dfrac{x}{z} - \dfrac{y}{z} = \dfrac{x - y}{z}$ for all $x, y$ in $S$ and all nonzero $z$ in $S$

(l) $\dfrac{x}{y} - \dfrac{u}{v} = \dfrac{x \cdot v - y \cdot u}{y \cdot v}$ for all $x, u$ in $S$ and all nonzero $y, v$ in $S$

3. Using various field theorems and results of Exercise 1, compute in two ways the value of the following in the field

$$\{W_7, (+ \bmod 7), (\times \bmod 7)\}$$

(a) $(-3)5$    (b) $(-3)(-5)$    (c) $\dfrac{3}{5}$

(d) $\dfrac{-3}{5}$    (e) $\dfrac{3}{-5}$    (f) $\dfrac{-3}{-5}$

(g) $\dfrac{3}{5} + \dfrac{4}{5}$    (h) $\dfrac{3}{5} + \dfrac{1}{4}$    (i) $\dfrac{2}{5} \times \dfrac{3}{4}$

(j) $\dfrac{3}{5} \div \dfrac{2}{5}$    (k) $\dfrac{4}{5} \div 2$    (l) $\dfrac{6}{5} \div 3$

(m) $\dfrac{2}{5} \div \dfrac{3}{4}$    (n) $4 \times \dfrac{5}{4}$    (o) $(3 - 2) + 5$

(p) $\dfrac{3}{5} - \dfrac{2}{5}$    (q) $\dfrac{2}{3} - \dfrac{1}{3}$    (r) $\dfrac{1}{3} - \dfrac{2}{3}$

4. Find the solution set for each of the following conditions if the solutions must come from the field $\{W_7, (+ \bmod 7), (\times \bmod 7)\}$. Try to find another way of obtaining your solutions.

(a) $3x = 4$    (b) $4x = 3$
(c) $x + 6 = 2$    (d) $x - 6 = 2$
(e) $x \div 2 = 6$    (f) $x \div 6 = 2$
(g) $2 \div x = 6$    (h) $2x + 3 = 1$
(i) $3x + 2 = 1$    (j) $x \cdot x = 1$

(k) $x \cdot x = 2$
(l) $x \cdot x = 3$
(m) $x \cdot x = 4$
(n) $x \cdot x = 5$
(o) $x \cdot x = 6$
(p) $x(x+1) = 0$
(q) $(x+1)(x+2) = 0$
(r) $x(x+1) = 1$
(s) $x(x+1) = -1$
(t) $x(x-1) = 2$
(u) $(x+1)(x+2) = 3$
(v) $x \cdot x = -1$
(w) $x \cdot x = -2$
(x) $x \cdot x = -3$
(y) $x \cdot x = -4$
(z) $x \cdot x = -5$

5. Let $\{S, +, \times\}$ be a field.
  (a) If $x \in S$ and $x \cdot x = 0$, prove that $x = 0$.
  (b) If $x \in S$ and $x + x = 0$, must $x = 0$? Why?
  (c) If $x \in S$ and $(x + x) + x = 0$, must $x = 0$? Why?

## 5.3 SOME FURTHER BASIC FIELD THEOREMS

The associativity of both addition and multiplication in any ring $\{S, +, \cdot\}$ assures us that for any elements $x, y, z$ in $S$

$$(x+y)+z = x+(y+z) \quad \text{and} \quad (x \cdot y) \cdot z = x \cdot (y \cdot z)$$

It is therefore unnecessary to use parentheses when writing the expressions in these equations. It is sufficient to use simpler expressions

$$x+y+z \quad \text{and} \quad x \cdot y \cdot z$$

to represent, respectively, the sums or products that appear. The same observation applies to expressions such as

$$\begin{Bmatrix} x+y+z+u \\ x+y+z+u+v \\ \text{etc.} \end{Bmatrix} \quad \text{and} \quad \begin{Bmatrix} x \cdot y \cdot z \cdot u \\ x \cdot y \cdot z \cdot u \cdot v \\ \text{etc.} \end{Bmatrix}$$

A further simplification is customary when the terms or factors are all the same. Expressions such as

$$\begin{Bmatrix} x+x \\ x+x+x \\ \text{etc.} \end{Bmatrix} \quad \text{and} \quad \begin{Bmatrix} x \cdot x \\ x \cdot x \cdot x \\ \text{etc.} \end{Bmatrix}$$

are abbreviated

$$\begin{Bmatrix} x+x = 2x \\ x+x+x = 3x \\ x+x+x+x = 4x \\ \text{etc.} \end{Bmatrix} \quad \text{and} \quad \begin{Bmatrix} x \cdot x = x^2 \\ x \cdot x \cdot x = x^3 \\ x \cdot x \cdot x \cdot x = x^4 \\ \text{etc.} \end{Bmatrix}$$

## 5.3 Some Further Basic Field Theorems

To maintain the pattern indicated here it is also customary to write

$$x = 1 \cdot x \quad \text{and} \quad x = x^1$$

Notice that the numerals "1," "2," "3," "4," etc., which appear as coefficients or as exponents do not necessarily name actual elements of the ring $\{S, +, \cdot\}$. For example, in the ring $\{W_2, +(\text{mod } 2), \times(\text{mod } 2)\}$ the members of the set $W_2$ are 0 and 1. When we write

$$\text{"}1 + 1 = 2 \cdot 1\text{,"}$$

the coefficient "2" merely tells us that 1 has been added to itself. "2" does not name an element of $W_2$.

On the other hand, in the ring $\{W_3, +(\text{mod } 3), \cdot(\text{mod } 3)\}$ the members of $W_3$ are 0, 1, and 2 and hence the expression "$2 \cdot 1$" can represent the product of two elements of $W_3$. Fortunately, this product has the same value as $1 + 1$ so it makes no difference which interpretation is used.

As another example of the distinction we are making here, consider the ring of even integers $\{E, +, \cdot\}$. The set $E$ contains the element 2 and hence also the element

$$2 + 2 + 2$$

In accordance with the convention mentioned, this element may also be designated by

$$3 \cdot 2$$

In this instance, the coefficient "3" does not name any element of $E$.

These examples show that one must exercise some care when dealing with expressions such as

$$1 \cdot x, \quad 2 \cdot x, \quad 3 \cdot x, \quad \ldots, \quad \text{etc.}$$

For some rings these expressions may represent products of actual elements of the system, but for other rings this may not be true. In any event, we may certainly interpret "$1 \cdot x$" to mean "$x$," "$2 \cdot x$" to mean "$x + x$," etc. Similarly, it is always correct to interpret "$3 \cdot x^2$," for example, to mean

$$(xx) + (xx) + (xx)$$

regardless of whether "3" and "2" do, or do not represent elements of the ring.

If a ring $\{S, +, \cdot\}$ happens to be a field, then $S$ certainly contains the two (distinct) elements 0 and 1. If the sum $1 + 1$ is distinct from both 0 and 1, then we can define

$$1 + 1 = 2$$

so $S$ will now contain 2 as an actual member. Similarly, $S$ must contain $1+1+1$, and if this is indeed a new element we can call it "3," that is, we can define
$$2+1 = 1+1+1 = 3$$
Thus in this case $S$ also contains 3. These definitions agree very nicely with the conventions introduced above because in any such field we have, for example,
$$2 \cdot x = x+x = 1x+1x = (1+1)x = 2x$$
$$3 \cdot x = x+x+x = 1x+1x+1x = (1+1+1)x = 3x$$
and so on.

We are now ready to develop some additional field theorems. (Many of them are also valid in any commutative ring.)

**THEOREM 5.8** (*Products of Binomials*) If $\{S, +, \cdot\}$ is a field (or more generally, a commutative ring), then

(a) $(x+y)^2 = x^2 + 2 \cdot xy + y^2$
(b) $(x-y)^2 = x^2 - 2 \cdot xy + y^2$
(c) $(x-y)(x+y) = x^2 - y^2$     for all $x, y$ in $S$
(d) $(x+y)^3 = x^3 + 3 \cdot x^2 y + 3 \cdot xy^2 + y^3$
(e) $(x-y)^3 = x^3 - 3 \cdot x^2 y + 3 \cdot xy^2 - y^3$
(f) $(x+y)(u+v) = xu + xv + yu + yv$ for all $x, y, u, v$, in $S$

To prove Theorem 5.8a we proceed as follows:

$$\begin{aligned}(x+y)(x+y) &= x(x+y) + y(x+y) && \text{(Why?)} \\ &= (xx+xy) + (yx+yy) && \text{(Why?)} \\ &= (x^2+xy) + (xy+y^2) && \text{(Why?)} \\ &= x^2 + (xy+xy) + y^2 && \text{(Why?)} \\ (x+y)^2 &= x^2 + 2 \cdot xy + y^2 && \text{(Why?)}\end{aligned}$$

To prove (b) we first observe that
$$(x-y)^2 = (x+(-y))^2$$
Then we apply (a) and simplify the result with the aid of previous theorems. The other parts of this theorem, which are proved by using similar techniques, are left as exercises.

**THEOREM 5.9** (*Perfect Squares*) Let $\{S, +, \times\}$ be a field.
(a) If $x^2 = y^2$, then $x = y$ or $x = -y$
(b) If $x = y$ or $x = -y$, then $x^2 = y^2$    for all $x, y$ in $S$

## 5.3 Some Further Basic Field Theorems

*Note:* This theorem is often abbreviated as follows:

> In any field, $x^2 = y^2$ if and only if $x = \pm y$.

However, it should be remembered that $x = \pm y$ is merely an abbreviation for $x = y$ or $x = -y$.

To prove Theorem 5.9a we observe that if $x^2 = y^2$, then

$$x^2 - y^2 = 0$$

Hence by Theorem 5.8c,

$$(x - y)(x + y) = 0$$

By applying Theorem 5.5a, this implies $x - y = 0$ or $x + y = 0$, that is,

$$x = y \quad \text{or} \quad x = -y$$

To prove Theorem 5.9b we observe that if $x = y$ or $x = -y$, then by Theorem 5.3b it follows that

$$xx = yy \quad \text{or} \quad xx = (-y)(-y)$$

and each of these implies that

$$x^2 = y^2$$

## EXERCISES 5.3

1. Compute each of the following for the fields

    (1) $\{W_2, +(\text{mod } 2), \times(\text{mod } 2)\}$

    and

    (2) $\{W_3, +(\text{mod } 3), \times(\text{mod } 3)\}$

    Check your result for $x = 1$.
    (a) $(x + 1)^2$  (b) $(x - 1)^2$  (c) $(x + 1)^3$
    (d) $(x - 1)^3$  (e) $(x + 1)^4$  (f) $(x - 1)^4$

2. Do Exercise 1 for the ring $\{W_4, +(\text{mod } 4), \times(\text{mod } 4)\}$.

3. If $x$ and $y$ are elements of a ring, compute each of the following:
    (a) $(x + y)^2$  (b) $(x - y)^2$  (c) $(x + y)^3$
    (d) $(x - y)^3$  (e) $(x + x)^2$  (f) $(x + x)^3$

4. Do Exercise 3 if $xy + yx = 0$ whenever $x, y$ are in the ring.

5. We have seen in Exercises 5.1, Exercise 4h, that

$$\{\{0, 1, a, b\}, +, \cdot\}$$

is a field where $+, \times$ are defined by the following tables:

| + | 0 | 1 | a | b |
|---|---|---|---|---|
| 0 | 0 | 1 | a | b |
| 1 | 1 | 0 | b | a |
| a | a | b | 0 | 1 |
| b | b | a | 1 | 0 |

| $\times$ | 0 | 1 | a | b |
|---|---|---|---|---|
| 0 | 0 | 0 | 0 | 0 |
| 1 | 0 | 1 | a | b |
| a | 0 | a | b | 1 |
| b | 0 | b | 1 | a |

(a) For this system compute:
- (1) $(a+1)^2$
- (2) $(a+b)^2$
- (3) $(a+1)^3$
- (4) $(a+b)^3$
- (5) $a(1+b)$ and $a+ab$
- (6) $b(a+b)$ and $ba+bb$
- (7) $\dfrac{1}{a} - \dfrac{1}{b}$ and $\dfrac{1}{ab}$
- (8) $\dfrac{1}{a} + \dfrac{1}{b}$ and $\dfrac{a+b}{ab}$
- (9) $\dfrac{-a}{b}, \ -\dfrac{a}{b},$ and $\dfrac{a}{-b}$
- (10) $\dfrac{1}{a} \div \dfrac{a}{b}, \ \dfrac{1}{a} \times \dfrac{b}{a},$ and $\dfrac{1 \div a}{a \div b}$

(b) For this system solve:
- (1) $x^2 = 0$
- (2) $x^2 = 1$
- (3) $x^2 = a$
- (4) $x^2 = b$
- (5) $x^2 = -1$
- (6) $x(x+1) = 0$
- (7) $(x+1)(x+a) = 0$
- (8) $x(x+1) = 1$
- (9) $x(x+1) = a$
- (10) $x^2 + ax = 0$
- (11) $x^2 + ax = 1$
- (12) $x^2 + bx = a$
- (13) $ax^2 + bx = 1$

**6.** Consider the following system $\{\{0, 1, a, b, c, d, e, f\}, +, \times\}$ where $+, \cdot$ are defined by the tables:

| + | 0 | 1 | a | b | c | d | e | f |
|---|---|---|---|---|---|---|---|---|
| 0 | 0 | 1 | a | b | c | d | e | f |
| 1 | 1 | 0 | c | f | a | e | d | b |
| a | a | c | 0 | d | 1 | b | f | e |
| b | b | f | d | 0 | e | a | c | 1 |
| c | c | a | 1 | e | 0 | f | b | d |
| d | d | e | b | a | f | 0 | 1 | c |
| e | e | d | f | c | b | 1 | 0 | a |
| f | f | b | e | 1 | d | c | a | 0 |

| $\cdot$ | 0 | 1 | a | b | c | d | e | f |
|---|---|---|---|---|---|---|---|---|
| 0 | 0 | 0 | 0 | 0 | 0 | 0 | 0 | 0 |
| 1 | 0 | 1 | a | b | c | d | e | f |
| a | 0 | a | b | c | d | e | f | 1 |
| b | 0 | b | c | d | e | f | 1 | a |
| c | 0 | c | d | e | f | 1 | a | b |
| d | 0 | d | e | f | 1 | a | b | c |
| e | 0 | e | f | 1 | a | b | c | d |
| f | 0 | f | 1 | a | b | c | d | e |

The following exercises are for the mathematical system defined above.

(a) Compute:
   (1) $(b + c) + e$ and $b + (c + e)$
   (2) $(b + d) + f$ and $b + (d + f)$
   (3) $(b \cdot c) \cdot e$ and $b \cdot (c \cdot e)$
   (4) $(b \cdot d) \cdot f$ and $b \cdot (d \cdot f)$
   (5) What do your results for Exercises (1)–(4) suggest?
   (6) $b \cdot (c + e)$ and $(b \cdot c) + (b \cdot e)$
   (7) $c \cdot (d + f)$ and $(c \cdot d) + (c \cdot f)$
   (8) What do your results for Exercises (6) and (7) suggest?
(b) Verify that the system is closed for both $+$ and $\cdot$ and that $+$ and $\cdot$ are commutative.
(c) Verify that $\{\{0, 1, a, b, c, d, e, f\}, +\}$ is a commutative group.
(d) Verify that $\{\{1, a, b, c, d, e, f\}, \cdot\}$ is a commutative group.
(e) Do you think that $\{\{0, 1, a, b, c, d, e, f\} +, \cdot\}$ is a field? What needs further checking to be certain?
(f) For this system compute each of the following:
   (1) $(b + f)^2$  (2) $(b - f)^2$
   (3) $(x + y)^2$  (4) $(x - y)^2$
   (5) $(x + y)^3$  (6) $(x - y)^3$
   (7) $3a + 4b - c$  (8) $\dfrac{1}{a} \cdot \dfrac{1}{b}$ and $\dfrac{1}{ab}$
   (9) $\dfrac{1}{a} \div \dfrac{1}{b}$ and $\dfrac{b}{a}$  (10) $\dfrac{1}{a} + \dfrac{1}{b}$ and $\dfrac{a+b}{ab}$
   (11) $\dfrac{1}{a} - \dfrac{1}{b}$ and $\dfrac{b-a}{ab}$  (12) $\dfrac{a}{b} \div \dfrac{c}{d}$, $\dfrac{ad}{bc}$, and $\dfrac{a \div c}{b \div d}$
   (13) $\dfrac{-a}{b}$, $-\dfrac{a}{b}$, and $\dfrac{a}{-b}$

(g) For this system find the solution set for each of the following:
   (1) $x^2 = 0$  (2) $x^2 = 1$
   (3) $x^2 = a$  (4) $x^2 = b$
   (5) $x^2 = c$  (6) $x(x - 1) = 0$
   (7) $x(x + 1) = 0$  (8) $(x - a)(x - b) = 0$
   (9) $(x + a)(x + b) = 0$  (10) $(x + 1)^2 = a$
   (11) $(x + 1)^2 = b$  (12) $(x + 1)^2 = c$
   (13) $(x + a)^2 = b$  (14) $(x + a)^2 = c$
   (15) $(x + c)^2 = a$  (16) $x^2 + bx + e = 0$
   (17) $bx^2 + cx + a = 0$  (18) $cx^2 + dx + e = 0$

7. If $a$ and $b$ are elements of a field, prove that
$$a \cdot b = 0 \quad \text{if and only if} \quad a = 0 \text{ or } b = 0.$$

8. Let $\{S, +, \cdot\}$ be a field for which $1 + 1 \neq 0$ and let $x$ be any element of $S$. Prove that

$$x + x = 0 \quad \text{if and only if} \quad x = 0$$

9. Show that in $\{W_2, +(\text{mod } 2), \cdot(\text{mod } 2)\}$ Theorems 5.8 a, b, c, d, and e may be expressed as follows:
   (a) $(x + y)^2 = x^2 + y^2$      for all $x, y$ in $W_2$
   (b) $(x - y)^2 = x^2 + y^2$      for all $x, y$ in $W_2$
   (c) $(x - y)(x + y) = x^2 + y^2$      for all $x, y$ in $W_2$
   (d) $(x + y)^3 = x^3 + x^2y + xy^2 + y^3$      for all $x, y$ in $W_2$
   (e) $(x - y)^3 = x^3 + x^2y + xy^2 + y^3$      for all $x, y$ in $W_2$

10. Express Theorems 5.8 a, b, c, d, and e in the field

$$\{W_3, +(\text{mod } 3), \cdot(\text{mod } 3)\}$$

## 5.4 SQUARE ROOTS AND THE QUADRATIC FORMULA

In the field $\{W_5, +(\text{mod } 5), \cdot(\text{mod } 5)\}$ the members of $W_5$ are 0, 1, 2, 3, and 4. Let us see what happens if we "square" each of these numbers, that is, if we multiply each element of $W_5$ by itself.

$$0^2 = 0 \cdot 0 = 0$$
$$1^2 = 1 \cdot 1 = 1$$
$$2^2 = 2 \cdot 2 = 4$$
$$3^2 = 3 \cdot 3 = 4$$
$$4^2 = 4 \cdot 4 = 1$$

We observe that although 0, 1, and 4 can each be obtained by squaring some member of $W_5$, the numbers 2 and 3 cannot be obtained in this manner. There is no number $x$ in $W_5$ such that $x^2 = 2$ or $x^2 = 3$, but there are, for example, two numbers in $W_5$ such that $x^2 = 4$, namely $x = 2$ or $x = 3$. The number 4 is therefore called a "**perfect square**" (or simply a "square") in $W_5$, and each of the numbers 2 or 3 is called a **square root** of 4 in $W_5$. Clearly, the numbers 0 and 1 are also squares in $W_5$, but the numbers 2 and 3 are not squares in $W_5$.

**Definition.** Let $\{S, +, \cdot\}$ be a field and let $c$ be any element of $S$. If there exists an element $x$ in $S$ such that $x^2 = c$, then $c$ is called a **perfect square** or simply a **square** in $S$, and $x$ is called a **square root** of $c$ in $S$.

**THEOREM 5.10** (*Square Roots*) $\{$Let $S, +, \cdot\}$ be a field and let $c$ be any element of $S$ which is a perfect square in $S$. If $x$ is a square root of $c$ in $S$, then $-x$ is also a square root of $c$ in $S$. Moreover, there are no other square roots of $c$ in $S$.

*Proof:* If $x$ is a square root of $c$, then by definition $x^2 = c$. But by Theorem 4.4,

$$(-x)^2 = (-x)(-x) = x \cdot x = x^2 = c$$

Hence $-x$ is also a square root of $c$. Moreover, if $y$ is any square root of $c$, then $y^2 = c$. Hence $y^2 = x^2$ and by Theorem 5.9b it follows that

$$y = x \quad \text{or} \quad y = -x$$

Theorem 5.10 does not guarantee that an element $c$ of a field $\{S, +, \cdot\}$ actually has any square roots in $S$. It merely asserts that whenever $c$ does have a square root, say $x$, then the additive inverse $-x$ is also a square root of $c$. Now $-x$ may or may not be distinct from $x$. If $-x \neq x$, then $c$ has two square roots in $S$. If $-x = x$, then $c$ has only one square root in $S$. Let us examine some further examples of these important ideas.

*Example 1.* In the field $\{W_2, +(\text{mod } 2), \times(\text{mod } 2)\}$, the element 0 has one square root, namely 0, and the element 1 also has one square root, namely 1. This is due to the fact that $-0 = 0$ and $-1 = 1$ in this field.

*Example 2.* In the field $\{W_3, +(\text{mod } 3), \times (\text{mod } 3)\}$, the element 0 has one square root, namely 0, the element 1 has two square roots, namely 1 and 2 (that is, 1 and $-1$), the element 2 has no square root in $W_3$ because there is no solution in $W_3$ for the equation $x^2 = 2$. (Verify this.)

*Example 3.* In the field $\{W_7, +(\text{mod } 7), \times(\text{mod } 7)\}$, the element 0 has one square root and each of the elements 1, 2, and 4 has two square roots (find them). The elements 3, 5, and 6 have no square roots in $W_7$.

*Example 4.* In the field of rational numbers $\{Q, +, \cdot\}$, 0 has one square root, and the negative rational numbers have no square roots in $Q$. The positive rational numbers, however, are of two types. Many positive rational numbers have two square roots in $Q$;

some examples are 1, 4, 9, $\frac{1}{4}$, and $\frac{16}{25}$ whose square roots are, respectively, $\pm 1$, $\pm 2$, $\pm 3$, $\pm\frac{1}{2}$, and $\pm\frac{4}{5}$. On the other hand, there are many positive rational numbers that have no square root in $Q$, for example, 2, 3, 5, 6, $\frac{1}{2}$, $\frac{2}{3}$.

In a sense, therefore, the rational number field is "incomplete." To obtain a number system within which every positive number has a square root, it is necessary to extend the rational number field. A complete discussion of how this can be done is beyond the scope of this text. (We shall, however, throw further light on this subject in Section 7.3.) It is sufficient to say that one such extension is the field of so-called **real numbers**, $\{R, +, \cdot\}$, where $R$ is a set that contains in addition to the rational numbers, new numbers called **irrational** numbers. In this more extensive real number field, every positive number $z$ has two square roots. If one of these square roots is $x$, that is, if $x^2 = c$, then by Theorem 5.10 the other square root is $-x$.

This new field $\{R, +, \cdot\}$, although more "complete" than the rational number field, is still "algebraically incomplete" in the sense that the negative real numbers do not have square roots within the real number field. To remedy this deficit a further extension of the real number field is necessary. This extension is, in fact, possible, although again the details are beyond the scope of this text. The resulting extension of the real number system is called the field of **complex numbers**. It is denoted by $\{C, +, \cdot\}$, where the set $C$ of complex numbers may be regarded as including the real numbers among its subsets.

If an element $c$ of a field has square roots, it is customary to select one of them and designate it by the symbol

$$\sqrt{c}$$

This means that if $x^2 = c$, then either $x = \sqrt{c}$ or $-x = \sqrt{c}$ so that $x = \pm\sqrt{c}$. If $c$ has more than one square root, the particular square root selected as $\sqrt{c}$ depends on the field being considered. Unfortunately, there is no general rule for selection that will apply to every field. In the field of real numbers, however, there is a definite convention. If $c$ is a positive real number, then we have seen that $c$ has two square roots, let us say $x$ and $-x$ (see Theorem 5.10). In this field one of these two numbers ($x$ or $-x$) must be positive and the other negative. (Positiveness and negativeness are more carefully discussed in Chapter 6.) In the field of real numbers $\sqrt{c}$ is always the positive one of these

## 5.4 Square Roots and the Quadratic Formula

two square roots. In particular, therefore,

$$\sqrt{1} = 1, \quad \sqrt{4} = 2, \quad \sqrt{\frac{9}{16}} = \frac{3}{4}, \quad \text{etc.}$$

In the case $c = 0$ there is only one square root to choose from, that is, $\sqrt{0} = 0$. Observe that $\sqrt{c}$ *is never a negative number.* Therefore we should never write $\sqrt{4} = \pm 2$ as uncritical students sometimes do.

It is convenient to rephrase Theorem 5.10 as follows.

**THEOREM 5.10** (*rephrased*) Let $\{S, +, \cdot\}$ be a field and let $c$ be any element of $S$ which is a perfect square in $S$. Let $\sqrt{c}$ be one of the square roots of $c$. The solution set of the equation $x^2 = c$ is

$$\{\sqrt{c}, -\sqrt{c}\}$$

This theorem is closely related to the problem of solving quadratic equations in a field. A **quadratic equation** is an equation of the form

$$ax^2 + bx + c = 0$$

where $a \neq 0$ and where $a$, $b$, $c$ are elements of a field. One procedure for solving quadratic equations is based on a technique known as **completing the square**. This consists of reducing the quadratic equation to an equivalent equation where one side represents a perfect square. Then a solution exists if and only if the other side also denotes a perfect square. Theorem 5.11 embodies this procedure.

**THEOREM 5.11** (*The Quadratic Formula*) Let $\{S, +, \cdot\}$ be any field in which $1 + 1 \neq 0$* and let $a$, $b$, $c$ be elements of $S$ with $a \neq 0$. The quadratic equation

$$ax^2 + bx + c = 0 \quad (a \neq 0)$$

has a solution in $\{S, +, \cdot\}$ if and only if $b^2 - 4ac$ is a perfect square in $S$. In this case the solution set of the quadratic equation is

$$\left\{\frac{-b + \sqrt{b^2 - 4ac}}{2a}, \frac{-b - \sqrt{b^2 - 4ac}}{2a}\right\}$$

---

* A field in which $1 + 1 = 0$ is $\{W_2, +, \cdot\}$. Other examples are mentioned in Exercises 5 and 6 of Section 5.3. For such fields, the quadratic formula does not apply, but fortunately these fields are rarely encountered except in more advanced studies.

where $\sqrt{b^2 - 4ac}$ is one of the square roots of $b^2 - 4ac$.*

*Proof:* Since $a \neq 0$, the equation
$$ax^2 + bx + c = 0$$
is equivalent† to
$$a \cdot (ax^2 + bx + c) = a \cdot 0$$
that is, to
$$a^2x^2 + abx + ac = 0$$

*Adding the left number to itself three times*, we obtain $4a^2x^2 + 4abx + 4ac = 0$, which is equivalent to
$$4a^2x^2 + 4abx = -4ac$$

Then adding $b^2$ to both sides we obtain
$$4a^2x^2 + 4abx + b^2 = b^2 - 4ac$$

With the aid of Theorem 5.8a and other field properties we can express this equation as
$$(2ax + b)^2 = b^2 - 4ac$$

If $b^2 - 4ac$ is a perfect square and if $\sqrt{b^2 - 4ac}$ is one of its square roots, then by Theorem 5.10 it follows that
$$2ax + b = \sqrt{b^2 - 4ac} \quad \text{or} \quad 2ax + b = -\sqrt{b^2 - 4ac}$$
that is,
$$2ax = -b + \sqrt{b^2 - 4ac} \quad \text{or} \quad 2ax = -b - \sqrt{b^2 - 4ac}$$
Now *if* $1 + 1 \neq 0$, then $(1 + 1)a \neq 0$ because $a \neq 0$. This yields $2a \neq 0$. Hence $2a$ has a reciprocal $\dfrac{1}{2a}$ in $S$, and it then follows that
$$x = \frac{-b + \sqrt{b^2 - 4ac}}{2a} \quad \text{or} \quad x = \frac{-b - \sqrt{b^2 - 4ac}}{2a}$$
that is, the solution set is
$$\left\{ \frac{-b + \sqrt{b^2 - 4ac}}{2a}, \frac{-b - \sqrt{b^2 - 4ac}}{2a} \right\}$$

* In the expression "$4ac$" we may always interpret "4" as "$1+1+1+1$".
† Equivalent equations have the same solution set.

## 5.4 Square Roots and the Quadratic Formula

Further observations concerning Theorem 5.11 and its proof are worthy of mention. In the quadratic formula for the solutions

$$\frac{-b + \sqrt{b^2 - 4ac}}{2a}, \quad \frac{-b - \sqrt{b^2 - 4ac}}{2a}$$

$a$, $b$, and $c$ are arbitrary elements in the field with $a \neq 0$, but the numerals "2" and "4" need not denote any such elements. As already pointed out, the expressions "$2a$" and "$4ac$" are abbreviations, respectively, for "$a + a$" and "$ac + ac + ac + ac$." It is only in certain fields such as $\{Q, +, \cdot\}$, $\{R, +, \cdot\}$, $\{W_5, +, \cdot\}$, and so on, that we may justifiably regard 2 and 4 as actual elements of the system. In the proof of Theorem 5.11, we therefore carefully avoided treating either 2 or 4 as an actual element in the field. This made the proof a little longer but more general. For fields that do contain 2 and 4 as actual elements (for example, in the field of real numbers) we can vary the initial steps in the proof somewhat by *multiplying* both sides of the equation $ax + bx + c = 0$ by 4a:

$$(4a) \cdot (ax^2) + (4a) \cdot (bx) + (4a) \cdot c = (4a) \cdot (0)$$

Then if 4 is an element in the field, it is legitimate to rewrite the equation as

$$4(a \cdot ax^2) + 4(abx) + 4(ac) = 0$$

that is,

$$4a^2x^2 + 4abx + 4ac = 0$$

Other variations of the proof are possible if special fields are considered. The proof we have adopted applies very generally to any field in which $1 + 1 \neq 0$.

Let us examine some examples of the use of the quadratic formula in various fields.

**Example 5.** Solve $x^2 + x + 2 = 0$ in the field $\{W_3, +(\text{mod } 3), \times(\text{mod } 3)\}$.

*Solution:* In this field

$$b^2 - 4ac = (1)^2 - 4(1)(2) = 1 - 2 = 1 + 1 = 2$$

Since 2 is not a perfect square in $W_3$, the equation has no solution.

**Example 6.** Solve $x^2 + x + 2 = 0$ in the field $\{W_7, +(\text{mod } 7), \cdot(\text{mod } 7)\}$.

*Solution:* In this field

$$b^2 - 4ac = (1)^2 - 4(1)(2) = 1 - 1 = 0$$

Since 0 is a perfect square, a solution exists in this field. In fact, all solutions are given by the quadratic formula

$$x = \frac{-b \pm \sqrt{b^2 - 4ac}}{2a} = \frac{-1 \pm \sqrt{0}}{2}$$

$$x = \frac{-1 \pm 0}{2} = \frac{-1}{2} = (-1) \times \frac{1}{2} = 6 \times 4 = 3$$

Hence the quadratic equation $x^2 + x + 2 = 0$ has exactly one solution in $W_7$, namely 3.

**Example 7.** Solve $x^2 + x + 2 = 0$ in the field $\{W_{11}, +(\text{mod } 11), \cdot(\text{mod } 11)\}$.

*Solution:* In this field

$$b^2 - 4ac = (1)^2 - (4)(1)(2) = 1 - 8 = 1 + 3 = 4$$

Since 4 is a perfect square, a solution exists and, in fact, all solutions are given by

$$x = \frac{-b \pm \sqrt{b^2 - 4ac}}{2a} = \frac{-1 \pm \sqrt{4}}{2} = \frac{-1 \pm 2}{2}$$

$$\therefore \quad x = \frac{-1 + 2}{2} \quad \text{or} \quad x = \frac{-1 - 2}{2}$$

$$\therefore \quad x = \frac{1}{2} \quad \text{or} \quad x = \frac{-3}{2} = \frac{8}{2}$$

$$\therefore \quad x = 6 \quad \text{or} \quad x = 4$$

Hence the quadratic equation $x^2 + x + 2 = 0$ has exactly two solutions in $W_{11}$, namely 4 and 6.

**Example 8.** Solve $x^2 + x + 2 = 0$ in the field of real numbers $\{R, +, \cdot\}$.

*Solution:* In this field

$$b^2 - 4ac = 1 - 4(1)(2) = 1 - 8 = -7$$

Since $-7$ is not a perfect square in the set of real numbers, there is no solution.

**Example 9.** Solve $x^2 + x - 2 = 0$ in the field of rational numbers $\{Q, +, \cdot\}$.

*Solution:* This time

$$b^2 - 4ac = 1 - 4(1)(-2) = 1 + 8 = 9$$

### 5.4 Square Roots and the Quadratic Formula

Since 9 is a perfect square in the field of rational numbers, a solution exists and, in fact, all solutions are given by

$$x = \frac{-b \pm \sqrt{b^2 - 4ac}}{2a} = \frac{-1 \pm \sqrt{9}}{2} = \frac{-1 \pm 3}{2}$$

$$\therefore \quad x = \frac{-1+3}{2} \quad \text{or} \quad x = \frac{-1-3}{2}$$

$$\therefore \quad x = \frac{2}{2} \quad \text{or} \quad x = \frac{-4}{2}$$

$$\therefore \quad x = 1 \quad \text{or} \quad x = -2$$

Consequently, the equation $x^2 + x - 2 = 0$ has two solutions in the field of rational numbers, namely 1 and $-2$.

We pause to remark that a quadratic equation can often be solved by *factoring* $ax^2 + bx + c$ and then making use of Theorem 5.5a. Thus in Example 9 we might write

$$x^2 + x - 2 = 0$$

or

$$(x-1)(x+2) = 0$$
$$\therefore \quad x-1 = 0 \quad \text{or} \quad x+2 = 0 \quad \text{(By Theorem 5.5a)}$$
$$\therefore \quad x = 1 \quad \text{or} \quad x = -2$$

In some fields, however, the factors may not be as obvious as they are in Example 9. For instance, in Example 7, the quadratic equation is

$$x^2 + x + 2 = 0$$

It requires a considerable amount of trial to see that this equation is equivalent to

$$(x+5)(x+7) = 0$$

so that

$$x+5 = 0 \quad \text{or} \quad x+7 = 0$$
$$\therefore \quad x = -5 \quad \text{or} \quad x = 7$$
$$\therefore \quad x = 6 \quad \text{or} \quad x = 4$$

Of course, now that we know the solutions we can also see that another factorization is given by

$$(x-6)(x-4) = 0$$
$$\therefore \quad x-6 = 0 \quad \text{or} \quad x-4 = 0$$
$$\therefore \quad x = 6 \quad \text{or} \quad x = 4$$

but it is unlikely that we would have noticed this in advance.

In general, unless a pair of factors is fairly obvious it is probably best to resort to the quadratic formula or to the method used in deriving the quadratic formula, completing the square. We illustrate this method for Example 7.

*Solution:*

$$x^2 + x + 2 = 0 \quad \text{in } \{W_{11}, +, \cdot\}$$
$$4x^2 + 4x + 8 = 0$$
$$4x^2 + 4x = -8$$
$$4x^2 + 4x + 1 = 1 - 8$$
$$(2x + 1)^2 = 1 + 3$$
$$(2x + 1)^2 = 4$$
$$2x + 1 = \pm 2$$
$$2x = -1 \pm 2$$

$2x = 1$ or $2x = -3$
$x = \tfrac{1}{2}$ or $2x = 8$
$x = 6$ or $x = 4$

## EXERCISES 5.4

1. Solve each of the following equations in the fields $\{W_7, +(\bmod 7), \times(\bmod 7)\}$ and $\{W_{11}, +(\bmod 11), \times(\bmod 11)\}$. Check your solutions for each field and compare the solutions obtained for one field with those obtained for the other field.
   (a) $x^2 = 4$
   (b) $(x + 1)^2 = 4$
   (c) $x^2 = -5$
   (d) $(x + 1)^2 = -5$
   (e) $3x^2 = 5$
   (f) $3(x - 4)^2 = 5$
   (g) $x^2 + 3x = 4$
   (h) $2x^2 + x + 1 = 0$
   (i) $2x^2 + x + 5 = 0$
   (j) $3x^2 + x + 3 = 0$
   (k) $3x^2 + x + 4 = 0$

2. Solve the equations of Exercise 1 in the fields $\{Q, +, \cdot\}$ and $\{R, +, \cdot\}$ and compare the results for these two fields.

3.

| + | 0 | 1 | a | b | c | d | e | f |
|---|---|---|---|---|---|---|---|---|
| 0 | 0 | 1 | a | b | c | d | e | f |
| 1 | 1 | 0 | c | f | a | e | d | b |
| a | a | c | 0 | d | 1 | b | f | e |
| b | b | f | d | 0 | e | a | c | 1 |
| c | c | a | 1 | e | 0 | f | b | d |
| d | d | e | b | a | f | 0 | 1 | c |
| e | e | d | f | c | b | 1 | 0 | a |
| f | f | b | e | 1 | d | c | a | 0 |

|   | 0 | 1 | a | b | c | d | e | f |
|---|---|---|---|---|---|---|---|---|
| 0 | 0 | 0 | 0 | 0 | 0 | 0 | 0 | 0 |
| 1 | 0 | 1 | a | b | c | d | e | f |
| a | 0 | a | b | c | d | e | f | 1 |
| b | 0 | b | c | d | e | f | 1 | a |
| c | 0 | c | d | e | f | 1 | a | b |
| d | 0 | d | e | f | 1 | a | b | c |
| e | 0 | e | f | 1 | a | b | c | d |
| f | 0 | f | 1 | a | b | c | d | e |

(a) Why is the quadratic formula not applicable for the field in Exercise 6, Exercises 5.3? (Tables are shown above.)

★(b) Try to find a technique for solving quadratic equations for this field and apply it to solving each of the following:
 (1) $x^2 + x = a$
 (2) $x^2 + x = b$
 (3) $x^2 + x = c$
 (4) $ax^2 + x = c$
 (5) $x^2 + ax = c$
 (6) $x^2 + ax = f$
 (7) $dx^2 + ex = c$

## 5.5 FURTHER APPLICATION OF FIELD THEOREMS TO SOLUTION OF EQUATIONS

In Section 5.4 we applied field theorems to the problem of solving quadratic equations in any given field. In this section we shall explore various other types of equations. In so doing we shall uncover some surprising, possibly even startling results. Nevertheless these results are all governed by the field theorems.

***Example 1.*** Solve the equation

$$\frac{1}{x} - \frac{2}{3} = \frac{1}{2}$$

in (a) the field of rational numbers $\{Q, +, \times\}$,
 (b) the field $\{W_5, +\text{(mod 5)}, \times\text{(mod 5)}\}$,
 (c) the field $\{W_7, +\text{(mod 7)}, \times\text{(mod 7)}\}$.

*Solution:* Our initial steps in solving this equation are valid in any field where $2 \neq 0$ and $3 \neq 0$. (*Note:* This excludes from consideration such fields as

$$\{W_2, +(\text{mod } 2), \times(\text{mod } 2)\}$$

and

$$\{W_3, +(\text{mod } 3), \times(\text{mod } 3)\}$$

In the former field the fraction $\frac{1}{2}$ is meaningless; in the latter field the fraction $\frac{2}{3}$ is meaningless.)

$$\frac{1}{x} = \frac{1}{2} + \frac{2}{3}$$

$$\frac{1}{x} = \frac{3(1) + 2(2)}{(2)(3)}$$

$$x = \frac{2(3)}{3 + 2(2)}$$

provided $3 + 2(2) \neq 0$.

(a) In the rational number field $\{Q, +, \times\}$ the solution becomes

$$\frac{6}{3+4} = \frac{6}{7}$$

This result is readily checked by ordinary arithmetic

$$\frac{1}{\frac{6}{7}} \stackrel{?}{=} \frac{1}{2} + \frac{2}{3}$$

$$\frac{7}{6} \stackrel{?}{=} \frac{3}{6} + \frac{4}{6}$$

$$\frac{7}{6} = \frac{7}{6}$$

(b) In the field $\{W_5, +(\text{mod } 5), \times(\text{mod } 5)\}$ the solution becomes

$$\frac{1}{3+4} = \frac{1}{2} = 3$$

To check this result we proceed as follows:

$$\frac{1}{3} \stackrel{?}{=} \frac{1}{2} + \frac{2}{3}$$

$$2 \stackrel{?}{=} 3 + 4$$

$$2 = 2$$

## 5.5 Further Application of Field Theorems to Solution of Equations

(c) In the field $\{W_7, +(\text{mod } 7), \times(\text{mod } 7)\}$,

$$3 + 2(2) = 3 + 4 = 0$$

Hence for this field the last step is not possible and there is therefore no solution.

**Example 2.** Solve

$$\frac{x^2 + x - 2}{x + 2} = 0$$

in the fields

(a) $\{W_2, +(\text{mod } 2), \times(\text{mod } 2)\}$
(b) $\{W_3, +(\text{mod } 3), \times(\text{mod } 3)\}$

*Solution:* Once again, the initial step is valid in any field.

$$\frac{(x+2)(x-1)}{(x+2)} = 0$$

(a) In the field $\{W_2, +, \times\}$, we interpret "2" as "0." Hence the equation is equivalent to

$$\frac{x(x-1)}{x} = 0$$

Now by Theorem 5.6a we may cancel the common factor $x$ provided that $x \neq 0$. Therefore this equation is equivalent to

$$x - 1 = 0$$

or

$$x = 1$$

So in this field the only solution is 1.

(b) In the field $\{W_3, +(\text{mod } 3), \times (\text{mod } 3)\}$, $2 = -1$. Hence the equation is equivalent to

$$\frac{(x-1)(x-1)}{x-1} = 0$$

Now by Theorem 5.6a we may cancel the common factor $x - 1$ provided that $x - 1 \neq 0$. Hence this equation is equivalent to

$$x - 1 = 0 \quad \text{provided that} \quad x - 1 \neq 0.$$

But this is clearly impossible because we cannot have both $x = 1$ and $x \neq 1$. Therefore in the field

$$\{W_3, +(\text{mod } 3), \times(\text{mod } 3)\}$$

the equation has no solution.

**Example 3.** Solve

$$x^3 + x^2 + x = 0$$

in each of the following fields:

(a) $\{W_2, +(\text{mod } 2), \times(\text{mod } 2)\}$
(b) $\{W_7, +(\text{mod } 7), \times(\text{mod } 7)\}$
(c) $\{R, +, \times\}$ (field of real numbers)
(d) $\{C, +, \times\}$ (field of complex numbers)

*Solution:* Our first two steps are valid in any field.

$$x(x^2 + x + 1) = 0$$
$$x = 0 \quad \text{or} \quad x^2 + x + 1 = 0$$

We see therefore that 0 is a solution in every field.

(a) In the field $\{W_2, +(\text{mod } 2), \times(\text{mod } 2)\}$ the only number other than 0 is 1. But 1 is clearly not a solution because

$$1^2 + 1 + 1 = 1 \neq 0$$

Hence 0 is the only solution in this field. (*Note:* The quadratic formula could not be applied in this field.)

(b) In the field $\{W_7, +(\text{mod } 7), \times(\text{mod } 7)\}$ let us test the quadratic equation

$$x^2 + x + 1 = 0$$

to see if it has any solutions.

$$b^2 - 4ac = 1^2 - 4(1)(1) = 1 - 4 = 1 + 3 = 4$$

Since 4 is a square in $W_7$, we see that further solutions do indeed exist. They are

$$\frac{-1 + \sqrt{4}}{2} = \frac{-1 + 2}{2} = \frac{1}{2} = 4$$

and

$$\frac{-1 - \sqrt{4}}{2} = \frac{-1 - 2}{2} = \frac{-3}{2} = \frac{4}{2} = 2$$

Hence in the field $\{W_7, +(\mathrm{mod}\ 7), \times(\mathrm{mod}\ 7)\}$ equation
$$x^3 + x^2 + x = 0$$
has three solutions, namely
$$0, 4, \text{ and } 2$$

With the "hindsight" thus obtained, we can now be clever and solve the original equation by factoring.

$$x^3 + x^2 + x = 0$$
$$x(x^2 + x + 1) = 0$$
$$x(x - 4)(x - 2) = 0 \quad \text{(Verify these factors.)}$$
$$\therefore x = 0, \quad \text{or} \quad x = 4, \quad \text{or} \quad x = 2$$

(c) In the field $\{R, +, \times\}$ the quadratic equation
$$x^2 + x + 1 = 0$$
has no solution because
$$b^2 - 4ac = 1^2 - 4(1)(1) = 1 - 4 = -3$$
There is no number in $R$ whose square is $-3$. Therefore the only solution in this field is 0.

(d) On the other hand, in the complex number field $\{C, +, \times\}$ there is a number whose square is $-3$. This number is usually expressed as $i\sqrt{3}$. Hence the equation
$$x^2 + x + 1 = 0$$
has two solutions in $C$,
$$\frac{-1 + i\sqrt{3}}{2} \quad \text{and} \quad \frac{-1 - i\sqrt{3}}{2}$$

Therefore the equation $x^3 + x^2 + x = 0$ has three solutions in the complex number field:
$$0, \quad -\frac{1}{2} + \frac{\sqrt{3}}{2}i, \quad \text{and} \quad -\frac{1}{2} - \frac{\sqrt{3}}{2}i$$

# EXERCISES 5.5

For each equation below find the solution set in each of the following fields.

(a) $\{W_2, +(\mathrm{mod}\ 2), \times(\mathrm{mod}\ 2)\}$
(b) $\{W_3, +(\mathrm{mod}\ 3), \times(\mathrm{mod}\ 3)\}$

**5 ‖ Fields**

(c) $\{W_5, +(\text{mod } 5), \times(\text{mod } 5)\}$
(d) $\{W_7, +(\text{mod } 7), \times(\text{mod } 7)\}$
(e) $\{W_{11}, +(\text{mod } 11), \times(\text{mod } 11)\}$
(f) $\{Q, +, \times\}$ (the field of rational numbers)
(g) $\{R, +, \times\}$ (the field of real numbers)
(h) $\{C, +, \times\}$ (the field of complex numbers)

*Note:* Where necessary, interpret the numerical coefficients appropriately in relation to the particular field. For example, interpret

$$12 = 5 \text{ in } W_7, \qquad 12 = 2 \text{ in } W_5, \qquad 12 = 0 \text{ in } W_3$$

1. $5 + 4x = x - 7$
2. $4 - 3x = 2x - 2 - 5x$
3. $2x^2 + 4x = 0$
4. $x^2 - 8x + 1 = 0$
5. $2x^2 + 5x - 9 = 3$
6. $\dfrac{3x - 7}{x^2 + 1} = 0$
7. $\dfrac{x^2 - 7x + 12}{x - 5} = 0$
8. $\dfrac{x^2 - 7x + 10}{x - 5} = 0$
9. $2x^3 - 4x^2 + 7x = 0$
10. $x^4 + 6x^2 + 5 = 0$
11. $\dfrac{x}{2} + \dfrac{x}{3} = 15$
12. $2x + 3 = \dfrac{2}{x}$
13. $\dfrac{2}{x - 3} + \dfrac{3}{x + 3} = 1$
14. $\dfrac{x^2 - 1}{x - 1} = 5$
15. $\dfrac{x + 7}{x^2 - 1} + \dfrac{x - 3}{x^2 - x} + \dfrac{x - 2}{x^2 + x} = 0$
16. $\dfrac{1}{x + 1} = 1 - \dfrac{x}{x + 1}$
17. $\dfrac{1}{x + 1} = 1 - \dfrac{1}{x + 1}$
18. $\dfrac{1}{x - 1} = 1 + \dfrac{1}{x - 1}$
19. $\dfrac{1}{x + 1} = \dfrac{1}{x(x + 1)}$
20. $\dfrac{1}{x - 1} = \dfrac{1}{x(x - 1)}$
21. $x^3 = 1$
22. $(x - 1)^3 = 1$
23. $x^3 = 2$

## REVIEW EXERCISES

In Exercises 1–10 find the solution set in the field $\{W_{11}, +(\text{mod } 11), \cdot(\text{mod } 11)\}$.

1. $x^2 = -2$. Find another field for which the solution set is not empty. Find a field for which the solution set is empty.
2. $x^2 = x$. Do solutions exist in every field? Why?
3. $x^2 = x + 1$. Specify a field in which this equation has exactly one solution.

4. $x^2 = x + 2$. Is there a field in which there is no solution? Why?

5. $\dfrac{2}{x+1} = x$

6. $1 + \dfrac{x}{x+1} = \dfrac{1}{x+1}$

7. $1 - \dfrac{x}{x+1} = \dfrac{1}{x+1}$

8. $1 + \dfrac{x}{x+1} = 1 - \dfrac{1}{x+1}$

9. $\dfrac{1}{x+2} = \dfrac{2}{x+3}$

★10. $x + \dfrac{2}{x+10} = \dfrac{2x}{x+10}$

11. Explain why the following are not fields.

(a) $\{S, +, \cdot\}$, where $S = \left\{ \dfrac{a}{b} \middle| a, b \text{ are integers}, b \text{ is odd} \right\}$.

(b) $\{S, +, \cdot\}$, where $S = \{x \mid x^2 \text{ is rational}\}$.

(c) $\{W_6, +(\text{mod } 6), \cdot(\text{mod } 6)\}$.

12. Let $S$ be a set consisting of the following nine elements:

$$0, \quad 1, \quad 2, \quad c, \quad c+1, \quad c+2, \quad 2c, \quad 2c+1, \quad 2c+2$$

Furthermore, let addition (+) and multiplication ($\cdot$) on $S$ obey the customary rules of algebra but with two additional restrictions: (1) operations involving $0, 1, 2$ are computed modulo 3 and (2) $c \cdot c = c + 1$. For example,

$$\begin{aligned}(c+1)(c+2) &= (c \cdot c) + (1 \cdot c) + (2 \cdot c) + (1 \cdot 2) \\ &= (c \cdot c) + (1 + 2)c + 2 \\ &= c + 1 + 0 \cdot c + 2 \\ &= c + 1 + 2 \\ &= c + 0 \\ &= c\end{aligned}$$

(a) Set up addition and multiplication tables for this system.

(b) Show that $\{S, +, \cdot\}$ is a field.

(c) Compute in two ways:

(1) $\dfrac{1}{c} + \dfrac{2}{c+1}$

(2) $\dfrac{1}{c} \cdot \dfrac{2}{c+1}$

(d) Solve for $y$:

(1) $cy + y = 2$

(2) $(c+1)y + 2y = c$
(3) $y^2 - y - 2 = 0$
(4) $y^2 - 2y - c = 0$
(5) $y^2 + 2cy + c + 1 = 0$

(e) Find all the subfields for this field.

## RESEARCH PROBLEM

The following is extracted from *The Role of Axiomatics and Problem Solving in Mathematics* compiled by the Conference Board of Mathematical Sciences (E. G. Begle, Chairman), issued by Ginn and Company in 1966 and credited to Leon Henkin.

**Axiom 1.** $\{P, +, \cdot\}$ *is a mathematical system with* $P \neq \emptyset$ *and* $+$ *and* $\cdot$ *binary operations on P.*

**Axiom 2.** *For all x, y, z in P we have*
$(x + y) + z = x + (y + z)$     ($+$ *is associative*)
$(x \cdot y) \cdot z = x \cdot (y \cdot z)$     ($\cdot$ *is associative*)

**Axiom 3.** *For all x, y, z in P we have*
$x \cdot (y + z) = (x \cdot y) + (x \cdot z)$     ($\cdot$ *distributes over* $+$
$(y + z) \cdot x = (y \cdot x) + (z \cdot x)$     *in both directions*)

**Axiom 4.** *For all a, b in P there is an x in P such that* $a \cdot x = b$ *and there is a y in P such that* $y \cdot a = b$.

With these axioms, prove the following theorems.

**THEOREM 1**  There is exactly one $e$ in $P$ such that $e \cdot x = x$ for every $x$ in $P$. Moreover, for this $e$, we also have $x \cdot e = x$ for all $x$ in $P$.

**THEOREM 2**  For each $a$ in $P$ there is exactly one $x$ in $P$ such that $a \cdot x = e$. Moreover, for this $x$ we also have $x \cdot a = e$. We shall denote this $x$ by our usual inverse notation, $a'$, and say that $a'$ is the multiplicative inverse of $a$.

**THEOREM 3**   For all $x$ and $y$ in $P$
(a) $(x')' = x$
(b) $(x \cdot y)' = y' \cdot x'$

**THEOREM 4**   Denote $e$ by "1," $a'$ by "$a^{-1}$" or "$\frac{1}{a}$," $a \cdot b'$ by "$\frac{a}{b}$," or "$a/b$." Then for all $x, y, z$ in $P$ we have

(a) $\dfrac{x}{x} = 1$

(b) $\left(\dfrac{x}{y}\right)^{-1} = \dfrac{y}{x}$

(c) $\dfrac{x \cdot z}{y \cdot z} = \dfrac{x}{y}$

(d) $\dfrac{x}{y} \cdot \dfrac{y}{z} = \dfrac{x}{z}$

(e) $\dfrac{x}{z} + \dfrac{y}{z} = \dfrac{x+y}{z}$

*Axiom 5.*   Let $G$ be any nonempty subset of $P$ which is closed under the operations of $+$ and $/$. In other words, for all $x, y$ in $G$, $x+y$ is in $G$ and $x/y$ is in $G$. It then must follow that $G = P$.

With this additional axiom, prove the following theorems.

**THEOREM 5**   For all $x, y$ in $P$, $x \cdot y = y \cdot x$.   ($\cdot$ is commutative)

**THEOREM 6**   For all $x, y, u, v$ in $P$

(a) $\left(\dfrac{x}{u}\right) \cdot \left(\dfrac{y}{v}\right) = \dfrac{x \cdot y}{u \cdot v}$

(b) $\dfrac{\left(\dfrac{x}{u}\right)}{\left(\dfrac{y}{v}\right)} = \dfrac{x \cdot v}{y \cdot u}$

(c) $\left(\dfrac{x}{u}\right) + \left(\dfrac{y}{v}\right) = \dfrac{(x \cdot v) + (u \cdot y)}{u \cdot v}$

(d) $\dfrac{x}{u} = \dfrac{y}{v}$ if and only if $x \cdot v = u \cdot y$

*Axiom 6.*   If $x$ and $y$ are any elements in $P$ such that $x+1 = y+1$, then $x = y$.

## 5 ‖ Fields

With this additional axiom prove the following additional theorems.

**THEOREM 7** If $x, y, z$ are any elements in $P$ such that $x + z = y + z$, then $x = y$.

**THEOREM 8** For all $x, y$ in $P$, $x + y = y + x$.  ($+$ is commutative)

**THEOREM 9** (*Actually a "Meta Theorem"*) Show that it is impossible to prove that $1 + 1 \neq 1$.

***Axiom 7.*** *There exist elements $x, y$ in $P$ such that $x \neq y$.*

Using this additional axiom, prove the following theorem.

**THEOREM 10** For all $x, y$ in $P$, $x + y \neq x$.

New York Public Library

*Giuseppe Peano (1858-1932)*

# 6 | Ordered Fields

> A discussion of order... has become essential to any understanding of mathematics.
>
> BERTRAND RUSSELL

## 6.1 AXIOMS FOR AN ORDERED FIELD

Thus far ten out of eleven of the field theorems considered express properties that are shared by all fields, and even the eleventh theorem (Theorem 5.11) applies quite generally since the restriction $1 + 1 \neq 0$ is valid for most of the fields that we encounter. Therefore, these theorems have wide applicability, but for that very reason they are of little help in pinpointing properties of special fields.

Among the special fields that are of considerable importance in mathematics are the field of **real numbers** $\{R, +, \cdot\}$ and the field of **rational numbers** $\{Q, +, \cdot\}$. These two fields possess a property which is not shared by many of the other fields studied. Unlike the finite fields such as $\{W_5, +(\text{mod } 5), \cdot(\text{mod } 5)\}$ and the complex number field $\{C, +, \cdot\}$, the fields of real numbers and rational numbers are both **ordered** fields. If we are given any two (distinct) real numbers $a$ and $b$, one will always be "greater" than the other. When, for example, $a$ is greater than $b$, we usually express that fact by an inequality

$$a > b$$

## 6 || Ordered Fields

Although the field axioms and field theorems up to this point are adequate for treating *equations,* which are mathematical sentences that express equality relationships, these axioms and theorems have limited applicability to *inequations* (or *inequalities*) which are sentences that involve the notion of order. Inequations (or inequalities) often arise in special fields such as $\{Q, +, \cdot\}$ and $\{R, +, \cdot\}$. The machinery for treating inequations is provided by introducing some additional axioms concerning the order of the numbers in these fields.

There are various ways to formulate these additional axioms. A particularly elegant approach is to single out the so-called *positive* elements of the ordered field by assuming that there is a set of elements $P$ which has three fairly simple and quite reasonable properties.

**Definition.** A field $\{S, +, \cdot\}$ is called an **ordered field** if and only if there exists a subset $P$ of $S$ having the following properties.

*Axioms of Order.*   *O*1. Zero is not a member of $P$.

*O*2. For every $x$ in $S$, other than zero, either $x$ is in $P$ or else*—$x$ is in $P$.

*O*3. Addition and multiplication are binary operations on $P$, that is, if $x$ and $y$ are members of $P$, then $x + y$ and $x \cdot y$ are also members of $P$. (*Note:* This axiom can also be expressed by saying that $P$ is closed with respect to addition and multiplication.)

Let us check to see which of these additional axioms are actually satisfied by some of the fields we have already studied.

*Example 1.*   Let $\{S, +, \cdot\}$ be the field $\{Q, +, \cdot\}$ of rational numbers. If we let $P$ be the set of all positive rational numbers, we observe the following:

1. Zero is not a positive rational number and hence is not a member of $P$.
2. Every rational number $x$ other than zero is either positive or else negative. If $x$ is positive, then $x$ is a member of $P$. If $x$ is negative, then $-x$ will be positive and hence $-x$ will be a member of $P$. In any case, if $x$ is not zero, then either $x$ is in $P$ or $-x$ is in $P$.

---
* "or else" is used here in the exclusive sense, that is, one or the other but not both possibilities must hold.

3. The sum and product of a pair of positive rational numbers will always be positive. Hence $P$ is closed under both addition and multiplication.

We see therefore that the field of rational numbers does indeed meet all requirements stipulated by the axioms for an ordered field. A very similar argument holds for the field of real numbers.

*Example 2.* Let $\{S, +, \cdot\}$ be the field $\{W_3, +(\text{mod } 3), \cdot(\text{mod } 3)\}$ where $W_3 = \{0, 1, 2\}$. We shall prove that this is not an ordered field by showing that there does not exist a subset $P$ satisfying the requirements of Axioms $\mathcal{O}1$, $\mathcal{O}2$, and $\mathcal{O}3$.

Since 0 may not be a member of $P$ (by Axiom $\mathcal{O}1$), let us examine only those subsets of $W_3$ that do not contain 0:

$$\{1, 2\}, \quad \{1\}, \quad \{2\}, \quad \varnothing*$$

Can $P = \{1, 2\}$? No, because $1 + 2 = 0$, which is not a member of $P$. Therefore Axiom $\mathcal{O}3$ does not hold for the set $\{1, 2\}$.

Can $P = \{1\}$? No, because $1 + 1 = 2$, which is not a member of $P$, thus violating Axiom $\mathcal{O}3$.

Can $P = \{2\}$? No, because $2 + 2 = 1$, which is not a member of $P$, thus violating Axiom $\mathcal{O}3$.

Finally, can $P = \varnothing$? No, because $P$ cannot be empty since Axiom $\mathcal{O}2$ requires that either 1 or $-1$ (namely 2) must belong to $P$.

Thus we see that none of the subsets of $W_3$ can qualify as the desired set $P$.

It is apparent that some fields are ordered fields, whereas others are not. Our object in this chapter is to study ordered fields. We shall do this by considering some of the consequences which stem from the additional axioms we have introduced for ordered fields.

Let us therefore examine once again our definition of an ordered field $\{S, +, \cdot\}$. The definition requires first that there be a subset $P$ of $S$ that satisfies Axioms $\mathcal{O}1$, $\mathcal{O}2$, and $\mathcal{O}3$.

**Definition.** The members of the subset $P$ are called the **positive elements** of $S$.

\* $\varnothing$ is the *null* set (*empty* set). Logically, the null set must be regarded as a subset of any set.

Axiom $\mathcal{O}1$ stipulates that 0 is not a positive element of $S$.

What about the remaining elements of $S$, those elements of $S$ which are neither zero nor positive? These remaining members form another subset of $S$ which we designate by $N$.

**Definition.** The members of the subset $N$ are called the **negative elements of $S$**. (The negative elements are defined to be those members of $S$ which are neither zero nor positive.)

The subsets $P$, $N$, and $\{0\}$ therefore form a *partition* of the set $S$. This means that no two of these three sets have members in common and the union of these three sets is $S$.

$$S = P \cup N \cup \{0\}$$

A brief way to express this idea is as follows:

> Every element in an ordered field is either positive, or else negative, or else zero.

To summarize, zero is neither positive nor negative and every nonzero element of $S$ is either positive or negative (not both).

It is also convenient to restate Axioms $\mathcal{O}1$, $\mathcal{O}2$, and $\mathcal{O}3$ as follows:

$\mathcal{O}1$. Zero is not positive.
$\mathcal{O}2$. Every nonzero element $x$ of $S$ is either positive or else its additive inverse $-x$ is positive.
$\mathcal{O}3$. Both the sum and the product of a pair of positive elements of $S$ are positive.

In Section 6.2 we shall begin to derive various useful theorems from these additional axioms, but first test and consolidate your understanding of these axioms by working the following exercises.

### EXERCISES 6.1

1. Show that the field $\{W_2, +(\text{mod } 2), \cdot (\text{mod } 2)\}$ is not an ordered field.
2. Show that the field $\{W_5, +(\text{mod } 5), \cdot (\text{mod } 5)\}$ is not an ordered field.

3. (a) Prove that any field in which $1 + 1 = 0$ cannot be an ordered field. (*Hint:* In any field $1 \neq 0$. If $1 + 1 = 0$, then $1 = -1$. If the field were ordered, then by Axiom $\mathcal{O}2$ either 1 is in $P$ or $-1$ is in $P$. Obtain a contradiction.)
   (b) Prove that a field $\{S, +, \cdot\}$ cannot be an ordered field if $S$ contains an element $a$ such that $a \neq 0$ and $a + a = 0$.
4. Prove that if an element $x$ of an ordered field is positive, then its additive inverse $-x$ must be negative. (*Hint:* Prove $-x \neq 0$ and $-x$ is not positive using $x + (-x) = 0$.)
5. Use the result of Exercise 4 and the fact that

$$(-1)(-1) = 1$$

to prove that, in any ordered field, 1 is positive and $-1$ is negative.

## 6.2 BASIC THEOREMS FOR ORDERED FIELDS

In this section we shall see that the notions of positiveness and negativeness defined in Section 6.1 can serve very nicely to define the order relations $>$ (greater than) and $<$ (less than), and that the order relations so defined do indeed possess certain basic properties that we would expect these order relations to have.

First, we must show that (additive) inverses of positive elements are negative and (additive) inverses of negative elements are positive.

**THEOREM 6.1** (*Additive Inverses*) Let $\{S, +, \cdot\}$ be an ordered field and let $x$ be an element of $S$.
    (a) If $x$ is positive, then $-x$ is negative.
    (b) If $x$ is negative, then $-x$ is positive.

*Proof:* (a) If $x$ is positive, then $x \neq 0$ (by Axiom $\mathcal{O}1$). Hence $-x \neq 0$ (see Exercise 1) and therefore $-x$ is either positive or negative. If $-x$ were positive, then, since $x$ is positive, it would follow (from Axiom $\mathcal{O}3$) that $x + (-x)$ is positive, that is, that 0 is positive. But that is impossible by Axiom $\mathcal{O}1$. Hence $-x$ cannot be positive and since $-x \neq 0$, the only remaining possibility is that $-x$ is negative.
(b) If $x$ is negative, then, by definition, $x \neq 0$ and $x$ is not in $P$. Hence by Axiom $\mathcal{O}2$, $-x$ must be in $P$, that is, $-x$ is positive.

We shall now use the notion of positiveness to define the order relations *greater than* (>) and *less than* (<) as follows.

**Definition.** Let $x, y$ be elements of an ordered field.
1. "$x > y$" means "$x - y$ is positive."
2. "$x < y$" means "$y > x$."

If we choose $y = 0$ in part 1 of this definition, we see that

$$x > 0 \quad \text{if and only if} \quad x - 0 \text{ is positive}$$

But $x - 0 = x$ (by Theorem 5.2c). Hence

$$x > 0 \quad \text{if and only if} \quad x \text{ is positive}$$

Similarly, if we choose $y = 0$ in part 2, we conclude that

$$x < 0 \quad \text{if and only if} \quad 0 > x$$

which, by part 1, means that

$$x < 0 \quad \text{if and only if} \quad 0 - x \text{ is positive}$$

Now since $0 - x = -x$ (see Theorem 5.2d), we obtain

$$x < 0 \quad \text{if and only if} \quad -x \text{ is positive}$$

But by Theorem 6.1, $-x$ is positive if and only if $-(-x)$, that is, $x$ is negative. Consequently,

$$x < 0 \quad \text{if and only if} \quad x \text{ is negative}$$

From the definition of ">" and "<" we have therefore deduced the following.

**THEOREM 6.2** (*Positiveness and Negativeness*) Let $x$ be any element of an ordered field.
(a) $x > 0$ if and only if $x$ is positive.
(b) $x < 0$ if and only if $x$ is negative.

The next few theorems show that the relations "greater than" and "less than" as defined here really do have the properties usually required for an order relation. Among these properties are the so-called *trichotomy* and *transitivity* requirements. Let us see what these are.

We have already observed that each element in an ordered field is either positive, or else negative, or else zero. These three possibilities are mutually exclusive and exactly one of them must be true. The Trichotomy Theorem states this fact in terms of the order relations in a slightly more general form.

**THEOREM 6.3** *(The Trichotomy Theorem)* For each pair $x, y$ of elements of an ordered field, exactly one of the following three possibilities must be true:
Either $x > y$, or else $x = y$, or else $x < y$.

*Proof:* Consider the element $x - y$. This element must be either positive, or else zero, or else negative. If $x - y$ is positive, then $x > y$ (by the definition of $>$). If $x - y = 0$, then it follows that $x = y$. If $x - y$ is negative, then by Theorem 6.1, its additive inverse $-(x - y)$ is positive, that is, $y - x$ is positive and hence $y > x$ (by the definition of $>$).

**THEOREM 6.4** *(The Transitivity Theorem)* Let $x, y, z$ be elements of an ordered field.
(a) If $x > y$ and $y > z$, then $x > z$.
(b) If $x < y$ and $y < z$, then $z < z$.

*Proof:* (a) $x > y$ means $x - y$ is positive. Similarly, $y > z$ means $y - z$ is positive. Hence by Axiom 3, $(x - y) + (y - z)$ is positive; that is, $x - z$ is positive, which means that $x > z$.
(b) By definition, $y < z$ means $z > y$; similarly, $x < y$ means $y > x$. Applying Theorem 6.4a to $z > y$ and $y > x$, we obtain $z > x$, which is equivalent to $x < z$ by definition.

*Note:* Whenever $x < y$ and $y < z$ we express these two facts more briefly by writing
$$x < y < z$$
Similarly, whenever $x > y$ and $y > z$ we use the abbreviation
$$x > y > z$$
In both cases we say "$y$ is *between* $x$ and $z$." The notion of *betweenness* is thus closely related to the notion of order in an ordered field.

The following theorems express some important facts of elementary mathematics which are seldom, if ever, actually proved when they are first encountered.

**THEOREM 6.5** (*Product of Negatives*) Let $x$, $y$ be elements of an ordered field. If $x < 0$ and $y < 0$, then $xy > 0$. (The product of a pair of negative elements is positive.)

*Proof:* By Theorems 6.1 and 6.2, if $x < 0$ and $y < 0$, then $-x > 0$ and $-y > 0$. Hence by Axiom $\mathcal{O}3$,
$$(-x)(-y) > 0,$$
and therefore (using Theorem 4.4), $xy > 0$.

Theorem 6.5 should not be confused with Theorem 4.4 which asserts that in *any ring*
$$(-x)(-y) = xy$$

*Theorem 4.4 actually has nothing to do with positiveness or negativeness!* It merely asserts that a product of additive inverses of a pair of elements is the same as the product of the elements themselves. Theorem 4.4 holds true in any ring and consequently applies to any field whether the field is ordered or not. Theorem 6.5, on the other hand, applies only to ordered fields. This theorem has no meaning unless the field has an order relation.

An important consequence of Theorem 6.5 (actually a special case of this theorem) is the following.

**THEOREM 6.6** (*Squares Never Negative*) If $x$ is any nonzero element of an ordered field, then $x^2 > 0$. (The square of any nonzero element of an ordered field is positive.)

*Proof:* If $x \neq 0$, then (by Theorem 6.3) either $x > 0$ or else $x < 0$.
If $x > 0$, then $x^2 = x \cdot x > 0$ by Axiom $\mathcal{O}3$.
If $x < 0$, then $x^2 = x \cdot x > 0$ by Theorem 6.5.
In both cases $x^2 > 0$.

We have already pointed out that the additive identity element, 0, is neither positive nor negative in an ordered field. How about the multiplicative identity 1? We know that it must be either positive or else negative because every element other than 0 must belong to either $P$ or else $N$. To which of these sets does 1 belong? Theorem 6.7 answers this question.

**THEOREM 6.7** In any ordered field
$$1 > 0 \quad \text{(that is, 1 is positive)}$$

## 6.2 Basic Theorems for Ordered Fields

*Proof:* By definition of a field $1 \neq 0$. Hence $1^2 > 0$ by Theorem 6.6. But
$$1^2 = 1 \cdot 1 = 1$$
Hence $1 > 0$.

## EXERCISES 6.2

1. (a) Prove that in a field if $x \neq 0$, then $-x \neq 0$.
   (b) Does this also hold for a ring? Explain.
   (c) What is the analogue for an abstract group? Prove it.
2. Prove that in an ordered field
   (a) $-1 < 0$.
   (b) if $x \in P$, then $-x < x$.
   (c) if $x \in N$ and $y \in P$, then $x < y$.
   (d) if $x \in P$, then $x + 1 \in P$.
   (e) if $x \in N$, then $x - 1 \in N$.
   (f) if $x \in P$, then $\dfrac{1}{x} \in P$.
   (g) if $x \in N$, then $\dfrac{1}{x} \in N$.
   (h) if $x$ and $xy$ are positive, then $y$ is positive.
   (i) if $x$ and $xy$ are negative, then $y$ is positive.
3. Discuss the following fallacious argument that Axiom $\mathcal{O}2$ is deducible from $\mathcal{O}1$ and $\mathcal{O}3$. By Axiom $\mathcal{O}3$, $P$ is closed under addition. Therefore $x$ and $-x$ cannot both be in $P$ because if they were, then $x + (-x)$ or $0$ would be in $P$, thus contradicting Axiom $\mathcal{O}1$. Hence either $x$ or $-x$ is in $P$, and we do not have to list this fact as a separate axiom.
4. Find a field that has a set $P$ which
   (a) satisfies $\mathcal{O}1$ and $\mathcal{O}2$ but not $\mathcal{O}3$.
   (b) satisfies $\mathcal{O}1$ and $\mathcal{O}3$ but not $\mathcal{O}2$.
   (c) satisfies $\mathcal{O}2$ and $\mathcal{O}3$ but not $\mathcal{O}1$.
5. Let $\{S, +, \cdot\}$ be an ordered field. Let $\{T, +, \cdot\}$ be a field where $T$ is a proper subset of $S$. Prove that $\{T, +, \cdot\}$ is an ordered field.
6. Why is the following field not an ordered field?

| + | 0 | 1 | a | b |
|---|---|---|---|---|
| 0 | 0 | 1 | a | b |
| 1 | 1 | 0 | b | a |
| a | a | b | 0 | 1 |
| b | b | a | 1 | 0 |

| × | 0 | 1 | a | b |
|---|---|---|---|---|
| 0 | 0 | 0 | 0 | 0 |
| 1 | 0 | 1 | a | b |
| a | 0 | a | b | 1 |
| b | 0 | b | 1 | a |

7. (a) Prove that any field in which
$$1 + 1 + 1 = 0$$
cannot be an ordered field.
   (b) Prove that a field $\{S, +, \cdot\}$ cannot be an ordered field if $S$ contains an element $a$ such that $a \neq 0$ and
$$a + a + a = 0$$
   (c) Generalize (a).
   (d) Generalize (b).
★(e) Use (c) to prove that any field $\{S, +, \cdot\}$ for which $S$ has a finite number of elements cannot be ordered.

## 6.3 EFFECT OF OPERATIONS ON ORDER AND APPLICATIONS TO SOLUTIONS OF INEQUATIONS

We know that equality is preserved under binary operations such as addition, subtraction, multiplication, and division (see Theorems 1.3 and 5.3). The next few theorems show that order relations are preserved under addition. Order relations are also preserved under multiplication by positive multipliers, but they are reversed under multiplication by negative multipliers.

**THEOREM 6.8** (*Preservation of Order under Addition*) Let $\{S, +, \cdot\}$ be an ordered field and let $x, y, z, u, v$ be any elements of $S$.
(a) $x > y$ if and only if $x + z > y + z$.
(b) $x < y$ if and only if $x + z < y + z$.
(c) If $x > y$ and $u > v$, then $x + u > y + v$.
(d) If $x < y$ and $u < v$, then $x + u < y + v$.

*Proof:* (a) By definition $x > y$ if and only if $x - y$ is positive. But
$$x - y = (x + z) - (y + z)$$
(See Exercise 1h of Exercises 5.2 on page 174.) Hence $x > y$ if and only if $(x + z) - (y + z)$ is in $P$, that is, if and only if
$$x + z > y + z$$

## 6.3 Effect of Operations on Order and Applications to Solutions of Inequations

(b) By definition $x < y$ means that $y > x$. But by Theorem 6.8a, $y > x$ if and only if $y + z > x + z$, that is, if and only if

$$x + z < y + z$$

(c) If $x > y$ and $u > v$, then $x - y$ is in $P$ and $u - v$ is in $P$. Consequently, by Axiom 3 it follows that

$$(x - y) + (u - v)$$

is in $P$. But

$$(x - y) + (u - v) = (x + u) - (y + v)$$

(See Exercise 1k on page 174.) Therefore, if $x > y$ and $u > v$, it follows that $(x + u) - (y + v)$ is in $P$, which means that

$$x + u > y + v$$

(d) Proof left for reader. (See Exercise 14a.)

**THEOREM 6.9** (*Preservation of Order under Multiplication by Positive Elements*) Let $\{S, +, \cdot\}$ be an ordered field and let $x, y, z, u, v$ be any elements of $S$.
(a) If $x > y$ and $z > 0$, then $zx > zy$.
(b) If $x < y$ and $z > 0$, then $zx < zy$.
(c) If $x > y > 0$ and $u > v > 0$, then $xu > yv$.
(d) If $0 < x < y$ and $0 < u < v$, then $xu < yv$.

*Proof:* (a) If $x > y$, then $x - y$ is in $P$ by definition. Moreover, if $z > 0$, then $z$ is in $P$ by Theorem 6.2. Therefore by Axiom 3, $z(x - y)$ is in $P$, that is, $zx - zy$ is in $P$, which simply means that

$$zx > zy$$

(b) Proof left for reader. (See Exercise 14b.)
(c) If $u > v > 0$, then by the Transitivity Theorem (Theorem 6.4) $u > 0$. Hence, since $x > y$, it follows by Theorem 6.9a that $xu > yu$. Similarly, because $y > 0$ and $u > v$, it follows that $yu > yv$. Therefore we now have

$$xu > yu \quad \text{and} \quad yu > yv$$

and it follows from the Transitivity Theorem that $xu > yv$.
(d) Proof left for reader. (See Exercise 14c.)

Parts a and b of Theorem 6.9 are often expressed, somewhat awkwardly, as follows:

> If unequals are multiplied by positive equals, the products remain unequal in the same sense.

A more elegant formulation is

> Multiplication of both members of an inequality by a positive element preserves the sense of the inequality.

With even greater brevity,

> Multiplication by a positive element preserves order.

Similarly, parts c and d of Theorem 6.9 are often expressed

> If positive unequals are multiplied by positive unequals in the same sense, the products remain unequal in the same sense.

The preservation of order under multiplication by a positive element has an important immediate consequence. Suppose $x$ is a positive element of an ordered field and suppose that $y$ is a negative element of this field. Since $y$ is negative, we have (by Theorem 6.2)

$$y < 0$$

If we multiply both members of this inequality by the positive element $x$, then (applying Theorem 6.9a)

$$x \cdot y < x \cdot 0$$

that is,

$$xy < 0$$

We express this result as follows.

## 6.3 Effect of Operations on Order and Applications to Solutions of Inequations

**THEOREM 6.10** (*Product of a Positive and a Negative Element*) Let $x, y$ be elements of an ordered field. If $x > 0$ and $y < 0$, then $xy < 0$. (The product of a positive element and a negative element is negative.)

Theorem 6.10 completes our list of multiplication properties for elements in an ordered field. Axiom $\mathcal{O}3$ stipulates that the product of a pair of positive elements is to be considered positive. Theorem 6.5 asserts that the product of a pair of negative elements is also positive. Theorem 6.10 now completes the story—it tells us that the product of a positive and a negative element is negative. (Of course, if either element is zero, then their product is zero.)

Theorem 6.10 should not be confused with Theorem 4.3 which asserts that in any ring

$$x \cdot (-y) = -(x \cdot y)$$

Theorem 4.3 actually has nothing to do with positiveness or negativeness. It is concerned solely with products in which one of the factors is an additive inverse. It holds true in every ring and hence in every field whether the field is ordered or not. On the other hand, Theorem 6.10 applies specifically to *ordered fields*.

Returning to Theorem 6.9, it is only natural to ask what happens to an inequality if both members are multiplied by a negative element. Theorem 6.11 answers this question.

**THEOREM 6.11** (*Reversal of Order under Multiplication by Negative Elements*) Let $\{S, +, \cdot\}$ be an ordered field and let $x, y, z$ be elements of $S$.
(a) If $x > y$ and $z < 0$, then $zx < zy$.
(b) If $x < y$ and $z < 0$, then $zx > zy$.

*Proof:* If $x > y$, then $x - y$ is positive. Hence if $z < 0$, that is, if $z$ is negative, then $z(x - y)$ is negative. Therefore

$$zx - zy < 0$$

Adding $zy$ to both members and applying Theorem 6.8b, we obtain

$$zx < zy$$

(b) Proof left for reader. (See Exercise 14d.)

Theorem 6.11 is readily remembered as follows:

> Multiplication of both members of an inequality by a negative element reverses the sense of the inequality.

Or even more briefly:

> Multiplication by a negative element reverses order.

Let us now examine how these axioms and theorems can be used in the manipulation and solution of inequations in an ordered field. This type of mathematical problem is becoming increasingly important in modern applications of mathematics. Therefore the student should make every effort to acquire a reasonable degree of skill and facility in handling inequalities.

In this connection it is worth reminding the student that solving an equation or inequation in an ordered field means finding the set of all members of the field that satisfy that equation or inequation. Depending on the particular problem, the solution set might consist of a single value, several values, or infinitely many values. It might also turn out to be the null set.

The procedure for solving inequations resembles that used for solving equations. Using one or more of the preceding Theorems 6.2 to 6.11, one replaces the given inequation by successively simpler inequations or combinations of inequations **equivalent** to the original (that is, which yield the same solution set). This process is continued until the complete solution set is readily apparent. The following examples illustrate this procedure. In each of these examples the underlying field is considered to be the field of real numbers.* The reader should make certain that he understands clearly which axioms or theorems are used at each stage of the solution.

To save space the symbol $\Leftrightarrow$ is used as an abbreviation for the phrase "is equivalent to." This symbol may also be read "if and only if" or "has the same solution set as."

***Example 1.*** Solve $2x - 5 < 7$ (in the field of real numbers).

*Solution:*

$2x - 5 < 7 \Leftrightarrow 2x < 7 + 5$     Add 5 to each member (Theorem 6.8b)

$\Leftrightarrow 2x < 12$     Replace "$7 + 5$" by "$12$"

$\Leftrightarrow x < 6$     Multiply each member by $\frac{1}{2}$ (Theorem 6.9b)

---

* However, it can readily be seen that the techniques used here apply equally well in any ordered field even though the solutions are interpreted as real numbers.

### 6.3 Effect of Operations on Order and Applications to Solutions of Inequations

Thus the solution consists of the set of all real numbers that are less than 6. This set may be denoted by $\{x < 6\}$. This solution set is conveniently depicted in the following way.

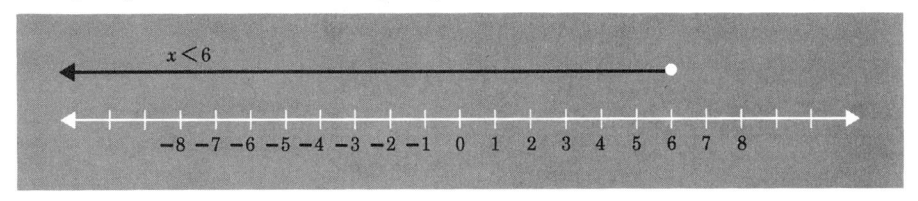

*Example 2.* Solve $4x > 15 + 7x$ (in the field of real numbers).

*Solution:* $4x > 15 + 7x \Leftrightarrow 4x - 7x > 15$  Add $-7x$ to each member (Theorem 6.8b)

$\Leftrightarrow \quad -3x > 15$  Replace "$4x - 7x$" by "$-3x$"

$\Leftrightarrow \quad x < -5$  Multiply both members by $(-\frac{1}{3})$ (Theorem 6.11b) Note reversal of inequality!

Therefore the solution consists of the set of all real numbers that are less than $-5$. This set may be depicted as follows:

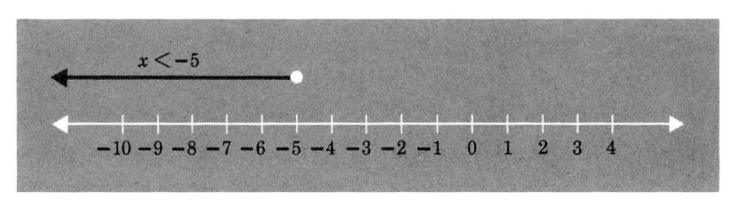

*Example 3.* Solve $-2 < x - 4 < 3$.

*Solution:* Observe first that the mathematical sentence

$$-2 < x - 4 < 3$$

is a conjunction of two simpler inequations. It means

$$-2 < x - 4 \quad and \quad x - 4 < 3$$

The desired solution set must consist of all real numbers that satisfy both of these inequalities. Consequently, we proceed as follows:

$$-2 < x-4 < 3 \Leftrightarrow \quad -2 < x-4 \text{ and } x-4 < 3$$
$$\Leftrightarrow -2+4 < x \quad \text{and} \quad x < 3+4$$
$$\text{by Theorem 6.8b}$$
$$\Leftrightarrow \quad 2 < x \quad \text{and} \quad x < 7$$
$$\Leftrightarrow \quad 2 < x < 7$$

where the last sentence $2 < x < 7$ is an abbreviation for the conjunction of the two simpler sentences $2 < x$ and $x < 7$. The desired solution set consists of all real numbers which are (both) greater than 2 and less than 7. It can also be described as the set of all real numbers between 2 and 7. (The word "between" is used in a noninclusive sense.) The solution set is conveniently depicted as follows:

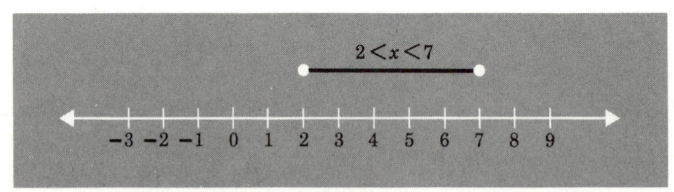

**Example 4.** Solve $-2x \geq x - 2$.

*Solution:* The sentence

$$-2x \geq x - 2$$

is actually an abbreviation for the longer sentence

$$-2x > x - 2 \quad \text{or} \quad -2x = x - 2$$

Hence we can proceed as follows:

$$-2x \geq x - 2 \Leftrightarrow -2x > x - 2 \quad \text{or} \quad -2x = x - 2$$
$$\Leftrightarrow -3x > -2 \quad \text{or} \quad -3x = -2$$
$$\Leftrightarrow \quad x < \frac{2}{3} \quad \text{or} \quad x = \frac{2}{3}$$
$$\text{by Theorem 6.11a}$$

Note the reversal of order in the last step resulting from dividing both sides by the negative number $-3$. If proper care is taken not to overlook this reversal of sense, the procedure can be shortened by operating simultaneously on both the equation and the inequality as follows:

### 6.3 Effect of Operations on Order and Applications to Solutions of Inequations

$$-2x \geq x - 2 \Leftrightarrow -3x \geq 2$$
$$\Leftrightarrow x \leq \frac{2}{3} \quad \text{Note reversal of the sense of the inequality}$$

The solution set consists of all real numbers that are either less than $\frac{2}{3}$ or equal to $\frac{2}{3}$, that is, all real numbers up to and including $\frac{2}{3}$

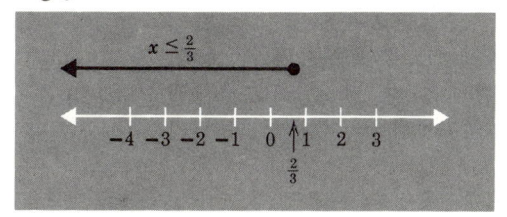

**Example 5.** Solve $-2 < 2x + 4 < x + 1$.

**Solution:** $-2 < 2x + 4 < x + 1 \Leftrightarrow -2 < 2x + 4$ and $2x + 4 < x + 1$
$$\Leftrightarrow -6 < 2x \quad \text{and} \quad x < -3$$
$$\Leftrightarrow x > -3 \quad \text{and} \quad x < -3$$

Since it is impossible to have both $x < -3$ and $x > -3$, there is no solution for this particular problem.

**Example 6.** Solve $-2 \leq 2x + 4 \leq x + 1$.

**Solution:** $-2 \leq 2x + 4 \leq x + 1 \Leftrightarrow -2 \leq 2x + 4$ and $2x + 4 \leq x + 1$
$$\Leftrightarrow -6 \leq 2x \quad \text{and} \quad x \leq -3$$
$$\Leftrightarrow x \geq -3 \quad \text{and} \quad x \leq -3$$
$$\Leftrightarrow x = -3$$

(*Note:* The solution is unique in this particular problem.)

## EXERCISES 6.3

Solve each inequality and determine its solution set for the field of real numbers. Sketch each solution set.

1. $3x - 2 < 13$
2. $2x - 3 > 1$
3. $4x + 1 \leq 9$
4. $x < 6 - x$
5. $2x - 3 \leq x + 4$
6. $3x > 2 + 5x$
7. $2x \leq 4x - 5$
8. $0 < x - 1 < 4$
9. $3 < x + 5 \leq 7$
10. $-1 \leq 2x + 5 \leq x + 8$
11. $-1 < 2x + 5 < x + 2$
12. $-1 \leq 2x + 5 \leq x + 2$

13. In the ordered field of rational numbers $\{Q, +, \cdot\}$ suppose it is given that $x > y$ and $y > 8$. By citing appropriate theorems prove the following:

   (a) $x + y > y + 8$
   (b) $x > 8$
   (c) $-x < -y$
   (d) $x - y > 0$
   (e) $y - x < 0$
   (f) $(x - y)(y - x) < 0$

14. (a) Prove that if $x, y, u, v$ are any elements in an ordered field such that $x < y$ and $u < v$, then $x + u < y + v$. (*Note:* This completes the proof of Theorem 6.8.)
    (b) Prove that in an ordered field if $x < y$ and $z > 0$, then $zx < zy$. (This proves Theorem 6.9b.)
    (c) Prove that in an ordered field if $0 < x < y$ and $0 < u < v$, then $xu < yv$. (This proves Theorem 6.9d.)
    (d) Prove that in an ordered field if $x < y$ and $z < 0$, then $zx > zy$. (This proves Theorem 6.11b.)

★15. Prove that in any ordered field
    (a) if $x < y$, then $x^3 < y^3$.
    (b) if $x^3 < y^3$, then $x < y$.
    (c) if $x^3 = y^3$, then $x = y$.

★16. Prove that for any ordered field
    (a) if $0 < x < 1$, then $0 < x^2 < 1$.
    (b) if $0 < x$ and $x^2 < 1$, then $x < 1$.
    (c) if $-1 < x < 0$, then $0 < x^2 < 1$.
    (d) if $x^2 < 1$, then $-1 < x$.

★17. Prove that the field of complex numbers is not an ordered field.

   (*Hint:* Both the numbers 0 and $i$ belong to the field of complex numbers. If this field were an ordered field, then since $i \neq 0$, we would have either $i > 0$ or $i < 0$. Prove that each of these possibilities yields a contradiction.)

## 6.4 THEOREMS CONCERNING RECIPROCALS AND PRODUCTS IN ORDERED FIELDS AND FURTHER APPLICATIONS

The theorems of Section 6.4 were adequate for the purpose of solving linear inequations. In this section we introduce several theorems which enable us to solve more complicated types of inequations.

### 6.4 Theorems Concerning Reciprocals and Products in Ordered Fields

**THEOREM 6.12** Let $x$ be a nonzero element of an ordered field.

(a) If $x > 0$, then $\dfrac{1}{x} > 0$.

(The reciprocal of a positive element is positive.)

(b) If $x < 0$, then $\dfrac{1}{x} < 0$.

(The reciprocal of a negative element is negative.)

*Proof:* (a) Since $\dfrac{1}{x} \neq 0$, therefore

$$\left(\frac{1}{x}\right)^2 > 0$$

by Theorem 6.6. Hence if $x > 0$, then by Axiom $\mathcal{O}3$, $x\left(\dfrac{1}{x}\right)^2 > 0$, that is, $\dfrac{1}{x} > 0$.

(b) Similarly, if $x < 0$, then by Theorem 6.11, $x\left(\dfrac{1}{x}\right)^2 < 0$, that is, $\dfrac{1}{x} < 0$.

**THEOREM 6.13** Let $x, y$ be nonzero elements of an ordered field. If $x > y$ and $xy > 0$, then

$$\frac{1}{x} < \frac{1}{y}$$

*Proof:* If $xy > 0$, then $\dfrac{1}{xy} > 0$ by Theorem 6.12. Therefore, if $x > y$, then $y < x$ (by definition) and hence

$$y\left(\frac{1}{xy}\right) < x\left(\frac{1}{xy}\right)$$

by Theorem 6.9b. Therefore

$$\frac{1}{x} < \frac{1}{y}$$

**THEOREM 6.14** Let $x, y$ be elements of an ordered field. If $xy > 0$, then either $x > 0$ and $y > 0$ or else* $x < 0$ and $y < 0$.

---

\* "or else" is used here in the exclusive sense, that is, one or the other but not both possibilities must hold.

*Proof:* If $xy > 0$, then neither $x$ nor $y$ can be 0, for otherwise $xy = 0$ by Theorem 4.2. This leaves four mutually exclusive possibilities:
(a) $x > 0$ and $y > 0$      (b) $x < 0$ and $y < 0$
(c) $x > 0$ and $y < 0$      (d) $x < 0$ and $y > 0$
However, (c) and (d) each imply $xy < 0$ by Theorem 6.10. This leaves (a) or else (b) as the only remaining possibilities.

**THEOREM 6.15**   Let $x$, $y$ be elements of an ordered field. If $xy < 0$, then either $x > 0$ and $y < 0$, or else $x < 0$ and $y > 0$.

*Proof:* Since $xy \neq 0$, we have the same four possibilities (a), (b), (c), (d) listed in the preceding proof. However, this time (a) and (b) are eliminated because $xy < 0$; this leaves (c) and (d) as the only remaining possibilities.

With the aid of these additional theorems we can enlarge considerably the variety of inequations we are now able to solve, Once again we interpret the solutions in the field of real numbers although the techniques used apply in any ordered field.

**Example 7.**   Solve $2 > \dfrac{1}{x}$.

*Solution:*

$$2 > \frac{1}{x} \Leftrightarrow 2 - \frac{1}{x} > 0$$

$$\Leftrightarrow \frac{2x-1}{x} > 0$$

$$\Leftrightarrow \frac{1}{x}(2x-1) > 0$$

$$\Leftrightarrow \left(\frac{1}{x} > 0 \text{ and } 2x-1 > 0\right) \text{ or else } \left(\frac{1}{x} < 0 \text{ and } 2x-1 < 0\right)$$

$$\Leftrightarrow (x > 0 \text{ and } 2x > 1) \quad \text{or else } (x < 0 \text{ and } 2x < 1)$$

$$\Leftrightarrow \left(x > 0 \text{ and } x > \frac{1}{2}\right) \quad \text{or else } \left(x < 0 \text{ and } x < \frac{1}{2}\right)$$

$$\Leftrightarrow \left(x > \frac{1}{2}\right) \qquad\qquad \text{or else } \ (x < 0)$$

### 6.4 Theorems Concerning Reciprocals and Products in Ordered Fields

The solution set thus consists of all real numbers that are either greater than $\frac{1}{2}$ or else less than 0. This set is depicted as follows.

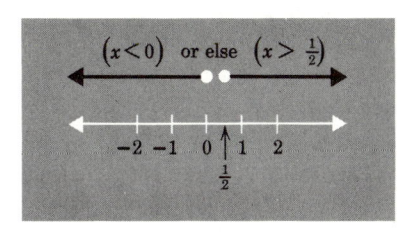

$(x < 0)$ or else $\left(x > \frac{1}{2}\right)$

$-2 \quad -1 \quad 0 \quad \underset{\frac{1}{2}}{\uparrow} \quad 1 \quad 2$

Notice that there are two **disjoint subsets** (two separate parts) that together make up the complete solution set. The solution set is the **union** of the two disjoint subsets.

*Alternate Solution to Example 7:* Clearly, $x$ may not be 0 because $\frac{1}{x}$ is not defined for $x = 0$. Hence by Theorem 3.2 (Trichotomy Theorem) we must have either $x > 0$ or else $x < 0$. We must consider these two possibilities as separate cases because multiplying both sides of the original inequality by $x$ preserves the sense of the inequality when $x > 0$ but reverses the sense of the inequality when $x < 0$. Consequently, we proceed as follows:

$$2 > \frac{1}{x} \Leftrightarrow (x > 0 \text{ and } 2x > 1) \text{ or else } (x < 0 \text{ and } 2x < 1)$$

$$\Leftrightarrow \left(x > 0 \text{ and } x > \frac{1}{2}\right) \text{ or else } \left(x < 0 \text{ and } x < \frac{1}{2}\right)$$

$$\Leftrightarrow \left(x > \frac{1}{2}\right) \text{ or else } (x < 0)$$

**Example 8.** Solve $2 < \frac{1}{x}$.

*Solution:* $2 < \frac{1}{x} \Leftrightarrow 2 - \frac{1}{x} < 0$

$$\Leftrightarrow \frac{2x - 1}{x} < 0$$

$$\Leftrightarrow \frac{1}{x}(2x - 1) < 0$$

$\Leftrightarrow \left(\dfrac{1}{x} > 0 \text{ and } 2x - 1 < 0\right)$ or else $\left(\dfrac{1}{x} < 0 \text{ and } 2x - 1 > 0\right)$

$\Leftrightarrow \left(x > 0 \text{ and } x < \dfrac{1}{2}\right)$ or else $\left(x < 0 \text{ and } x > \dfrac{1}{2}\right)$

$\Leftrightarrow \left(x > 0 \text{ and } x < \dfrac{1}{2}\right)$

(Because the second alternative is impossible)

$\Leftrightarrow 0 < x < \dfrac{1}{2}$

The solution set consists of all real numbers between 0 and $\tfrac{1}{2}$.

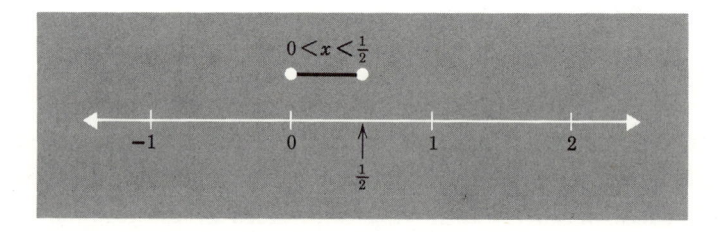

*Alternate Solution to Example 8:* Clearly, $x$ may not be 0. Hence by Theorem 3.2 (Trichotomy Theorem) we must have either $x > 0$ or else $x < 0$. We must treat these two possibilities separately because multiplying both sides by $x$ preserves the sense of the inequality when $x > 0$ but reverses the sense of the inequality when $x < 0$. Consequently, we proceed as follows:

$2 < \dfrac{1}{x} \Leftrightarrow (x > 0 \text{ and } 2x < 1)$ or else $(x < 0 \text{ and } 2x > 1)$

$\Leftrightarrow \left(x > 0 \text{ and } x < \dfrac{1}{2}\right)$ or else $\left(x < 0 \text{ and } x > \dfrac{1}{2}\right)$

$\Leftrightarrow \left(x > 0 \text{ and } x < \dfrac{1}{2}\right)$ (Because the second alternative is impossible)

$\Leftrightarrow 0 < x < \dfrac{1}{2}$

Notice that Examples 7 and 8 were readily solved by expressing them either in the form $ab > 0$ or $ab < 0$. Many other inequalities are of this "factorable" type and can be treated essentially the same way.

### 6.4 Theorems Concerning Reciprocals and Products in Ordered Fields

**Example 9.** Solve $x^2 > 3x + 10$.

*Solution:*

$x^2 > 3x + 10$
$\Leftrightarrow x^2 - 3x - 10 > 0$
$\Leftrightarrow (x+2)(x-5) > 0$
$\Leftrightarrow (x+2 > 0 \text{ and } x-5 > 0)$ or else $(x+2 < 0 \text{ and } x-5 < 0)$
$\Leftrightarrow (x > -2 \text{ and } x > 5)$ or else $(x < -2 \text{ and } x < 5)$
$\Leftrightarrow (x > 5)$ or else $(x < -2)$

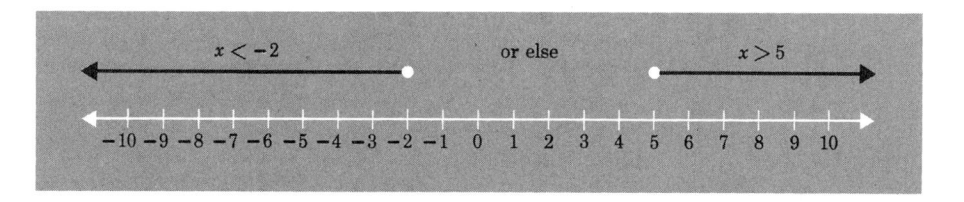

**Example 10.** Solve $x^2 < 3x$.

*Solution:*

$x^2 < 3x \Leftrightarrow x^2 - 3x < 0$
$\Leftrightarrow x(x-3) < 0$
$\Leftrightarrow (x > 0 \text{ and } x - 3 < 0)$ or else $(x < 0 \text{ and } x - 3 > 0)$
$\Leftrightarrow (x > 0 \text{ and } x < 3)$ or else $(x < 0 \text{ and } x > 3)$
$\Leftrightarrow (x > 0 \text{ and } x < 3)$
   (Because the other alternative is impossible)
$\Leftrightarrow 0 < x < 3$

The solution set consists of all real numbers between 0 and 3.

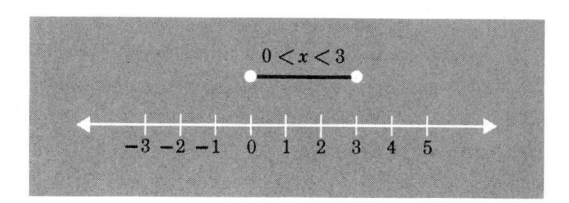

Of course, not all quadratic inequalities are factorable in this manner. We shall presently discuss a more general method for treating any quadratic inequality regardless of whether it is factorable or not. We shall do this, in Section 6.6 after introducing the important notion of absolute value of a real number.

## EXERCISES 6.4

*Solve each inequality. Determine its solution set for the field of real numbers. Sketch the solution set.*

1. $2 > \dfrac{8}{x}$
2. $3 < \dfrac{9}{x}$
3. $1 + \dfrac{5}{x} > 0$
4. $1 + \dfrac{5}{x} \geq 0$
5. $1 + \dfrac{5}{x} \leq 0$
6. $x^2 + 2x > 8$
7. $x^2 < x + 2$
8. $x^2 \leq 5x$
9. $1 < x^2$
10. $x^2 < 1$
11. $x < x^2$
12. $x^2 < x$
13. $1 < (x + 1)^2$
14. $(x + 1)^2 < 1$
15. $(x + 1)(x - 1) < 1$
16. $(x + 2)(x - 3) < 1$
17. $1 < \dfrac{1}{x}$
18. $\dfrac{1}{x} < 1$
19. $x < \dfrac{1}{x}$
20. $\dfrac{1}{x} < x$
21. $x + 1 < \dfrac{2}{x}$
22. $x - 1 < \dfrac{2}{x}$
23. $1 - x < \dfrac{2}{x}$
24. $x^2 < \dfrac{1}{x}$
25. $x < \dfrac{1}{x^2}$
26. $\dfrac{1}{x} < \dfrac{1}{x+1}$

*Prove that for any ordered field* $\{S, +, \cdot\}$ *with a and b in S*

27. if $a < b$, then $a^5 < b^5$.
28. if $a^5 < b^5$, then $a < b$.

*Sketch a graph for each sentence.*

29. $y = x$; $y > x$; $y < x$
30. $y = \dfrac{12}{x}$; $y > \dfrac{12}{x}$; $y < \dfrac{12}{x}$
31. $y = (x + 1)(x - 2)$; $y > (x + 1)(x - 2)$; $y < (x + 2)(x - 2)$
32. $y = \dfrac{x}{x - 1}$; $y < \dfrac{x}{x - 1}$; $y > \dfrac{x}{x - 1}$
33. $y = x^2$; $y > x^2$; $y < x^2$

## 6.5 SOME MORE ADVANCED THEOREMS FOR ORDERED FIELDS

Our next theorem expresses a rather simple, almost obvious relation among elements of an ordered field. Yet this theorem turns out to have a surprisingly large number of diverse applications. Its usefulness extends into many advanced areas of modern mathematics.

**THEOREM 6.16** Let $\{S, +, \cdot\}$ be an ordered field and let $x, y$ be any elements of $S$. Then
(a) $x^2 + y^2 \geq 2xy$ and
(b) if $x \neq y$, then $x^2 + y^2 > 2xy$.

*Proof:* By Theorems 4.2 and 6.6 $(x-y)^2 \geq 0$ where equality holds if and only if $x = y$. But

$$(x-y)^2 \geq 0 \Leftrightarrow x^2 - 2xy + y^2 \geq 0 \quad \text{(see Theorem 5.8b)}$$
$$\Leftrightarrow x^2 + y^2 \geq 2xy \quad \text{(see Theorem 6.8)}$$

where equality holds if and only if $x = y$.

**THEOREM 6.17** Let $\{S, +, \cdot\}$ be an ordered field in which every positive element has a square root. Then for all positive $x, y$ in $S$ (that is, for all $x, y$, in $P$)
(a) $\dfrac{x+y}{2} \geq \sqrt{xy}$
(b) if $x \neq y$, then $\dfrac{x+y}{2} > \sqrt{xy}$

*Proof:* In Theorem 6.16 replace $x$ by $\sqrt{x}$ and $y$ by $\sqrt{y}$. This is legitimate because $\sqrt{x}$ and $\sqrt{y}$ exist for all positive $x, y$. Theorem 6.16 then yields
(a) $x + y \geq 2\sqrt{x}\sqrt{y}$
(b) if $x \neq y$, then $x + y > 2\sqrt{x}\sqrt{y}$
Now $(\sqrt{x} \cdot \sqrt{y})^2 = (\sqrt{x})^2(\sqrt{y})^2 = x \cdot y$; but since $xy$ is positive, $xy$ has a unique positive square root. Hence

$$\sqrt{xy} = \sqrt{x}\sqrt{y}$$

and the theorem now follows.

$\dfrac{x+y}{2}$ is called the **average** or **arithmetic mean** of $x$ and $y$. $\sqrt{xy}$ is called the **geometric mean** of $x$ and $y$. Theorem 6.17 states the following.

> The **arithmetic mean** of two (distinct) positive real elements is always greater than their **geometric mean**.

**THEOREM 6.18** Let $\{S, +, \cdot\}$ be an ordered field. Then for all positive $x, y$ in $S$

(a) $\dfrac{x+y}{2} \geq \dfrac{2xy}{x+y}$ and

(b) if $x \neq y$, then $\dfrac{x+y}{2} > \dfrac{2xy}{x+y}$

*Proof:* By Theorem 6.17

$$\dfrac{x+y}{2} \geq \sqrt{xy}$$

where equality holds if and only if $x = y$. Consequently,

$$\dfrac{(x+y)^2}{4} \geq xy$$

Then multiplying both members of this inequality by the positive element $\dfrac{2}{x+y}$, we obtain

$$\dfrac{x+y}{2} \geq \dfrac{2xy}{x+y}$$

where equality holds if and only if $x = y$.

$\dfrac{2xy}{x+y}$ is called the **harmonic mean** of $x$ and $y$. It can also be expressed as

$$\dfrac{1}{\dfrac{\dfrac{1}{x}+\dfrac{1}{y}}{2}}$$

and hence represents the reciprocal of the arithmetic mean of the reciprocals of $x$ and $y$.

## 6.5 Some more Advanced Theorems for Ordered Fields

Theorem 6.18 asserts that the arithmetic mean of two (distinct) positive elements is always greater than their harmonic mean. (*Note:* This theorem also holds under the more general requirement, $x + y > 0$. We leave it to the reader to prove this for himself.)

A theorem which has far-reaching consequences in analysis and in linear algebra is the following.

**THEOREM 6.19** (*The Cauchy-Schwarz Inequality*) Let $\{S, +, \cdot\}$ be an ordered field. Then for all $x, y, u, v$ in $S$

$$(x^2 + y^2)(u^2 + v^2) \geq (xu + yv)^2$$

where equality holds if and only if $xv = yu$.

*Proof:*

$$(x^2 + y^2)(u^2 + v^2) = x^2u^2 + y^2v^2 + x^2v^2 + y^2u^2$$

But Theorem 6.17 implies that

$$x^2v^2 + y^2u^2 \geq 2 \cdot (xv) \cdot (yu)$$

where equality holds if and only if $xv = yu$. Hence

$$(x^2 + y^2)(u^2 + v^2) \geq x^2u^2 + y^2v^2 + 2xuyv$$

that is,

$$(x^2 + y^2)(u^2 + v^2) \geq (xu + yv)^2$$

where equality holds if and only if $xv = yu$.

## EXERCISES 6.5

1. Prove that for all positive $x$ in $R$

$$x + \frac{1}{x} \geq 2$$

where equality holds if and only if $x = 1$.

2. Prove that for all $x, y, z$ in $R$

$$x^2 + y^2 + z^2 \geq xy + yz + xz$$

where equality holds if and only if $x = y = z$.

3. Prove that for all $x, y$ in $R$ for which $xy > 0$

$$x^3y + yx^3 \geq 2x^2y^2$$

where equality holds if and only if $x = y$.

4. Prove that the geometric mean of two (distinct) positive real numbers is always greater than their harmonic mean.

★5. Prove that the Cauchy-Schwarz Inequality also holds in "three dimensions," that is, prove that for all $x, y, z, u, v, w$ in $R$

$$(x^2 + y^2 + z^2)(u^2 + v^2 + w^2) \geq (xu + yv + zw)^2$$

where equality holds if and only if

$$xv = yu, \quad yw = zv, \quad \text{and} \quad zu = xw$$

6. If $x$ and $y$ are positive elements of an ordered field, prove

(a) $\dfrac{x^2 + y^2}{2} \geq \left(\dfrac{x+y}{2}\right)^2$

(b) $\dfrac{x^3 + y^3}{2} \geq \left(\dfrac{x+y}{2}\right)^3$

(c) $\dfrac{x^4 + y^4}{2} \geq \left(\dfrac{x+y}{2}\right)^4$

(d) What generalization is suggested by (a), (b), and (c)?

★7. Prove that for all real numbers $x, y, z, w$ such that

$$x > y \quad \text{and} \quad z > w$$

we have

(a) $xz + yw > xw + yz$ and
(b) $(x + z)(y + w) < (x + w)(y + z)$

★8. Try to extend the result for Exercise 7 to the case where $x > y$, $z > w$, $u > v$, and all numbers are nonnegative.

★9. Let $\{S, +, \cdot\}$ be a field. Prove that if there exist elements $a$ and $b$ in $S$ such that

$$\frac{1}{a} + \frac{1}{b} = \frac{1}{a+b}$$

then $\{S, +, \cdot\}$ cannot be an ordered field.

## 6.6 ABSOLUTE VALUE IN ORDERED FIELDS

Suppose that $p$ is any positive element of an ordered field. Then by Theorem 6.1 $-p$ is a negative element. Furthermore, if $n$ is any negative element, then $-n$ is a positive element. In short, if $x$ is any element other than 0, then of the two elements $x$ and $-x$ exactly one must be positive and the other negative. Of these two elements the positive one is called the **absolute value** of either of them. For example, the absolute value of either 3 or $-3$ is simply 3; the absolute value of $2-7$ is the same as the absolute value of $7-2$, namely 5. It is convenient to define the absolute value of 0 to be 0 itself. By this convention every element $x$ of an ordered field has an absolute value which is denoted by the symbol $|x|$.

In order to prove theorems about absolute value, it is necessary to formulate a precise definition that embodies the ideas just described.

**Definition.** Let $\{S, +, \cdot\}$ be an ordered field. Then for all $x$ in $S$

$$|x| = x \text{ if } x \geq 0 \quad \text{and} \quad |x| = -x \text{ if } x < 0$$

From this definition, together with the axioms and theorems already established, we can proceed to develop most of the properties that make absolute value a useful mathematical tool.

**THEOREM 6.20** Let $\{S, +, \cdot\}$ be an ordered field. Then for all $x$ in $S$, $|x| \geq 0$.

*Proof:* Either $x \geq 0$ or else $x < 0$ by the Trichotomy Theorem (Theorem 6.3).
If $x \geq 0$, then by definition

$$|x| = x \geq 0$$

If $x < 0$, then by definition

$$|x| = -x$$

and by Theorem 6.1b, $-x > 0$.
Hence in this case $|x| > 0$. Thus in all cases we have $|x| \geq 0$.

**THEOREM 6.21** Let $\{S, +, \cdot\}$ be an ordered field. Then for all $x$ in $S$, $|x| \geq x$.

*Proof:* If $x \geq 0$, then $|x| = x$ by definition. If $x < 0$, then since $0 \leq |x|$ (by Theorem 6.20), it follows that $x < |x|$. Thus in all cases we have $x \leq |x|$, that is,

$$|x| \geq x$$

**THEOREM 6.22** Let $\{S, +, \cdot\}$ be an ordered field. Then for all $x$, $y$, in $S$

$$|xy| = |x| \cdot |y|$$

*Proof:* Either $x \geq 0$ or else $x < 0$. With each of these two possibilities for $x$, there are two possibilities for $y$, namely $y \geq 0$ or else $y < 0$. There are therefore four cases to consider.

(a) If $x \geq 0$ and $y \geq 0$, then $xy \geq 0$ and, applying our definition, we obtain
$$|xy| = xy = |x| \cdot |y|$$

(b) If $x \geq 0$ and $y < 0$, then $xy \leq 0$ and, applying our definition, we obtain
$$|xy| = -(xy) = x(-y) = |x| \cdot |y|$$

(c) If $x < 0$ and $y \geq 0$, then $xy \leq 0$ and, applying our definition, we obtain
$$|xy| = -(xy) = (-x)y = |x| \cdot |y|$$

(d) If $x < 0$ and $y < 0$, then $xy > 0$ and, applying our definition, we obtain
$$|xy| = xy = (-x)(-y) = |x| \cdot |y|$$

Many important properties of absolute value stem from Theorem 6.22.

**THEOREM 6.23** Let $\{S, +, \cdot\}$ be an ordered field. Then for all $x$ in $S$
$$|x|^2 = |x^2| = x^2$$

*Proof:* By Theorem 6.22
$$|x|^2 = |x| \cdot |x| = |x \cdot x| = |x^2| = x^2$$
where the last equality follows from the fact that $x^2 \geq 0$ for all $x$ in $S$.

The following theorem is essentially a restatement of Theorem 5.9.

**THEOREM 6.24** Let $\{S, +, \cdot\}$ be an ordered field. Then for all $x, y$ in $S$
(a) $|x| = |y|$ if and only if $x^2 = y^2$
(b) $|x| = |y|$ if and only if $x = \pm |y|$

*Proof:* (a) If $|x| = |y|$, then $|x|^2 = |y|^2$ which by Theorem 6.23 is equivalent to $x^2 = y^2$. Conversely, if $x^2 = y^2$, then by Theorem 5.9, $x = y$ or $x = -y$; either of these two cases implies $|x| = |y|$.

(b) $|x| = |y| \Leftrightarrow x^2 = y^2$      by Theorem 6.24a
$\phantom{|x| = |y|} \Leftrightarrow x^2 = |y|^2$      by Theorem 6.23
$\phantom{|x| = |y|} \Leftrightarrow x = \pm|y|$      by Theorem 5.9

An immediate corollary and important special case of Theorem 6.24b is the following.

**THEOREM 6.25** Let $\{S, +, \cdot\}$ be an ordered field. Then for all $x$ in $S$

$$|x| = 0 \quad \text{if and only if} \quad x = 0$$

*Proof:* Let $|y| = 0$ in Theorem 6.24b

We turn now to some theorems that involve both inequality and absolute value.

**THEOREM 6.26** Let $\{S, +, \cdot\}$ be an ordered field. Then for all $x, y$ in $S$

$$|x| < |y| \quad \text{if and only if} \quad x^2 < y^2$$

*Proof:* Observe first that under either hypotheses ($|x| < |y|$ or $x^2 < y^2$) we must have $0 < |y|$. (See Exercises 26c and d.) By Theorem 6.23 $x^2 < y^2$ is equivalent to $|x|^2 < |y|^2$, which is in turn equivalent to

$$|y|^2 - |x|^2 > 0$$

that is, to

$$(|y| + |x|) \cdot (|y| - |x|) > 0$$

But since $|y| > 0$ and $|x| \geq 0$, then (by Theorem 6.3)

$$|y| + |x| > 0$$

It now follows (by Axiom $\mathcal{O}$3 and Theorem 6.3) that

$(|y| + |x|)(|y| - |x|) > 0$ if and only if $|y| - |x| > 0$

which is equivalent to $|x| < |y|$.

With the aid of Theorem 6.26 we are able to devise a neat proof of the following important theorem (sometimes called the *triangle inequality*).

**THEOREM 6.27** Let $\{S, +, \cdot\}$ be an ordered field. Then for all $x, y$ in $S$

(a) $|x+y| \leq |x| + |y|$  (b) $|x-y| \geq |x| - |y|$

*Proof:* (a) By Theorem 6.21 $xy \leq |xy|$. Hence by Theorem 6.22

$$xy \leq |x| \cdot |y|$$

But

$$x^2 + y^2 = |x|^2 + |y|^2$$

Thus it follows that

$$x^2 + 2xy + y^2 \leq |x|^2 + 2|x| \cdot |y| + |y|^2$$

that is,

$$(x+y)^2 \leq (|x| + |y|)^2$$

Applying Theorems 6.24 and 6.26 to this result, we obtain

$$|x+y| \leq ||x| + |y||$$

But

$$||x| + |y|| = |x| + |y|$$

because $|x| + |y| \geq 0$. Hence

$$|x+y| \leq |x| + |y|$$

(b) To prove this, replace $x$ by $x-y$ in (a). (See Exercise 26a.)

The following two theorems are often convenient when solving inequations involving absolute values.

**THEOREM 6.28** Let $\{S, +, \cdot\}$ be an ordered field. Then for all $x, y$ in $S$

$$|x| < |y| \quad \text{if and only if} \quad -|y| < x < |y|$$

*Proof:* By Theorem 6.26 $|x| < |y|$ is equivalent to $x^2 < y^2$, which by Theorem 6.23 is equivalent to $x^2 < |y|^2$, that is, to

$$x^2 - |y|^2 < 0$$

This is in turn equivalent to

$$(x - |y|)(x + |y|) < 0$$

and by Theorem 6.15 this can be true if and only if

$$\text{either} \quad (x - |y| < 0 \text{ and } x + |y| > 0)$$
$$\text{or else} \quad (x - |y| > 0 \text{ and } x + |y| < 0)$$

that is, if

$$\text{either} \quad (x < |y| \text{ and } x > -|y|)$$
$$\text{or else} \quad (x > |y| \text{ and } x < -|y|)$$

But since $|y| \geq 0$ (by Theorem 6.20), therefore $-|y| \leq 0$ (by Theorem 6.1), and hence it is impossible to have both $x > |y|$ and $x < -|y|$. Therefore $|x| < |y|$ if and only if $x < |y|$ and $x > -|y|$, that is, if and only if

$$-|y| < x < |y|$$

**THEOREM 6.29** Let $\{S, +, \cdot\}$ be an ordered field. Then for all $x$, $y$ in $S$

$$|x| > |y| \quad \text{if and only if} \quad x > |y| \text{ or } x < -|y|$$

*Proof:* By Theorem 6.26 $|x| > |y|$ is equivalent to $x^2 > y^2$, which by Theorem 6.23 is equivalent to $x^2 > |y|^2$, that is, to

$$x^2 - |y|^2 > 0$$

This in turn is equivalent to

$$(x - |y|)(x + |y|) > 0$$

and by Theorem 6.14 this can be true if and only if

$$\text{either} \quad (x - |y| > 0 \text{ and } x + |y| > 0)$$
$$\text{or else} \quad (x - |y| < 0 \text{ and } x + |y| < 0)$$

that is,

$$\text{either} \quad (x > |y| \text{ and } x > -|y|)$$
$$\text{or else} \quad (x < |y| \text{ and } x < -|y|)$$

But $|y| \geq 0$ (by Theorem 6.20) and therefore $-|y| \leq 0$ (by Theorem 6.1). Hence $x > |y|$ implies $x > -|y|$; so $(x > |y|$ and $x > -|y|)$ is equivalent simply to $x > |y|$. Similarly, $(x < |y|$ and $x < -|y|)$ is equivalent simply to $x < -|y|$.

We are now in a position to solve many new types of equations and inequations that arise frequently in mathematics. In the following examples we again interpret the results in the field of real numbers although the techniques used are for the most part valid in any ordered field in which the necessary square roots exist.

***Example 1.*** Solve $|x| = 5$.

***Solution (a):*** By Theorem 6.24a the equation $|x| = 5$ is equivalent to $x^2 = 25$. The solutions are therefore $\pm 5$ by Theorem 5.9a.

***Solution (b):*** By Theorem 6.24b letting $|y| = 5$, the solutions are $\pm |5|$, that is, $\pm 5$.

***Example 2.*** Solve $|2x - 1| = 7$.

***Solution:*** Using Theorem 6.24b,
$$2x - 1 = \pm 7$$
$$\therefore 2x = 1 \pm 7$$
$$x = \frac{1 \pm 7}{2}$$
that is, $x = 4$ or $x = -3$.

***Example 3.*** Solve $|2x - 1| = -7$.

***Solution:*** By Theorem 6.20 it is impossible for $|2x - 1|$ to be negative. Hence there is no solution.

***Example 4.*** Solve $|x| < 5$.

***Solution:*** Use Theorem 6.28 with $|y| = 5$. The solution of $|x| < 5$ is then
$$-5 < x < 5$$

***Example 5.*** Solve $|2x - 1| < 7$.

***Solution:*** By Theorem 6.28 $-7 < 2x - 1 < 7$. Adding 1 to each member of this inequality, we obtain
$$-6 < 2x < 8$$
Dividing each member by 2 gives
$$-3 < x < 4$$

## 6.6 Absolute Value in Ordered Fields

**Example 6.** Solve $(x-2)^2 < 7$.

**Solution:**
$$(x-2)^2 < 7 \Leftrightarrow |x-2| < \sqrt{7}$$
$$\Leftrightarrow -\sqrt{7} < x-2 < \sqrt{7}$$
$$\Leftrightarrow 2-\sqrt{7} < x < 2+\sqrt{7}$$

**Example 7.** Solve $|2x-1| > 7$.

**Solution:** By Theorem 6.29 either

$$2x-1 < -7 \quad \text{or} \quad 2x-1 > 7$$

that is, either

$$2x < -6 \quad \text{or} \quad 2x > 8$$

that is, either

$$x < -3 \quad \text{or} \quad x > 4$$

**Example 8.** Solve $4 - |2x-1| \geq -1$.

**Solution:**
$$4 - |2x-1| \geq -1 \Leftrightarrow 4+1 \geq |2x-1|$$
$$\Leftrightarrow |2x-1| \leq 5$$
$$\Leftrightarrow -5 \leq 2x-1 \leq 5$$
$$\Leftrightarrow -5+1 \leq 2x \leq 5+1$$
$$\Leftrightarrow -4 \leq 2x \leq 6$$
$$\Leftrightarrow -2 \leq x \leq 3$$

**Example 9.** Solve $1 < |2x-1| < 5$.

**Solution:** By the Trichotomy Theorem (Theorem 6.3) either $2x-1 \geq 0$ or else $2x-1 < 0$. Hence $1 < |2x-1| < 5$ is equivalent to $(2x \geq 1$ and $1 < 2x-1 < 5)$ or else $(2x < 1$ and $1 < 1-2x < 5)$, that is, $(2x \geq 1$ and $2 < 2x < 6)$ or else $(2x < 1$ and $0 < -2x < 4)$, that is, $(x \geq \frac{1}{2}$ and $1 < x < 3)$ or else $(x < \frac{1}{2}$ and $0 > x > -2)$, that is, $(1 < x < 3)$ or else $(-2 < x < 0)$.

**Example 10.** Solve $\left|\dfrac{x-2}{2}\right| = 5 + x$.

Solution: $\left|\dfrac{x-2}{2}\right| = 5+x \Leftrightarrow |x-2| = 10+2x$

$\Leftrightarrow (x-2 \geq 0 \text{ and } x-2 = 10+2x)$
or else $(x-2 < 0 \text{ and } 2-x = 10+2x)$

$\Leftrightarrow \underbrace{(x \geq 2 \text{ and } -12 = x)}_{\text{Impossible}}$

or else $\underbrace{(x < 2 \text{ and } -8 = 3x)}_{\text{Equiv. to } x = \dfrac{-8}{3}}$.

$\Leftrightarrow x = \dfrac{-8}{3}$

**Example 11.** Solve $\left|\dfrac{x-2}{2}\right| < 5+x$.

Solution: $\left|\dfrac{x-2}{2}\right| < 5+x \Leftrightarrow |x-2| < 10+2x$

$\Leftrightarrow (x-2 \geq 0 \text{ and } x-2 < 10+2x)$
or else $(x-2 < 0 \text{ and } 2-x < 10+2x)$

$\Leftrightarrow (x \geq 2 \text{ and } -12 < x)$
or else $(x < 2 \text{ and } -8 < 3x)$

$\Leftrightarrow (x \geq 2 \text{ and } x > -12)$
or else $\left(x < 2 \text{ and } x > -\dfrac{8}{3}\right)$

$\Leftrightarrow (x \geq 2)$ or else $\left(-\dfrac{8}{3} < x < 2\right)$

$\Leftrightarrow x > -\dfrac{8}{3}$

**Example 12.** Solve $x^2 - 4x < 3$.

Solution: $x^2 - 4x < 3 \Leftrightarrow x^2 - 4x + 4 < 3 + 4$

$\Leftrightarrow (x-2)^2 < 7$
$\Leftrightarrow |x-2| < \sqrt{7}$
$\Leftrightarrow -\sqrt{7} < x-2 < \sqrt{7}$
$\Leftrightarrow 2 - \sqrt{7} < x < 2 + \sqrt{7}$

## 6.6 Absolute Value in Ordered Fields

**Example 13.** Solve $x^2 + 10 < 6x$.

Solution: $x^2 + 10 < 6x \Leftrightarrow x^2 - 6x < -10$
$\Leftrightarrow x^2 - 6x + 9 < 9 - 10$
$\Leftrightarrow (x-3)^2 < -1$

But $(x-3)^2 \geq 0$ for all $x$. Hence the last condition is impossible. Therefore there is no solution.

**Example 14.** Solve $x^2 + 10 > 6x$.

Solution: $x^2 + 10 > 6x \Leftrightarrow x^2 - 6x > -10$
$\Leftrightarrow x^2 - 6x + 9 > 9 - 10$
$\Leftrightarrow (x-3)^2 > -1$

But $(x-3)^2 \geq 0$ for all $x$. And since $0 > -1$,

$$(x-3)^2 > -1$$

for all $x$.
The solution set is the set of all real numbers.

**Example 15.** Solve $x^2 > 4x + 2$.

Solution: $x^2 > 4x + 2 \Leftrightarrow x^2 - 4x > 2$
$\Leftrightarrow x^2 - 4x + 4 > 2 + 4$
$\Leftrightarrow (x-2)^2 > 6$
$\Leftrightarrow |x-2| > \sqrt{6}$
$\Leftrightarrow x - 2 > \sqrt{6}$ or else $x - 2 < -\sqrt{6}$
$\Leftrightarrow x > 2 + \sqrt{6}$ or else $x < 2 - \sqrt{6}$

## EXERCISES 6.6

Solve 1–25 in the set of real numbers.

1. $|x| = 5$
2. $|x| + 2 = 0$
3. $|x + 3| < 2$
4. $|x - 5| < 7$
5. $|2x + 1| = 9$
6. $5 + |1 - 2x| = 0$
7. $5 - |1 - 2x| = 0$
8. $5 + |1 - 2x| > 0$
9. $4 + 3|2x - 1| \leq 13$
10. $8 - 2|x + 3| \geq 10$
11. $8 - 2|x + 3| < 10$
12. $2 < |x| < 5$

13. $-2 < |x| < 5$
14. $|x| + x = 0$
15. $\left|\dfrac{x-1}{2}\right| + x = 7$
16. $\left|\dfrac{x-1}{2}\right| + x = -7$
17. $\left|\dfrac{x-1}{2}\right| + x < 7$
18. $\left|\dfrac{x-1}{2}\right| + x < -7$
19. $x^2 + 3 \leq 4x$
20. $x^2 - 12 < 0$
21. $x^2 - 12 \geq 0$
22. $2x^2 + 4x + 3 < 0$
23. $2x^2 + 4x - 3 < 0$
24. $x^2 + 12x + 37 \geq 0$
25. $5x^2 > 3(x+1)$
26. Prove the following for real $x$ and $y$.
    
    (a) $|x| - |y| \leq |x - y|$
    (b) $\left|\dfrac{1}{y}\right| = \dfrac{1}{|y|}$, $y \neq 0$
    (c) $\left|\dfrac{x}{y}\right| = \dfrac{|x|}{|y|}$, $y \neq 0$
    (d) If $|x| < |y|$, then $0 < |y|$
    (e) If $x^2 < y^2$, then $0 < |y|$
27. Interpret on the real number line $0 < |x - a| < b$.
28. Prove that in any ordered field, $x \leq y$ and $-x \leq y$ if and only if $|x| \leq y$.
29. (a) Let $a < b$. Under what circumstances will $\dfrac{1}{a} < \dfrac{1}{b}$ be true?
    
    Under what circumstances will $\dfrac{1}{a} < \dfrac{1}{b}$ be false?
    
    ★(b) Let $a < b$ and $c < d$. Under what circumstances will $ac < bd$ be true? Under what circumstances will $ac < bd$ be false?
30. The graph for $y = |x|$ is given below.

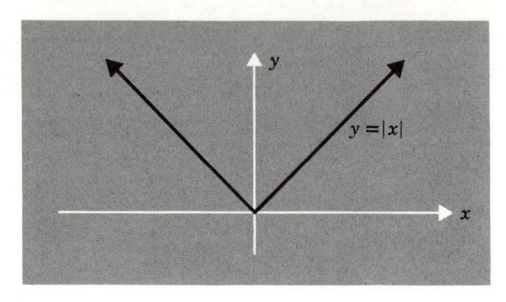

    (a) Indicate the region in the plane for which $y > |x|$.
    (b) Indicate the region in the plane for which $y < |x|$.
31. Sketch a graph in the $xy$-plane for each of the following.
    (a) $y = -|x|$, $y > -|x|$, $y < -|x|$
    (b) $|y| = x$, $|y| > x$, $|y| < x$

(c) $|y| = -x$, $|y| > -x$, $|y| < -x$
(d) $y = |x-2|$, $y > |x-2|$, $y < |x-2|$
(e) $y = |x+2|$, $y > |x+2|$, $y < |x+2|$
(f) $|y| = |x|$, $|y| > |x|$, $|y| < |x|$
(g) $|y| = |x-2|$, $|y| > |x-2|$, $|y| < |x-2|$
(h) $|y| = |x+2|$, $|y| > |x+2|$, $|y| < |x+2|$
(i) $y = x + |x|$, $y > x + |x|$, $y < x + |x|$
(j) $y = |x| - x$, $y > |x| - x$, $y < |x| - x$
(k) $y = x - |x|$, $y > x - |x|$, $y < x - |x|$
(l) $|x+y| = 4$, $|x+y| > 4$, $|x+y| < 4$
(m) $|x| + |y| = 4$, $|x| + |y| > 4$, $|x| + |y| < 4$
(n) $|x| - |y| = 4$, $|x| - |y| > 4$, $|x| - |y| < 4$
(o) $|y| - |x| = 4$, $|y| - |x| > 4$, $|x| - |x| < 4$
(p) $2|x| + y = 3$, $2|x| + y > 3$, $2|x| + y < 3$
(q) $|x| + 2|y| = 3$, $|x| + 2|y| > 3$, $|x| + 2|y| < 3$
(r) $y = ||x| - 3|$
(s) $y = |x^2 - 4|$
(t) $y = |x^3|$
(u) $|x| + |y| = 2x^2$
(v) $x < y < |x|$ (*Hint:* Consider the points common to the graphs of $x < y$ and $y < |x|$.)
(w) $x < y < |x| + 1$
(x) $x < |y| < |x|$
(y) $x < |y| < |x| + 1$

## 6.7 MATHEMATICAL INDUCTION—THE INTEGERS OF AN ORDERED FIELD

We have already proved (see Theorem 6.7) that in any ordered field $\{F, +, \cdot\}$ the additive identity is less than the multiplicative identity

$$0 < 1$$

Consequently, it follows that $0 + 0 < 1 + 0 < 1 + 1$, that is,

$$0 < 1 < (1+1)$$

We can continue this argument to prove that

$$0 < 1 < (1+1) < (1+1+1) < (1+1+1+1) < \cdots$$

showing that in any ordered field the elements $1, (1+1), (1+1+1), \ldots$ form an endless sequence of distinct* elements all of which are greater than 0,

---

* That they are distinct follows from Exercise 7 of Section 6.2.

that is, positive. It seems quite natural to designate these elements

$$1, 2, 3, \ldots,$$

and to call them **natural numbers** or **positive integers.** Moreover, since the members of this set are all positive, it follows from Theorem 6.1 that their additive inverses,

$$-1, -2, -3, \ldots,$$

are negative elements of $F$, and so it is equally tempting to call these the **negative integers** of $F$.

At this point, however, it is possible to raise several objections before proceeding. First, the description of the set we wish to call the positive integers is still quite vague. Precisely which subset of $F$ are we selecting when we write

$$1, \quad 1+1, \quad 1+1+1, \quad \ldots ?$$

The three dots indicating "and so on" can hardly be accepted as an accurate mathematical definition.

Second, even if we succeed in making this definition precise, how can we be sure that the resulting elements of $F$, $1, 2, 3, \ldots$ and their additive inverses $-1, -2, -3, \ldots$, actually deserve to be called integers? This is not an academic question because we have already assigned various properties to integers. In Chapter 2 we showed that many of these properties depend on the very fundamental Well-Ordering Axiom. At the very least, we should expect that the positive integers of $F$ ought to be a well-ordered set.

We are confronted therefore with the following task: Given an ordered field $\{F, +, \cdot\}$, we want to formulate a precise definition which will select from $F$ a subset $N$ that we can legitimately identify as the positive integers. This means that we want $N$ to possess the properties that we have already associated with these integers including (among others) the well-ordering property. We want to do all this in such a way that we make precise the assertion "$N$ contains the integers $1, 1+1, 1+1+1, \ldots$."

The clue to accomplishing all this is based on a very simple observation. The set $N$ we are seeking to define

(1) must contain the integer 1 and
(2) whenever $N$ contains an integer $x$, then $N$ must also contain the next integer $x+1$.

In order to capitalize on this observation and also to express it more precisely we set up the following definition.

## 6.7 Mathematical Induction—The Integers of an Ordered Field

**Definition.** Let $\{F, +, \cdot\}$ be an ordered field.
1. If $x$ is any element of $F$, we shall call $x + 1$ the **successor** of $x$.
2. If $S$ is a subset of $F$, we shall call $S$ a **successor subset** if and only if both of the following are true:
   (a) $1 \in S$ and
   (b) whenever $x \in S$, then $(x + 1) \in S$.

Clearly, the entire set $F$ is itself a successor subset. Furthermore, the set $P$ of all positive elements of $F$ is also a successor subset. Thus $F$ certainly includes at least two successor subsets (and presumably many more).

Now let us observe that

> If $S_1$ and $S_2$ are successor subsets, then their intersection $S_1 \cap S_2$ is again a successor subset.

This is proved as follows.

1. $1 \in S_1$ and $1 \in S_2$; therefore $1 \in S_1 \cap S_2$.
2. If $x \in S_1 \cap S_2$, then $x \in S_1$ and $x \in S_2$.
   Hence $(x + 1) \in S_1$ and $(x + 1) \in S_2$.
   Therefore $(x + 1) \in S_1 \cap S_2$.

This argument clearly applies to any collection of successor subsets.

> The intersection of any collection of successor subsets is again a successor subset.

In particular, the intersection of all successor subsets of $F$ is also a successor subset of $F$. We make use of this particular successor subset and define it to be the set $N$ that we have been seeking.

**Definition.** Let $\{F, +, \cdot\}$ be an ordered field and let $N$ be the intersection of all successor subsets of $F$. The members of set $N$ are called the **positive integers** of $F$.

Although we have now formulated a precise definition for the set $N$ of positive integers of $F$, we must still show that the elements of $N$ actually have the properties that we expect the positive integers to possess. We start this part of our task by observing that the set $N$ as defined is the "smallest" of all successor subsets of $F$. By "smallest" we mean that $N$ is included in every successor subset of $F$, that is, $N \subseteq S$ for every successor subset $S$ of $F$. To prove this we need merely recall that the intersection of any collection of sets is by definition a subset of each of these sets. Since $N$ is the intersection of all successor subsets of $F$, $N$ is included in every successor subset of $F$. Thus there can be no successor subsets that are smaller than $N$. This *minimality* property of $N$ leads immediately to the following simple, yet far-reaching theorem.

**THEOREM 6.30** (*Principle of Mathematical Induction*) Let $\{F, +, \cdot\}$ be an ordered field and let $N$ be the set of positive integers of $F$. If $A \subseteq N$ and $A$ is a successor subset of $F$, then $A = N$.

*Proof:* Since $A$ is given to be a successor subset of $F$, it follows that $N \subseteq A$. But it is also given that $A \subseteq N$. Consequently, $A = N$.

This deceptively simple theorem has rather surprising ramifications throughout mathematics. With its aid we can complete the task we set out to accomplish. We shall use Theorem 6.30 repeatedly to prove that the elements of $N$ actually do possess various properties that we would intuitively expect the positive integers to have. We select for detailed proof several obvious properties leading to a proof that the set $N$ is well-ordered. Other properties are included within the Exercises.

The reader should study carefully the manner in which the Principle of Mathematical Induction (Theorem 6.30) is used in these proofs. Although Theorem 6.30 is easy to prove, it is not always easy to use. Skill in applying Mathematical Induction is, however, worth acquiring. (See Section 6.8.)

**THEOREM 6.31** If $\{F, +, \cdot\}$ is an ordered field, then every positive integer in $F$ is equal to or greater than 1.

*Proof:* Let $A$ be the set of all positive integers in F which are equal to or greater than 1. Then certainly $A \subseteq N$. Furthermore, $1 \in A$ and if $x \in A$, then since $x \geq 1$, it follows that

$$x + 1 \geq 1 + 1 > 1$$

Hence $(x+1) \in A$. Therefore $A$ is a successor subset of $F$ and since $A \subseteq N$, it follows (from Theorem 6.30) that $A = N$.

**THEOREM 6.32** If $\{F, +, \cdot\}$ is an ordered field, then there is no positive integer in $F$ between 1 and $(1+1)$ (that is, there is no positive integer between 1 and 2).

*Proof:* Let $A$ be the set consisting of 1 together with all those elements of $N$ that are equal to or greater than $(1+1)$. Clearly, no element of this set $A$ is between 1 and $(1+1)$. We shall prove that $A = N$.

By definition $A \subseteq N$ and $1 \in A$. Now if $x \in A$, then (by Theorem 6.31), $x \geq 1$ and hence

$$x + 1 \geq 1 + 1$$

Thus

$$(x+1) \in A$$

We see therefore that $A$ is a successor subset of $F$, and since $A \subseteq N$, it follows (by Theorem 6.30) that $A = N$.

**THEOREM 6.33** Let $\{F, +, \cdot\}$ be an ordered field and let $x$ be a positive integer in $F$. If $x \neq 1$, then $(x-1)$ is a positive integer in $F$.*

*Proof:* Let $A$ be the set consisting of 1 together with all those elements $x \in N$ for which

$$x - 1 \in N$$

Then $A \subseteq N$ and certainly $1 \in A$. Moreover, if $x \in A$, this means that $x \in N$ and

$$\text{either} \quad x = 1 \quad \text{or} \quad x - 1 \in N$$

It follows that

$$(x+1) \in N \text{ and (in either case) } (x+1) - 1 = x \in N$$

---

* This theorem asserts that every positive integer except 1 has an "immediate predecessor."

which simply means that $x+1 \in A$. Therefore $A$ is a successor subset of $F$ and since $A \subseteq N$, it follows (by Theorem 6.30) that $A = N$.

**THEOREM 6.34** Let $\{F, +, \cdot\}$ be an ordered field. If $x$ is any positive integer in $F$, then there is no positive integer in $F$ between $x$ and $(x+1)$.

*Proof:* Let $A$ be the set of all those positive integers $x$ in $F$ for which there is no positive integer in $F$ between $x$ and $(x+1)$. We shall prove that $A = N$.

Clearly, $A \subseteq N$ and (by Theorem 6.32) $1 \in A$. We shall now prove that if $x \in A$, then $(x+1) \in A$. Suppose there were a positive integer $x$ such that $x \in A$ but $(x+1) \notin A$. This means
(a) there is no positive integer in $F$ between $x$ and $x+1$
and
(b) there is a positive integer $y$ such that

$$x+1 < y < (x+1)+1$$

Now since $x+1 < y$, it follows (from Theorem 6.31) that $y \neq 1$ and hence (by Theorem 6.33) that $y-1$ is a positive integer in $F$. But from (b) it follows that

$$x < y-1 < x+1$$

which contradicts (a), namely, that there is no positive integer between $x$ and $x+1$. This contradiction proves the above supposition false.

Summarizing, we have proved that $1 \in A$ and if $x \in A$, then $x+1 \in A$. Therefore $A$ is a successor subset of $F$ and since $A \subseteq N$, it follows (by Theorem 6.30) that $A = N$.

**THEOREM 6.35** Let $\{F, +, \cdot\}$ be an ordered field and let $x, y$ be positive integers in $F$. If $x < y$, then $x+1 \leq y$.

*Proof:* If $x+1 \leq y$ were false, then (by the Trichotomy Theorem) we would have $y < x+1$ and since $x < y$,

$$x < y < x+1$$

### 6.7 Mathematical Induction—The Integers of an Ordered Field

But this is impossible (by Theorem 6.34). Consequently, $x+1 \leq y$.

Many additional properties of positive integers can be derived from Theorem 6.30 in a manner similar to that used in Theorems 6.31 through 6.35. However, we now have enough properties available to prove that the set $N$ is well-ordered.

**THEOREM 6.36** *(The Well-Ordering Theorem).* Let $\{F, +, \cdot\}$ be an ordered field. If $N$ is the set of all positive integers in $F$, then $N$ is well-ordered (that is, every nonempty subset of $N$ has a least element).

*Proof:* Let $E$ be any nonempty subset of $N$. If $1 \in E$, there is nothing more to prove because (by Theorem 6.31) 1 is the least element of $N$ and hence would be the least element of $E$. We may suppose therefore that $1 \notin E$, that is, that every element of $E$ is $>1$. Now let $A$ be the set of all those elements of $N$ which are less than every element of set $E$. Clearly, $A \subseteq N$. Furthermore, $1 \in A$ because every element of $E$ is greater than 1.

Now if $x \in A$, then for all $y \in E$, $x < y$, and hence (by Theorem 6.35) $x+1 \leq y$. There are now two cases to consider.

CASE 1. Suppose $x+1 < y$ for all $x \in A$ and all $y \in E$. This means that $x+1 \in A$ whenever $x \in A$ and since $1 \in A$, we see that $A$ is a successor subset of $F$. Then because $A \subseteq N$, it follows (by Theorem 6.30) that $A = N$. But recalling that $E$ is a nonempty subset of $N$, $E$ contains at least one element $n$ of $N$. This element $n$ cannot belong to $A$ because every element of $A$ is less than every element of $E$. Since $n \notin A$, it is impossible that $A = N$. Case 1 is therefore impossible.

CASE 2. We now know that $x+1 = y$ for some $x \in A$ and some $y \in E$. Therefore, if $z$ is any element of $E$ and since $x < z$, we have (by Theorem 6.35) $x+1 \leq z$, that is, $y \leq z$. This means that $y$ is the least element of $E$. This completes the proof.

We digress to remark that from our definition of $N$ together with Theorems 6.30 and 6.31 we can easily obtain the following five properties known as *Peano's Axioms* (because they were first used by the Italian mathematician Peano to deduce all the known properties of the integers).

***Peano's Axioms.***
1. $1 \in N$.
2. If $x \in N$, then $x + 1 \in N$.
3. If $A \subseteq N$ such that $1 \in A$ and $x + 1 \in A$ whenever $x \in A$, then $A = N$.
4. $x + 1 \neq 1$ for all $x \in N$.
5. If $x \in N$ and $y \in N$ and if $x + 1 = y + 1$, then $x = y$.

Observe that Axioms 1 and 2 are immediate consequences of the definition of $N$. Axiom 3 is Theorem 6.30, Axiom 4 follows immediately from Theorem 6.31, and Axiom 5 follows from Theorem 2.5 (Cancellation Theorem for a Group).

Having now produced convincing evidence that the set $N$ is a worthy candidate for the name *positive integers* of $F$, we proceed to define the *integers* of $F$.

**Definition.** Let $\{F, +, \cdot\}$ be an ordered field and let $N$ be the set of positive integers of $F$. An element $x$ of $F$ is called a **negative integer** of $F$ if and only if $-x$ is a positive integer of $F$. The set of all negative integers of $F$ is denoted by $-N$. The union of $N$, $-N$, and $\{0\}$ is called the set of **integers** of $F$. The set of integers is denoted by $Z$.

$$Z = N \cup (-N) \cup \{0\}$$

Thus the set of integers $Z$ contains the element 0, the positive integers 1, 2, 3, ..., and the negative integers $-1, -2, -3, \ldots$. However, the above definition achieves our objective of defining $Z$ without resorting to the vague expression "...".

We can go even further and define the rational numbers of $F$.

**Definition.** Let $\{F, +, \cdot\}$ be an ordered field and let $Z$ be the set of integers of $F$. An element $x$ of $F$ is called a **rational number** of $F$ if and only if there are **integers** $p$ and $q$ in $F$ such that $x = \frac{p}{q}$. The set of all rational numbers of $F$ is denoted by $Q$.

It is a straightforward task to prove that the system $\{Q, +, \cdot\}$ is also a field, in fact, an ordered field. The operations $+$, $\cdot$, and the order relation $<$ are those which apply to $\{F, +, \cdot\}$ but which are restricted to elements of the subset $Q$. The rational numbers may therefore be viewed as an ordered **subfield** of any ordered field.* For example, the field of real numbers is an ordered field which includes the field of rational numbers among its ordered subfields. Another ordered subfield of the field of real numbers is the system $\{S, +, \cdot\}$ where $S$ is the set of all real numbers of the form $p + q\sqrt{2}$ where $p$ and $q$ are rational numbers. Observe that this field contains the rational number field as an ordered subfield (let $q = 0$).

## EXERCISES 6.7

1. Let $\{R, +, \cdot\}$ be the ordered field of real numbers. Which of the following subsets of $R$ are successor sets?
   (a) The set of rational numbers.
   (b) The set of positive real numbers.
   (c) The set of negative real numbers.
   (d) The set of positive rational numbers.
   (e) The set of negative rational numbers.
   (f) The set of rational numbers equal to or greater than 1.
   (g) The set of rational numbers equal to or greater than 0.
   (h) The set of rational numbers equal or to greater than 2.
   (i) The set of odd numbers equal to or greater than 1.
   (j) The set of even numbers equal to or greater than 2.
   (k) The set of rational numbers expressible as a quotient of two integers $\frac{p}{q}$ with $q$ an odd integer.
2. Let $S_1$ and $S_2$ be successor sets. Prove that
   (a) $S_1 \cup S_2$ is a successor set and
   (b) the union of any collection of successor sets is also a successor set.
3. Prove that there is no integer between 0 and 1.
4. Prove that the following sets are well-ordered (under the less-than relation).
   (a) The set of whole numbers $\{0\} \cup N$.

---

* Strictly speaking, there is in $F$ a subset $Q_F$ such that when the field operations $+$ and $\cdot$ are *restricted* to elements in $Q_F$, the resulting subsystem $\{Q_F, +, \cdot\}$ becomes an ordered field *isomorphic* to the field $\{Q, +, \cdot\}$ of rational numbers. The isomorphism preserves not only sums and products but order of elements as well.

(b) The set of integers equal to or greater than a given integer $m$. (We denote this set by $Z_m$.)

5. Prove that any subset of a well-ordered set is also well-ordered under the same order relation.
6. Prove that if $W_1$ and $W_2$ are two well-ordered subsets of the real numbers under the less-than relation, then
    (a) $W_1 \cap W_2$ is also well-ordered.
    (b) $W_1 \cup W_2$ is also well-ordered.
7. Show that every nonempty subset of each of the following sets has a greatest member.
    (a) The set of negative integers $(-N)$.
    (b) The set of integers less than a given integer $m$ (that is, $Z - Z_m$, where $Z_m$ is the set defined in Exercise 4b).
8. Why is the set of rational numbers not well-ordered under the less-than relation? (Describe a nonempty subset that does not have a least element.)
9. If a set has a well-ordered subset, show that the original set need not be well-ordered.
10. Show that the subfield of rational numbers of any ordered field is itself an ordered field.

## 6.8 APPLICATIONS OF MATHEMATICAL INDUCTION

The power and versatility of Theorem 6.30 (the Principle of Mathematical Induction) can hardly be overemphasized. In this section we get another glimpse of the variety of problems that this principle is capable of handling.

*Example 1.* As a first illustration consider the following sums of consecutive odd integers

$$\begin{aligned} 1 &= 1 \\ 1 + 3 &= 4 \\ 1 + 3 + 5 &= 9 \\ 1 + 3 + 5 + 7 &= 16 \end{aligned}$$

It requires little effort to recognize a pattern and to formulate the following conjecture.

If $n$ is any positive integer, then
$$1 + 3 + 5 + \cdots + (2n - 1) = n^2$$

We can, of course, verify the conclusion for additional values of $n$, say for $n = 5$, $n = 6$, etc. Such special verifications, however, even for hundreds of values of $n$, cannot be considered a proof because the conjecture requires that the equality

(1) $$1 + 3 + 5 + \cdots + (2n - 1) = n^2$$

prove to be a true statement for all positive integer values of $n$.

The Principle of Mathematical Induction (Theorem 6.30) provides the necessary general method for proving that a sentence such as (1) is indeed true for all positive integers. As in the proofs of Theorems 6.30 through 6.36, the method begins by introducing a suitable set $A$, which is a subset of $N$, and proving that $A$ is a successor set.

In the present example we let $A$ be the set of all those positive integers $n$ for which (1) is a true statement. Clearly, $1 \in A$ because $1 = 1^2$. Next, we suppose that $x \in A$ and proceed to prove that

$$(x + 1) \in A$$

This is done as follows: $x \in A$ implies that

$$1 + 3 + 5 + \cdots + (2x - 1) = x^2$$

If we add $2x + 1$ to both members of this equality, we obtain

$$1 + 3 + 5 + \cdots + (2x - 1) + (2x + 1) = x^2 + (2x + 1)$$

which implies that

$$1 + 3 + 5 + \cdots + (2x + 1) = (x + 1)^2$$

that is, that $(x + 1) \in A$. Since $A$ satisfies both requirements for a successor set and since $A \subseteq N$, it follows by the Principle of Mathematical Induction that $A = N$. This means that (1) is a true statement for all $n \in N$.

*Example 2.* Let $a$, $b$, and $m$ be integers and let $n$ be any positive integer. Prove: If $a \equiv b \pmod{m}$, then $a^n \equiv b^n \pmod{m}$.

*Proof:* (by Mathematical Induction) Let $A$ be the set of all those positive integers $n$ for which

$$a^n \equiv b^n \pmod{m}$$

Then surely $1 \in A$ because we are given that

(2) $$a \equiv b \pmod{m}$$

Now if we suppose that $x \in A$, this implies that

(3) $$a^x \equiv b^x \pmod{m}$$

Applying Theorem 3.10c to the congruences (2) and (3), we obtain

$$a^x \cdot a \equiv b^x \cdot b \pmod{m}$$

that is,

$$a^{x+1} \equiv b^{x+1} \pmod{m}$$

which means that $(x+1) \in A$. We see therefore that $A$ is a successor set and since $A \subseteq N$, it follows by the principle of Mathematical Induction that $A = N$, that is,

$$a^n \equiv b^n \pmod{m}$$

for all $n \in N$.

**Example 3.** Let $n$ be any positive integer. Prove: If $n$ circles are drawn in a plane so that each pair of distinct circles intersects in two points and no three circles have a point in common, then the plane is partitioned into $n^2 - n + 2$ distinct regions. (For example, in the case $n = 3$, there are $3^2 - 3 + 2 = 8$ regions.)

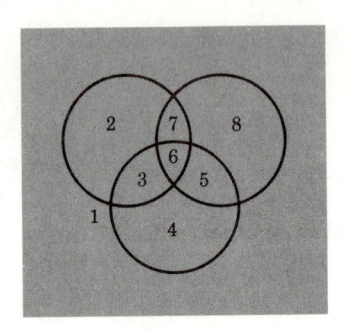

*Proof:* (by Mathematical Induction) Let $A$ be the set of all those positive integers $n$ for which $n$ circles, drawn as specified, yield $n^2 - n + 2$ regions. Clearly, $1 \in A$ because one circle yields

$$1^2 - 1 + 2 = 2$$

distinct regions. Next suppose that $x \in A$. This implies that $x$ circles drawn as specified partition the plane into

$$x^2 - x + 2$$

regions. Now if an additional circle is drawn intersecting each of the $x$ circles in two *new* points (because no three circles may have a point in common), this creates $2x$ new points of intersection all lying on the new circle and therefore subdividing this circle into $2x$ arcs. Each of these arcs cuts across one of the regions already present, thus producing $2x$ additional regions. Hence if $x$ circles yield $x^2 - x + 2$ regions, then $x + 1$ circles will yield

$$(x^2 - x + 2) + (2x)$$

regions. We may rewrite this number of regions as

$$(x^2 + 2x + 1) - (x + 1) + 2 \quad \text{regions}$$

that is, as

$$(x + 1)^2 - (x + 1) + 2 \quad \text{regions}$$

showing that $(x + 1) \in A$ whenever $x \in A$. This completes the proof that $A$ is a successor set and since $A \subseteq N$, it follows by Mathematical Induction that $A = N$.

# EXERCISES 6.8

*Solve each of the following problems by Mathematical Induction using the solutions of Examples 1 to 3 in Section 6.8 as models. Prove each of the following for all positive integers n.*

1. $1 + 2 + 3 + \cdots + n = \dfrac{n(n + 1)}{2}$

2. $9 + 13 + 17 + \cdots + (5 + 4n) = n(2n + 7)$

3. $1^2 + 2^2 + 3^2 + \cdots + n^2 = \dfrac{n(n + 1)(2n + 1)}{6}$

4. $1^3 + 2^3 + 3^3 + \cdots + n^3 = \dfrac{n^2(n + 1)^2}{4}$

5. $1 \cdot 2 + 2 \cdot 3 + 3 \cdot 4 + \cdots + n(n+1) = \dfrac{n(n+1)(n+2)}{3}$

6. $\dfrac{1}{2} + \dfrac{1}{4} + \dfrac{1}{8} + \cdots + \dfrac{1}{2^n} = 1 - \dfrac{1}{2^n}$

7. $\dfrac{1}{3} + \dfrac{1}{9} + \dfrac{1}{27} + \cdots + \dfrac{1}{3^n} = \dfrac{1}{2}\left(1 - \dfrac{1}{3^n}\right)$

8. $2^n > n$

★9. $2^n \geq n^2 - 1$

10. $3^n > 2^n$

11. $\dfrac{n(n+1)}{2}$ is an integer.

12. $\dfrac{n(n+1)(n+2)}{6}$ is an integer.

13. $\dfrac{5^n - 2^n}{3}$ is an integer.

14. $\dfrac{8^n - 3^n}{5}$ is an integer.

★15. $\dfrac{n^5 - n}{5}$ is an integer.

16. $\dfrac{n(n+1)(2n+1)}{6}$ is an integer.

17. $n^2 \geq n$

18. $2^2 - 3^2 + 4^2 - 5^2 + \cdots + (2n)^2 = 2n^2 + n + 1$
    (*Note:* For each $n$, the left member of this equation terminates with $(2n)^2$. This expression $(2n)^2$ does not generate the series.)

19. $1^2 - 3^2 + 5^2 - 7^2 + \cdots + (4n-3)^2 = 8n(n-1) + 1$
    (*Note:* For each $n$, $(4n-3)^2$ is the last term but not the generating term of the series.)

20. $1^2 + 3^2 + 5^2 + \cdots + (2n-1)^2 = \dfrac{n(2n-1)(2n+1)}{3}$

21. The number of diagonals of a convex polygon of $(n+2)$ sides is $\dfrac{(n+2)(n-1)}{2}$.

22. $\dfrac{1}{1 \cdot 2} + \dfrac{1}{2 \cdot 3} + \dfrac{1}{3 \cdot 4} + \cdots + \dfrac{1}{n(n+1)} = \dfrac{n}{n+1}$

23. $\dfrac{1}{1 \cdot 3} + \dfrac{1}{3 \cdot 5} + \dfrac{1}{5 \cdot 7} + \cdots + \dfrac{1}{(2n-1)(2n+1)} = \dfrac{n}{2n+1}$

24. The sum of the angles of a convex polygon of $(n+2)$ sides is $n \cdot 180$ degrees.

★25. The maximum number of regions into which $n$ chords of a circle can partition the interior is $\frac{1}{2}(n^2 + n + 2)$. (*Note:* Assume that the maximum number of regions is obtained when each pair of distinct chords intersect but no three are concurrent.)

★26. The number of chords joining $n$ points of a circle in every possible way is $\dfrac{n(n-1)}{2}$.

★27. Suppose there are $n$ points on a circle. The number of triangles with vertices from these points is $\dfrac{n(n-1)(n-2)}{6}$.

28. $\left(1+\dfrac{3}{1}\right)\left(1+\dfrac{5}{4}\right)\left(1+\dfrac{7}{9}\right)\cdots\left(1+\dfrac{2n+1}{n^2}\right) = (n+1)^2$

★29. $(1.1)^{n+1} > 1 + .1(n+1)$

★30. $(1.2)^{n+1} > 1 + .2(n+1)$

## REVIEW EXERCISES

1. For each of the following, determine the solution set in the field of real numbers. Sketch each solution set.
    (a) $-2 < 2x - 3 \le 3$
    (b) $-2 < x^2 - 2 \le 3$
    ★(c) $-2 < x^2 - 3x \le 3$
    (d) $(x-2)(x+3)(x+5) < 0$
    (e) $\dfrac{3}{x} \le x + 2$
    (f) $|x - 3| < \tfrac{1}{2}$
    (g) $|x - 3| < x$
    (h) $x < |x + 3|$

2. Let $\{S, +, \cdot\}$ be a field and let $a \in S$. Prove that this field is not an ordered field if
    (a) $a^2 + 1 = 0$
    (b) $a^2 + a + 1 = 0$
    (c) $a^2 + a + 2 = 0$
    (d) $ra^2 + sa + t = 0,\ s^2 - 4rt = -b^2,\ r, s, t, b \in S,\ b \ne 0$

3. Prove each of the following by Mathematical Induction.
    (a) $2^n > 3n$ for all integers $> 3$
    (b) $\dfrac{7^n - 2^n}{5}$ is an integer for all integers $\ge 0$
    ★(c) $\dfrac{n^5 - n}{30}$ is an integer for all integers $\ge 0$
    (d) $1 + 2 + 2^2 + 2^3 + \cdots + 2^n = 2^{n+1} - 1$ for all integers $\ge 0$

256    6 ‖ Ordered Fields

(e) $1 + 3 + 3^2 + 3^3 + \cdots + 3^n = \dfrac{3^{n+1} - 1}{2}$ for all integers $\geq 0$

★(f) The sum of the exterior angles of a convex polygon of $(n + 2)$ sides is 360° for all integers $n \geq 1$.

★(g) The number of ways of electing a president and a vice-president in a class of $n$ students is $n(n - 1)$. Assume that the same student may not hold both offices at the same time.

4. Let $a, b, c, d$ be positive integers and $\dfrac{a}{b} < \dfrac{c}{d}$.

Prove $\quad \dfrac{a}{b} < \dfrac{a+c}{b+d} < \dfrac{c}{d}$

5. Try to remove some of the restrictions in Exercise 4; for example,
   (a) must all the integers be positive?
   (b) must they be integers?

# RESEARCH PROBLEM

Prove: $\{Q(\sqrt{2}), +, \cdot\}$ is an ordered field assuming:

$\{Q(\sqrt{2}), +, \cdot\}$ is a field (see Exercises 5.1, Exercise 4c).

$\{Q, +, \cdot\}$ is an ordered field.

$a + b\sqrt{2} = 0 \iff a = 0$ and $b = 0$

$a + b\sqrt{2} > 0 \iff$ one of the following holds:

   1. $a \geq 0, b \geq 0,$ and $a + b > 0$
   2. $a > 0, b < 0,$ and $a^2 > 2b^2$
   3. $a < 0, b > 0,$ and $2b^2 > a^2$

$(a + b\sqrt{2}) + (r + s\sqrt{2}) = (a + r) + (b + s)\sqrt{2}$

$(a + b\sqrt{2}) \cdot (r + s\sqrt{2}) = (ar + 2bs) + (as + br)\sqrt{2}$

"Courtesy of Paul J. Cohen"

*Paul J. Cohen (1934-    )*

# 7 | Complete Ordered Fields

> I could be bounded in a nutshell and count myself a king of infinite space.
>
> WILLIAM SHAKESPEARE
> (*Hamlet*, Act II, Scene 2)

## 7.1 LOWER BOUNDS AND UPPER BOUNDS

We have already observed that in any ordered field $\{S, +, \cdot\}$ there are endless sequences of ever-increasing elements. For example, if $c$ is any element of $S$, then $c, c+1, c+2, \ldots$ is an endless increasing sequence because

(1) $$c < c+1 < c+2 < \cdots$$

Similarly, there are endless sequences of ever-decreasing elements

(2) $$c > c-1 > c-2 > \cdots$$

From (1) it is clear that none of the elements of the set

$$\{c, c+1, c+2, \ldots\}$$

is smaller than $c$ and from (2) it is clear that none of the elements of the set

$$\{c, c-1, c-2, \ldots\}$$

is greater than $c$. Consequently, $c$ is called a **lower bound** for the set of elements $\{c, c+1, c+2, \ldots\}$ and $c$ is called an **upper bound** for the set $\{c, c-1, c-2, \ldots\}$. This simple example illustrates the following more general definition.

**Definition.** Let $\{S, +, \cdot\}$ be an ordered field, let $A$ be any subset of $S$, and let $b$ be an element of $S$.

    1. $b$ is called a **lower bound** for (or of) set $A$ if and only if no element of $A$ is less than $b$. (Alternatively, $b$ is called a lower bound for set $A$ if and only if $x \geq b$ for all $x$ in $A$.)

    2. $b$ is called an **upper bound** for (or of) set $A$ if and only if no element of $A$ is greater than $b$. (Alternatively, $b$ is called an upper bound of set $A$ if and only if $x \leq b$ for all $x$ in $A$.)

Observe that this definition does not require that a lower bound $b$ actually belong to set $A$ although it may do so as in our example above (where $A = \{c, c+1, c+2, \ldots\}$). A similar remark holds for upper bounds.

Observe also that a subset $A$ of an ordered field $\{S, +, \cdot\}$ need not have either a lower bound or an upper bound. For example, the set $S$ itself is such a subset. To prove this, suppose that there were an element $b$ in $S$ which is an upper bound for $S$. By definition this would mean that $b \geq x$ for all $x$ in $S$. But since $b \in S$, therefore

$$b + 1 \in S$$

Choosing $x = b+1$ we now obtain $b \geq b+1$ and (subtracting $b$) therefore

$$0 \geq 1$$

This, however, contradicts Theorem 6.7 ($1 > 0$). This contradiction shows that there can be no upper bound for $S$. Similarly, we can prove that $S$ does not have a lower bound either. (See Exercise 5a.)

**THEOREM 7.1** If $\{S, +, \cdot\}$ is an ordered field, then $S$ has neither an upper bound nor a lower bound.

In contrast with $S$ itself which has neither upper nor lower bounds there are, of course, subsets which have both upper bounds and lower bounds. This must be true, for example, of any nonempty *finite* subset of $S$ because the elements of any such subset can be arranged in an increasing sequence

$$c_1 < c_2 < c_3 < \cdots < c_n$$

in which case $c_1$ (or any element less than $c_1$) can serve as a lower bound and $c_n$ (or any element greater than $c_n$) can serve as an upper bound for the subset.

## 7.1 Lower Bounds and Upper Bounds

As another example consider the ordered field of rational numbers $\{Q, +, \cdot\}$ and let $A$ be the subset of all positive rational numbers less than 5, that is,

$$A = \{x \in Q \mid 0 < x < 5\}$$

Set $A$ clearly has both lower and upper bounds even though there are infinitely many positive rational numbers in set $A$.

In an ordered field, subsets that have both a lower bound and an upper bound are called **bounded** sets.

### EXERCISES 7.1

1. Let $\{Q, +, \cdot\}$ be the ordered field of rational numbers. For each of the following subsets of $Q$, find (if they exist) an upper bound in $Q$ and a lower bound in $Q$.
   (a) $\{9, 9.9, 9.99, 9.999, \ldots\}$
   (b) $\{1, .1, .01, .001, \ldots\}$
   (c) $\{1, 1.1, 1.11, 1.111, \ldots\}$
   (d) $\{2, 2.2, 2.22, 2.222, \ldots\}$
   (e) $\left\{\dfrac{1}{2}, \dfrac{2}{3}, \dfrac{3}{4}, \dfrac{4}{5}, \ldots\right\}$
   (f) $\{9, -9.9, 9.99, -9.999, \ldots\}$
   (g) $\{1, -.1, .01, -.001, \ldots\}$.
   (h) $\left\{\dfrac{1}{2}, -\dfrac{2}{3}, \dfrac{3}{4}, -\dfrac{4}{5}, \ldots\right\}$
   (i) $\{n \in Q \mid n < 2\}$
   (j) $\{n^2 \mid n \in Q, n < 2\}$
   (k) $\{n \mid n \in Q, n^2 < 2\}$
   (l) $\{n \mid n \in Q, n^2 > 2\}$
   (m) $\{n \mid n \in Q, n^2 < 3\}$
   (n) $\{n \mid n \in Q, n^2 > 3\}$
   (o) $\left\{\dfrac{1}{9}, \dfrac{1}{9.9}, \dfrac{1}{9.99}, \dfrac{1}{9.999}, \ldots\right\}$
   (p) $\left\{\dfrac{1}{1}, \dfrac{1}{.1}, \dfrac{1}{.01}, \dfrac{1}{.001}, \ldots\right\}$
   (q) $\left\{\dfrac{2}{1}, \dfrac{3}{2}, \dfrac{4}{3}, \dfrac{5}{4}, \dfrac{6}{5}, \ldots\right\}$
   (r) $\left\{\dfrac{2}{1}, -\dfrac{3}{2}, \dfrac{4}{3}, -\dfrac{5}{4}, \dfrac{6}{5}, \ldots\right\}$

(s) $\left\{1, 1+\dfrac{1}{2}, 1+\dfrac{1}{2}+\dfrac{1}{4}, 1+\dfrac{1}{2}+\dfrac{1}{4}+\dfrac{1}{8}, \ldots\right\}$

(t) $\left\{1, 1-\dfrac{1}{2}, 1-\dfrac{1}{2}+\dfrac{1}{4}, 1-\dfrac{1}{2}+\dfrac{1}{4}-\dfrac{1}{8}, \ldots\right\}$

★(u) $\left\{1, 1+\dfrac{1}{2}, 1+\dfrac{1}{2}+\dfrac{1}{3}, 1+\dfrac{1}{2}+\dfrac{1}{3}+\dfrac{1}{4}, \ldots\right\}$

★(v) $\left\{1, 1-\dfrac{1}{2}, 1-\dfrac{1}{2}+\dfrac{1}{3}, 1-\dfrac{1}{2}+\dfrac{1}{3}-\dfrac{1}{4}, \ldots\right\}$

★(w) $\left\{1, 1+\dfrac{1}{3}, 1+\dfrac{1}{3}+\dfrac{1}{5}, 1+\dfrac{1}{3}+\dfrac{1}{5}+\dfrac{1}{7}, \ldots\right\}$

2. Let $\{S, +, \cdot\}$ be an ordered field and let $A$ be a subset of $S$ such that $A$ has an upper bound in $S$. Prove the following:
   (a) If a fixed element of $S$ is added to each member of $A$, then the new set which results also has an upper bound in $S$.
   (b) If each member of $A$ is multiplied by a fixed positive element of $S$, then the new set which results has an upper bound in $S$.
   (c) If each member of $A$ is multiplied by a fixed negative element of $S$, then the new set which results has a lower bound in $S$.

3. Let $A$ be a set of positive elements having an upper bound in an ordered field $\{S, +, \cdot\}$. Let $B$ be the set of reciprocals of elements of $A$, that is,

$$B = \left\{\dfrac{1}{x} \;\middle|\; x \in A\right\}$$

   Prove or disprove:
   (a) Set $B$ has an upper bound.
   (b) Set $B$ has a lower bound.

4. We showed that in an ordered field every nonempty finite subset is bounded and, in fact, has a least element as well as a greatest element. Cite examples from the field of rational numbers to show that a bounded infinite subset may or may not have a least element; a greatest element.

5. (a) Using an argument similar to that given in Section 7.1, prove that if $\{S, +, \cdot\}$ is an ordered field, then set $S$ has no lower bound.
   (b) Let $\{S, +, \cdot\}$ be an ordered field and let $A$ be a subset of $S$. Define $-A$ to be the set consisting of all the additive inverses of the elements of $A$, that is,

$$-A = \{-x \mid x \in A\}$$

   Prove that $A$ has an upper bound in $S$ if and only if $-A$ has a lower bound in $S$.

(c) Combine the first part of Theorem 7.1 with the result proved in (b) to obtain still another proof of (a).

6. By imitating the argument used in Section 7.1 show that if $\{S, +, \cdot\}$ is an ordered field, there is a proper subset of $S$ which has neither a lower bound nor an upper bound in $S$. (*Hint:* The set of integers $\{0, \pm 1, \pm 2, \ldots\}$ is an example of such a proper subset.)

★7. Let $\{S, +, \cdot\}$ be an ordered field.
 (a) Show that $\emptyset$ (the null set) has every element of $S$ as an upper bound.
 (b) Show that $\emptyset$ has every element of $S$ as a lower bound.

## 7.2 LEAST UPPER BOUND AND GREATEST LOWER BOUND

We have already pointed out that the field of rational numbers $\{Q, +, \cdot\}$ is an ordered field. It follows therefore (by Theorem 6.30) that $Q$ has neither an upper bound nor a lower bound.

On the other hand, if $Q^-$ is the subset of all negative rational numbers, then clearly $Q^-$ has many upper bounds. In fact, any positive rational number such as $\frac{1}{2}$, $\frac{3}{4}$, 1, $2\frac{7}{8}$, etc., is an upper bound for $Q^-$ because each negative number is less than any given positive number. Moreover, zero is also an upper bound for $Q^-$ because $x < 0$ for each $x \in Q^-$.

A natural question now arises: Are there any upper bounds for $Q^-$ in addition to positive rationals and zero? Suppose $n$ is any negative rational number. Then $\frac{n}{2}$ (that is, $n \times \frac{1}{2}$) will also be a negative number. But $n < \frac{n}{2}$. (Proof: $2 > 1$; therefore $\frac{1}{2} < \frac{1}{1}$; therefore $n \cdot \frac{1}{2} > n \cdot 1$, that is, $\frac{n}{2} > n$.) This shows that if $n$ is any negative number, there is always a greater negative number, namely $\frac{n}{2}$; thus no negative number can be an upper bound for $Q^-$.

We now see that the set of all rational upper bounds for $Q^-$ consists of zero together with all positive rational numbers. Since zero is less than any positive number, zero is called a **least upper bound** for $Q^-$.

As another example consider the sequence of positive rational numbers

$$1.9, \quad 1.99, \quad 1.999, \quad \ldots$$

where each numeral of this sequence after the first is obtained by appending an additional digit 9 to the preceding numeral. If we rename this sequence as follows:

$$2 - \frac{1}{10}, \quad 2 - \frac{1}{100}, \quad 2 - \frac{1}{1000}, \quad \ldots$$

we immediately see that each of the numbers in the set is less than 2. Therefore 2 is certainly an upper bound for this (infinite) set of numbers. Furthermore, if we choose any rational number $q$ less than 2, then $q$ will not be an upper bound for the set of numbers. To see this, we observe that if $q < 2$, then $2 - q > 0$. However, it is a property of the rational numbers that the sequence

$$\frac{1}{10}, \quad \frac{1}{100}, \quad \frac{1}{1000}, \quad \ldots, \quad \frac{1}{10^n}, \quad \ldots$$

ultimately becomes less than any given positive rational number. This means that corresponding to each rational number $r > 0$ we can choose an integer $n$ so that $\frac{1}{10^n} < r$. (The integer $n$ will, of course, depend on $r$.) Therefore, since $2 - q > 0$, let us choose $n$ so that

$$\frac{1}{10^n} < 2 - q$$

Then it follows that

$$q < 2 - \frac{1}{10^n}$$

Since $q$ is less than $2 - \frac{1}{10^n}$, $q$ cannot be an upper bound for the above set of numbers. This implies that 2 is the least upper bound for the set

$$\left\{ 2 - \frac{1}{10}, \quad 2 - \frac{1}{100}, \quad 2 - \frac{1}{1000}, \quad \ldots, \quad 2 - \frac{1}{10^n}, \quad \ldots \right\}$$

These two examples lead us, naturally, to the following general definition of a least upper bound (abbreviated LUB).

**Definition.** Let $\{S, +, \cdot\}$ be an ordered field and let $A$ be any subset of $S$. An element $l$ of $S$ is called a **least upper bound** for set $A$ if and only if
1. $l$ is an upper bound for $A$ in $S$;
2. whenever $b$ is any upper bound for $A$, then $l \leq b$.

In a similar manner we define a **greatest lower bound** (abbreviated GLB) for a subset of $S$.

**Definition.** Let $\{S, +, \cdot\}$ be an ordered field and let $A$ be any subset of $S$. An element $g$ of $S$ is called a **greatest lower bound** for set $A$ if and only if
1. $g$ is a lower bound for $A$;
2. whenever $c$ is any lower bound for $A$, then $g \geq c$.

For example, in the ordered field of rational numbers $\{Q, +, \cdot\}$, if $Q^+$ is the subset of all positive rational numbers, then zero is a greatest lower bound for $Q^+$. (See Exercise 5.)

Observe that a least upper bound or a greatest lower bound for a set $A$ may or may not be a member of $A$. For example, if $A$ is the set

$$\left\{1.9, \quad 1.99, \quad 1.999, \quad \ldots, \quad 2 - \frac{1}{10^n}, \quad \ldots\right\}$$

then the least upper bound for $A$, namely 2, is not a member of $A$. On the other hand, the greatest lower bound for $A$ is 1.9, and this number is a member of $A$.

It is also possible for a set to have upper bounds in an ordered field without having a least upper bound in this field. Likewise, a set may have lower bounds in an ordered field without having a greatest lower bound in that field. We discuss this important phenomenon in Section 7.3.

Although a bounded set may not have any least upper bound in an ordered field, it is legitimate to inquire whether a bounded set might possibly have more than one. The following theorem settles this question.

**THEOREM 7.2** If $\{S, +, \cdot\}$ is an ordered field and if $A$ is a subset of $S$, then

    (a) $A$ can have at most one least upper bound;

    (b) $A$ can have at most one greatest lower bound.

*Proof:* (a) Suppose $l_1$ is a least upper bound for $A$ and $l_2$ is also a least upper bound for $A$. Then $l_2$ is certainly an upper bound for $A$ by part 1 of the definition of least upper bound. Then, applying part 2 of the definition of least upper bound to $l_1$, we see that

$$l_1 \leq l_2$$

But by the same reasoning we also establish that

$$l_2 \leq l_1$$

Hence by the Trichotomy Theorem (Theorem 6.3) we obtain

$$l_1 = l_2$$

This completes the proof of (a).

(b)  We leave this proof for the reader. (See Exercise 4.)

In view of Theorem 7.2 we may speak of *the* least upper bound (if one exists) rather than *a* least upper bound and, similarly, we may refer to *the* greatest lower bound rather than *a* greatest lower bound.

# EXERCISES 7.2

1. Let $\{Q, +, \cdot\}$ be the ordered field of rational numbers. For each of the following subsets of $Q$, find (if they exist) a least upper bound and a greatest lower bound in $Q$.

    (a)  $\{9, 9.9, 9.99, 9.999, \ldots\}$

    (b)  $\{1, .1, .01, .001, \ldots\}$

    (c)  $\{1, 1.1, 1.11, 1.111, \ldots\}$

    (d)  $\{2, 2.2, 2.22, 2.222, \ldots\}$

    (e)  $\left\{\dfrac{1}{2}, \dfrac{2}{3}, \dfrac{3}{4}, \dfrac{4}{5}, \ldots\right\}$

    (f)  $\{9, -9.9, 9.99, -9.999, \ldots\}$

    (g)  $\{1, -.1, .01, -.001, \ldots\}$

    (h)  $\left\{\dfrac{1}{2}, -\dfrac{2}{3}, \dfrac{3}{4}, -\dfrac{4}{5}, \ldots\right\}$

    (i)  $\{n \in Q \mid n < 2\}$

    (j)  $\{n^2 \mid n \in Q \text{ and } n < 2\}$

    (k)  $\left\{\dfrac{1}{9}, \dfrac{1}{9.9}, \dfrac{1}{9.99}, \dfrac{1}{9.999}, \ldots\right\}$

    (l)  $\left\{\dfrac{1}{1}, \dfrac{1}{.1}, \dfrac{1}{.01}, \dfrac{1}{.001}, \ldots\right\}$

    (m)  $\left\{\dfrac{2}{1}, \dfrac{3}{2}, \dfrac{4}{3}, \dfrac{5}{4}, \ldots\right\}$

    (n)  $\left\{\dfrac{2}{1}, -\dfrac{3}{2}, \dfrac{4}{3}, -\dfrac{5}{4}, \ldots\right\}$

(o) $\left\{1, \dfrac{1}{2}, \dfrac{1}{4}, \dfrac{1}{8}, \ldots\right\}$

(p) $\left\{1, -\dfrac{1}{2}, \dfrac{1}{4}, -\dfrac{1}{8}, \ldots\right\}$

(q) $\left\{1, 1 + \dfrac{1}{2}, 1 + \dfrac{1}{2} + \dfrac{1}{4}, 1 + \dfrac{1}{2} + \dfrac{1}{4} + \dfrac{1}{8}, \ldots\right\}$

(r) $\left\{1, 1 - \dfrac{1}{2}, 1 - \dfrac{1}{2} + \dfrac{1}{4}, 1 - \dfrac{1}{2} + \dfrac{1}{4} - \dfrac{1}{8}, \ldots\right\}$

2. Let $\{S, +, \cdot\}$ be an ordered field and let $A$ be a subset of $S$ such that $A$ has a least upper bound in $S$. Prove the following:
   (a) If a fixed element of $S$ is added to each member of $A$, then the new set which results has a least upper bound in $S$.
   (b) If each member of $A$ is multiplied by a fixed positive element of $S$, then the new set which results has a least upper bound in $S$.
   (c) If each member of $A$ is multiplied by a fixed negative element of $S$, then the new set which results has a greatest lower bound in $S$.

3. Let $A$ be a set of positive elements having a least upper bound in an ordered field $\{S, +, \cdot\}$. Let $B$ be the set of reciprocals of elements of $A$, that is,

$$B = \left\{\dfrac{1}{x} \ \Big| \ x \in A\right\}$$

Prove or disprove:
   (a) Set $B$ has a least upper bound.
   (b) Set $B$ has a greatest lower bound.

4. Prove part b of Theorem 7.2, that is, a set $A$ can have at most one greatest lower bound in an ordered field.

5. (a) Prove that if $l$ is the least upper bound for set $A$ and if $l$ is an element of $A$, then $l$ must be the greatest element of set $A$.
   (b) State and prove a conclusion analogous to (a) concerning the greatest lower bound of set $A$.
   (c) Prove that if $l$ is the least upper bound for set $A$ and if $l$ is not an element of $A$, then there cannot be a greatest element in set $A$.
   (d) State and prove a conclusion analogous to (c) concerning the greatest lower bound of set $A$.

6. Let $\{S, +, \cdot\}$ be an ordered field and let $S^+$ be the set of all positive elements of $S$. Prove that 0 is the greatest lower bound for $S^+$ in $S$.

## 7.3 THE COMPLETENESS PROPERTY

In Chapter 5 we mentioned that the field of rational numbers $\{Q, +, \cdot\}$ is incomplete because there are positive rational numbers that do not have a square root in $Q$. (See page 184.) As a specific illustration let us prove the following theorem.

**THEOREM 7.3** There is no rational number whose square is 2.

*Proof:* Suppose that there were a rational number $r$ such that $r^2 = 2$. Since every rational number can be obtained as a quotient of integers, let $r = \dfrac{a}{b}$ where $a$ and $b$ are relatively prime integers. (By use of Theorem 5.6a, any quotient of integers can be reduced to a quotient of relatively prime integers.) Our supposition that $r^2 = 2$ implies that

$$a^2 = 2b^2$$

Hence the prime number 2 is a divisor of $a^2$ and therefore, by Theorem 3.6, 2 must be a divisor of $a$. Letting $a = 2c$ (where $c$ is some integer), we now obtain

$$4c^2 = 2b^2$$
$$\therefore 2c^2 = b^2$$

Applying Theorem 3.6 again, we see that 2 is also a divisor of $b$. But this is impossible because $a$ and $b$ are relatively prime and therefore both cannot be divisible by 2. Our supposition that there is a rational number whose square is 2 must therefore be false. This completes the proof.

In view of Theorem 7.3 the square of any rational number must either be less than 2 or else greater than 2 (Trichotomy Theorem for ordered fields). The set $Q^+$ of all positive rational numbers may therefore be partitioned into two disjoint subsets $A$ and $B$ where

$A = \{$ all positive rational numbers whose squares are less than 2 $\}$;
$B = \{$ all positive rational numbers whose squares are greater than 2 $\}$.

It is quite easy to prove that every member of $B$ is an upper bound for set $A$. In fact, if $x$ is any member of $A$ and $y$ is any member of $B$, then

$x^2 < 2$ and $2 < y^2$. Hence, by transitivity (Theorem 6.4), $x^2 < y^2$, and it follows (by Theorem 6.26) that

$$|x| < |y|$$

that is, $x < y$ (because $x$ and $y$ are positive). This shows that $y > x$ for every $x$ in $A$; thus $y$ must be an upper bound for set $A$.

Although set $A$ has infinitely many upper bounds in $Q$, it is a curious fact that set $A$ has no *least* upper bound in $Q$. To prove this we first observe that if $A$ had a least upper bound $l$ in $Q$, then $l$ would have to be either a member of set $A$ or a member of set $B$. If $l$ were a member of $A$, then $l$ would have to be the greatest element in $A$ (because $l \geq x$ for all $x$ in $A$). If $l$ were a member of set $B$, then $l$ would have to be the *least* element in $B$ (because each element $y$ in $B$ is an upper bound for $A$ and hence $l \leq y$ for all $y$ in $B$). Therefore, if we can prove that there is neither a greatest element in $A$ nor a least element in $B$, it will follow that there is no least upper bound in $Q$ for set $A$.

We now establish these facts by proving the following theorem.

**THEOREM 7.4**  If $a$ and $b$ are positive rational numbers with $a < b$, then there exists a positive rational number $r$ whose square is between $a$ and $b$, that is,

$$a < r^2 < b$$

*Proof:*  (*Note:* The proof is lengthy. Those who find it difficult may omit it on a first reading.)

There is certainly some integer $n > 1$ such that $n > a$. (For example, if $a = \dfrac{p}{q}$ where $p$ and $q$ are positive integers, then $q \geq 1$ and hence $qa \geq a$, so that

$$qa + 1 > a$$

Therefore $qa + 1$ ($= p + 1$) is such an integer $n$. Now, since $n > 1$, therefore

$$n^2 > n > a$$

This shows that the set of all integers whose squares are greater than $a$ is nonempty. Therefore, by the well-ordering property of the positive integers, there is a least integer $g$ such that $a < g^2$. If $g^2 < b$, then we have

$$a < g^2 < b$$

and there is nothing more to prove. Hence we suppose that $b \leq g^2$.

Because $g$ is the least integer such that $a < g^2$, it follows that

(1) $$(g-1)^2 \leq a < b \leq g^2$$

Therefore $g^2 - b \geq 0$ and $a - (g-1)^2 \geq 0$ and it follows that

$$0 < b - a \leq (b-a) + (g^2 - b) + a - (g-1)^2$$

that is, $\quad 0 < b - a \leq g^2 - (g-1)^2$

Hence $\quad 0 < b - a \leq 2g - 1 < 2g$

and therefore $\quad 1 < \dfrac{2g}{b-a}$

Using the well-ordering axiom again, we let $m$ be the least integer greater than $\dfrac{2g}{b-a}$. Then

$$1 < \dfrac{2g}{b-a} < m$$

and hence

(2) $$\dfrac{2g}{m} < b - a$$

Now consider the set of rational numbers

$$\left\{ g-1, g-1+\dfrac{1}{m}, g-1+\dfrac{2}{m}, \ldots, \right.$$
$$\left. g-1+\dfrac{j}{m}, \ldots, g-1+\dfrac{m}{m} \right\}$$

As $j$ ranges from 0 to $m$ these numbers form an increasing sequence which ranges from $g-1$ to $g$ and hence whose squares range from $(g-1)^2$ to $g^2$. Keeping in mind that $a < g^2$, let $k$ be the least integer such that

(3) $$a < \left( g - 1 + \dfrac{k}{m} \right)^2$$

### 7.3 The Completeness Property

(This is our third use of the well-ordering axiom in this proof.) For this value of $k$,

(4) $$\left(g - 1 + \frac{k-1}{m}\right)^2 \leq a$$

and furthermore,

(5) $$\left(g - 1 + \frac{k}{m}\right)^2 - \left(g - 1 + \frac{k-1}{m}\right)^2$$

$$= \frac{(mg - m + k)^2 - (mg - m + k - 1)^2}{m^2}$$

$$= \frac{2(mg - m + k) - 1}{m^2} < \frac{2(mg - m + k)}{m^2}$$

$$\leq \frac{2(mg - m + m)}{m^2} = \frac{2g}{m} < b - a \qquad \text{(from (2))}$$

Now let $r = g - 1 + \frac{k}{m}$. Then from (3)

$$a < r^2$$

and from (4) and (5)

$$r^2 - a \leq r^2 - \left(g - 1 + \frac{k-1}{m}\right)^2 < b - a$$

$$\therefore r^2 < b$$

Thus the rational number $r = \frac{mg - m + k}{m}$

has the desired property $a < r^2 < b$.

This completes the proof of Theorem 7.4.

Applying Theorem 7.4 to set $A$ we see that if $x$ is any element of $A$, that is, if $x^2 < 2$, then there exists a rational number $r$ with

$$x^2 < r^2 < 2$$

Since $r^2 < 2$, therefore $r$ is an element of $A$ and since $x^2 < r^2$, therefore $x < r$. This shows that for each $x$ in $A$ there always exists another member of $A$, namely $r$, which is greater than $x$. Hence there is no greatest element in $A$. Similarly, we prove there is no least element in $B$. (See Exercise 9.)

Finally, since there is neither a greatest element in set $A$ nor a least element in set $B$, there is no least upper bound for set $A$ in $Q$.

The fact that the rational numbers $Q$ have a bounded subset such as $A$ which has no least upper bound emphasizes further the *incompleteness* of the rational number system $\{Q, +, \cdot\}$. It appears desirable to attempt an extension or expansion of this system to a more inclusive ordered field that will be *complete* in the sense that every nonempty subset having an upper bound must perforce have a least upper bound. In such a complete system set $A$, discussed above, would, for example, have a least upper bound $l$, and since $l$ belongs neither to $A$ nor to $B$, that is, since $l^2$ is neither less than 2 nor greater than 2, it would follow (by the Trichotomy Theorem) that $l^2 = 2$. In other words, $\sqrt{2}$ would exist in this more extensive ordered field along with many other irrational numbers as we shall see later.

Let us therefore introduce the following definition.

**Definition.** An ordered field $\{S, +, \cdot\}$ is called a **complete ordered field** if and only if every nonempty subset of $S$ which has an upper bound (in $S$) has a least upper bound (in $S$).

We have already observed that in every ordered field $\{S, +, \cdot\}$ the set $S$ contains a subset which for all practical purposes can be regarded as the set $Q$ of rational numbers.* Hence any complete ordered field can be regarded as an extension of the field of rational numbers which "fills the gaps" in the rational number system.

It can also be shown† that any two complete ordered fields are isomorphic in the following sense: Their elements can be matched in a one-to-one correspondence in such a way that sums, products, and order of elements are all preserved. From this point of view all complete ordered fields are structurally alike and all may be regarded as identical with the field of real numbers $\{R, +, \cdot\}$.

Although a full discussion is beyond our present scope, we shall indicate here how several fundamental properties of the real numbers can be derived from the assumption that the real number system is a complete ordered field.

---

* Strictly speaking, $S$ contains a subset $Q_s$ such that when the field operations $+$ and $\cdot$ are restricted to elements in $Q_s$, the resulting subsystem $\{Q_s, +, \cdot\}$ becomes an ordered field isomorphic to the field $\{Q, +, \cdot\}$ of rational numbers. The isomorphism preserves not only sums and products but order of elements as well.

† See, for example, Richard Dean, *Elements of Abstract Algebra*, John Wiley and Sons, page 124.

**THEOREM 7.5** (*Archimedean Property*) Let $\{S, +, \cdot\}$ be a complete ordered field and let $a, b$ be any two positive elements of $S$. Then there exists a positive integer $n$ such that $na > b$.

*Proof:* Suppose that there were no such positive integer $n$, that is, suppose that $na \leq b$ for all positive integers $n$. Then $b$ would be an upper bound in $S$ for the set of all such multiples of $a$ and, consequently, there would be a least upper bound $b_0$ in $S$ for this set. Then certainly $b_0 \geq ma$ for every positive integer $m \geq 2$; and it follows that

$$b_0 - a \geq (m-1)a$$

for every positive integer $m \geq 2$, or letting $n = m - 1$,

$$b_0 - a \geq na$$

for every positive integer $n$. But this is impossible because

$$b_0 - a < b_0$$

which was supposed to be the *least* upper bound for all such multiples $na$.

**THEOREM 7.6** (*Condition of Eudoxus*) Let $\{S, +, \cdot\}$ be a complete ordered field and let $a$ and $b$ be any two elements of $S$ with $a < b$. Then there exists a rational element $r = \dfrac{m}{n}$ (where $m$ and $n$ are integers) such that $a < \dfrac{m}{n} < b$.

*Proof:* There is no loss of generality in assuming both $a > 0$ and $b > 0$ because if, for example, $a < 0$ and $b > 0$, then $a < 0 < b$, so $r = 0$ fulfills the requirements; if both $a < 0$ and $b < 0$, then $-a > 0$ and $-b > 0$, and since $a < b$, then $-b < -a$; if we determine $r$ so that

$$-b < r < -a$$

then $a < -r < b$, so $(-r)$ fulfills the requirements. We suppose then that $0 < a < b$. Hence $b - a > 0$; thus by the Archimedian Property (Theorem 7.5) there is a positive integer $n$ such that $n(b - a)$ is certainly greater than 1, that is

$$n(b - a) > 1$$

Therefore

(1) $$\frac{1}{n} < b - a$$

Now since $a > 0$, $na$ is also $> 0$ and it follows (again by the Archimedian Property) that there exists a positive integer $m$ such that

$$m \cdot 1 > na$$

By the well-ordering axiom there is a least such integer; so if we choose $m$ to be this least integer, it follows that

$$m \cdot 1 > na, \quad \text{but} \quad (m-1) \cdot 1 \leq na$$

From the first of these inequalities we see immediately that

(2) $$a < \frac{m}{n}$$

and from the second inequality we have

(3) $$\frac{m-1}{n} \leq a$$

If we write

$$\frac{m}{n} = \frac{(m-1)+1}{n} = \frac{m-1}{n} + \frac{1}{n}$$

and make use of both (1) and (3), we obtain

$$\frac{m}{n} < a + (b-a)$$

that is,

(4) $$\frac{m}{n} < b$$

Combining (2) and (4), we see that

$$a < \frac{m}{n} < b$$

as desired. This completes the proof.

## EXERCISES 7.3

1. An alternate proof that the square of no rational number is 2 depends on the number of primes in the prime factorization of $a^2$ and $2b^2$ in Theorem 7.3. Try to complete such a proof.

2. Prove that the square of no rational number is 3 using a method similar to
   (a) the proof given in Theorem 7.3;
   (b) the proof suggested in Exercise 1.

3. Prove that the square of no rational number is a prime using the method suggested in Exercise 1.

4. Extend the proof in Exercise 3 to show that if a whole number is not the square of some whole number, then it is not the square of any rational number.

5. Prove that each of the following is not the cube of any rational number.
   (a) 2 (b) 3 (c) Any prime $p$

6. For the given values of $a$ and $b$, find a rational number $r$ such that $a < r^2 < b$.
   (a) $a = 6, b = 7$
   (b) $a = 6, b = 6.2$
   (c) $a = 6, b = 6.1$
   (d) $a = 6, b = 6.01$

7. For the given values of $a$ and $b$, find a rational number $r$ such that $a < r^3 < b$.
   (a) $a = 9, b = 10$
   (b) $a = 9, b = 9.5$
   (c) $a = 9, b = 9.1$

★8. Prove that if $a$ and $b$ are positive rational numbers, then there exists a rational number $r$ whose cube is between $a$ and $b$, that is,
$$a < r^3 < b$$

9. Let set $B$ be defined by
$$B = \{x \in Q \mid x^2 > 2\}$$
Prove that $B$ has no least element.

## REVIEW EXERCISES

1. For each of the following sets find, if possible,
   (a) the least upper bound
   (b) the greatest lower bound
   (1) $\{.3, .33, .333, \ldots\}$
   (2) $\{.3, .03, .003, \ldots\}$
   (3) $\{2.3, 2.03, 2.003, \ldots\}$
   (4) $\{2.3, -2.03, 2.003, -2.0003, \ldots\}$
   (5) $\left\{1, 1 - \frac{1}{3}, 1 - \frac{1}{3} + \frac{1}{5}, 1 - \frac{1}{3} + \frac{1}{5} - \frac{1}{7}, \ldots\right\}$
   (6) $\left\{1, 1 + \frac{1}{3}, 1 + \frac{1}{3} + \frac{1}{5}, 1 + \frac{1}{3} + \frac{1}{5} + \frac{1}{7}, \ldots\right\}$

2. Let $B = \{x \mid x \text{ is a rational number } x^2 > 3\}$. Prove that $B$ has no least element.

3. Let $r$ be a positive real number. Prove that there exists a positive integer $p$ such that $\frac{1}{p} < r$.

4. If $r$ and $s$ are any two real numbers, prove that there are two rational numbers $q_1$ and $q_2$ between $r$ and $s$.

5. For each of the conditions listed below find an infinite subset of $Q$ (rational numbers) satisfying that condition.
   (a) Has 7 for its greatest lower bound.
   (b) Has 7 for its least upper bound.
   (c) Has $\sqrt{7}$ for its greatest lower bound.
   (d) Has $\sqrt{7}$ for its least upper bound.
   (e) Has $\sqrt{7}$ for its greatest lower bound and 7 for its least upper bound.
   (f) Has no greatest lower bound and no least upper bound in $Q$.
   (g) Has a greatest lower bound but no least upper bound in $Q$.
   (h) Has no greatest lower bound but has a least upper bound in $Q$.

*David Hilbert (1862-1943)*

# 8 | Vector Spaces

> Probably no aspect of twentieth century mathematics stands out more clearly than does the ever greater degree of generalization and abstraction. From the time of Hilbert and Frechet the notions of abstract set and abstract space have been fundamental in research.
>
> CARL BOYER

## 8.1 GENERALIZING THE NOTION OF AN OPERATION

Until now we have considered (almost exclusively) operations which are binary operations in (or on) some specified set $S$. A binary operation $\circ$ always "acts" on a pair of elements of $S$, say $u$ and $v$, and associates with this pair of elements an element $w$ that also belongs to $S$. Thus when writing

$$(u, v) \xrightarrow{\circ} w \quad \text{or} \quad u \circ v = w$$

we have always stipulated that $u$, $v$, and $w$ are elements of one and the same set $S$.

For the purposes of this chapter, however, we need a somewhat more general type of operation. Instead of confining $u$, $v$, and $w$ to the same set $S$, let us assume that the first operand $u$ belongs to a set $U$ and the second operand $v$ belongs to a set $V$ which may be different from $U$. Let us further assume that the operation $\circ$ associates with the ordered pair $(u, v)$ an element $w$ that

279

belongs to a set $W$ that may (but need not) be different from $U$ or $V$. Thus although we shall still write

$$(u, v) \overset{\circ}{\to} w \quad \text{or} \quad u \circ v = w$$

we do not assume that $u, v, w$ all necessarily belong to the same set.

***Example 1.*** Suppose that
$U = \{\text{All integer multiples of } \sqrt{2}\}$
$V = \{\text{All integer multiples of } \sqrt{3}\}$
and suppose that $\cdot$ denotes ordinary multiplication of real numbers. Then if $u \in U$ and $v \in V$ and if

$$w = u \cdot v$$

the product $w$ will belong to the set

$$W = \{\text{Integer multiples of } \sqrt{6}\}$$

In this case, the set $W$ is different from either $U$ or $V$.

***Example 2.*** Suppose that $U = Z$ (the set of all integers), that $V = Q$ (the set of all rational numbers), and again $\cdot$ is ordinary multiplication. If $u \in U$, $v \in V$, and if

$$w = u \cdot v$$

then the product $w$ will certainly be a rational number, although not necessarily an integer. Thus, if $W$ is the set of all possible values of $w$, then $W \subseteq V$. Moreover, choosing $u = 1$, we see that every rational number $v$, that is, every member of $V$, is a member of $W$, that is, $V \subseteq W$. Therefore, in this case, $W = V$. The set $W$ is identical with set $V$.

***Example 3.*** Suppose that $U = Q$ (the set of rational numbers) and $V$ is the set of all ordered pairs of rational numbers. If $u \in U$, then $u$ is simply a rational number. On the other hand, if $v \in V$, then $v$ is an ordered pair of rational numbers, that is,

$$v = (q_1, q_2)$$

where $q_1$ and $q_2$ are each rational numbers (members of $Q$). Now let us define the product $u \circ v$ as

$$u \circ v = u \circ (q_1, q_2) = (u \cdot q_1, u \cdot q_2)$$

where $u \cdot q_1$ and $u \cdot q_2$ are ordinary products of rational numbers. The new product $u \circ v$ is therefore an ordered pair $(u \cdot q_1, u \cdot q_2)$ of rational numbers and hence a member of $V$. In fact, in this example we readily see that the set $W$ consisting of all products $u \circ v$, where $u \in U$ and $v \in V$, is identical with the set $V$.

*Example 4.* Suppose that $U = R$ (the set of real numbers) and $V = C$ (the set of complex numbers). If $u \in U$ and $v \in V$ and if we define

$$w = u \circ v$$

to be simply the ordinary product $u \cdot v$ (the real number $u$ multiplied by the complex number $v$), then the set $W$ of all possible values of $w$ is readily seen to be identical with set $V$.

## EXERCISES 8.1

1. Suppose that $U = \{\text{Integer multiples of } \sqrt{5}\}$
   $V = \{\text{Integer multiples of } \sqrt{2}\}$
   Describe all ordinary products $u \cdot v$ if $u \in U$, $v \in V$.

2. Suppose that $U = \{\text{Integers}\}$
   $V = \{\text{Even integers}\}$
   Describe all products $u \circ v$ if $u \in U$, $v \in V$, and
   $$u \circ v = (u, v)$$

3. Suppose that $U = \{\text{Integers}\}$
   $V = \{\text{Integer multiples of } \sqrt{2}\}$
   Describe all products $u \circ v$ where $u \in U$, $v \in V$ if
   $u \circ v = u + v$    (ordinary addition of $u$ and $v$)

4. Suppose that $U = \{\text{Integer powers of 2}\}$
   $V = \{\text{Integer powers of 3}\}$
   Describe all products $u \circ v$ if $u \in U$, $v \in V$, and
   $u \circ v = u \cdot v$    (ordinary multiplication of $u$ and $v$)

5. Suppose that $U = \{0, 1, 2\}$
   $V = \{0, 1, 2, 3, 4\}$
   Describe all products $u \circ v$ with $u \in U$, $v \in V$ where
   $$u \circ v = u \cdot v \pmod 5.$$

   (Make a multiplication table.)

**6.** Suppose that $U = \{0, 1, 2, 3, 4\}$
$$V = \{(x, y) \mid x \in A \text{ and } y \in A\}$$
where $A = \{0, 1, 2\}$.

Describe all products $u \circ v$ where $u \in U$, $v \in V$, and
$$u \circ v = u \circ (x, y) = (ux, uy)$$
the products $ux$ and $uy$ being calculated modulo 5.

## 8.2 DEFINITION OF A VECTOR SPACE

In Examples 1 to 4 of Section 8.1 a pair of mathematical systems, one involving a set $U$ and the other a set $V$, were brought into relationship with each other by means of an external operation. An element of set $U$ was multiplied by an element of set $V$ to produce an element of a set $W$ which, in several of the examples, was the same as set $V$. Clearly, a great variety of more or less complicated algebraic structures can be defined in this manner. It turns out that several of these structures are particularly important in mathematics and its applications. One such fundamental structure is the **vector space.**

We can gain some insight into how the notion of a vector space arises by studying Examples 3 and 4 of Section 8.1 in greater detail. To conform with notation to be introduced shortly we rename the first set $S$ instead of $U$, but we retain the symbol $V$ for the second set.

In Example 3 each element of set $V$ is an ordered pair of rational numbers. One way of looking at these ordered pairs is to think of them as representing points in a plane. If the ordered pair is $(x, y)$, we can think of the rational numbers $x$ and $y$ as coordinates of a point. We locate this point in the usual manner of coordinate geometry by using two (fixed) intersecting number lines as axes. The axes need not be perpendicular, although they may be so in special cases. Figure 8.1 depicts a few such points $A$, $B$, $C$, and $D$.

If we connect each of the points $A$, $B$, $C$, and $D$ with the origin $O$, we obtain the directed segments $\overrightarrow{OA}$, $\overrightarrow{OB}$, $\overrightarrow{OC}$, and $\overrightarrow{OD}$, as shown in Figure 8.2.

Directed segments such as these are often called **geometric vectors** or simply **vectors**. When a pair of axes has been chosen, each ordered pair in set $V$ corresponds to a unique geometric vector on the graph. To remind us of this pictorial representation we designate the elements of set $V$ by symbols such as $\vec{u}$ and $\vec{v}$.

## 8.2 Definition of a Vector Space

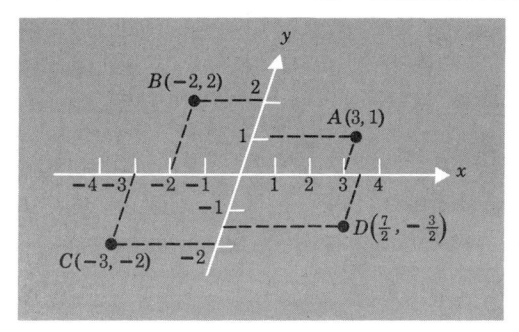

**Figure 8.1**

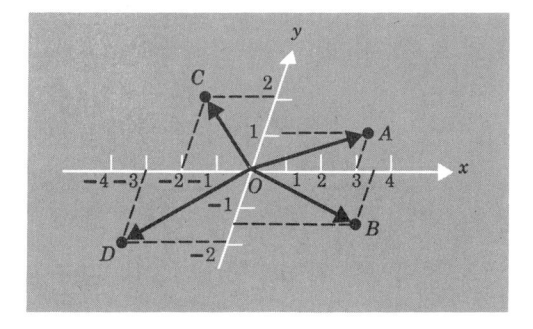

**Figure 8.2**

Suppose that $S$ is the set of all rational numbers (this set is named $U$ in Example 3). Let us choose an element $a \in S$, and element $\vec{v} \in V$, and let us form the product $a\vec{v}$ as in Example 3. For example, if $a = 2$ and $\vec{v} = (3, 1)$, then

$$a\vec{v} = 2(3, 1) = (6, 2)$$

Geometrically, this product corresponds to a directed segment in the same direction as $\overrightarrow{OA}$ but twice as long. (See Figure 8.3.)

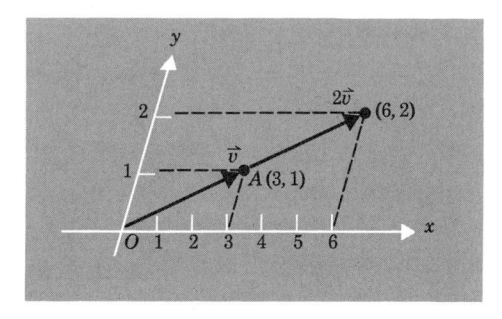

**Figure 8.3**

If the ordered pair $\vec{v} = (3, 1)$ is multiplied by $(-2)$, we obtain
$$(-2)\vec{v} = (-2)(3, 1) = (-6, -2)$$
Geometrically, this product corresponds to a directed segment in the opposite direction as $\overrightarrow{OA}$ and again twice as long. (See Figure 8.4.)

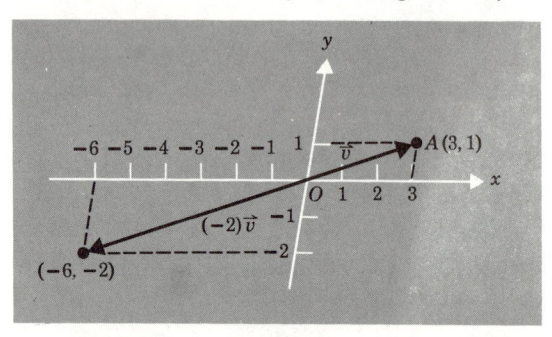

**Figure 8.4**

We have therefore a useful geometric interpretation for the operation of multiplying an element $\vec{v}$ from set $V$ by an element $a$ from set $S$. In this illustration the product $a\vec{v}$ is again an element of set $V$ and is associated with a geometric vector $\vec{w}$ whose length is $|a|$ times the length of the original vector $\vec{v}$. Moreover, when $a > 0$, the product vector $a\vec{v}$ has the same direction as the original vector, but when $a < 0$, the product vector is oppositely directed.

There is still another compelling reason for calling the members of set $V$ vectors. If $\vec{u}$ and $\vec{v}$ are elements of $V$, say
$$\vec{u} = (x_1, y_1) \quad \text{and} \quad \vec{v} = (x_2, y_2)$$
then it is only natural to add $\vec{u}$ and $\vec{v}$ as follows:
$$\vec{u} + \vec{v} = (x_1 + x_2, y_1 + y_2)$$

The addition defined here is clearly a *binary operation* on $V$ and it is left to the reader to show that the system $\{V, +\}$ is a commutative group (see Exercise 1 of this section).

Let us now examine the geometrical significance of this new addition operation. Suppose that $\vec{u} = (3, 1)$ and $\vec{v} = (-2, 2)$. Then $\vec{u}$ and $\vec{v}$ correspond, respectively, to the geometric vectors $\overrightarrow{OA}$ and $\overrightarrow{OB}$ in Figure 8.5. If we add $\vec{u}$ and $\vec{v}$, we obtain
$$\vec{u} + \vec{v} = (3, 1) + (-2, 2) = (3 + (-2), \ 1 + 2)$$
that is,
$$\vec{u} + \vec{v} = (1, 3)$$

## 8.2 Definition of a Vector Space

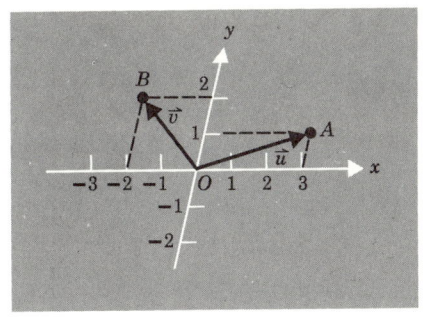

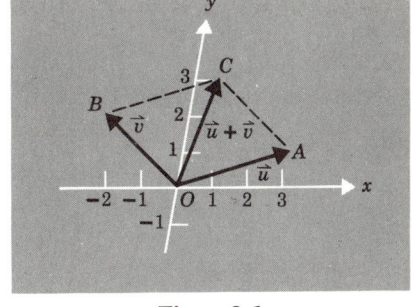

**Figure 8.5**

**Figure 8.6**

The sum $\vec{u} + \vec{v}$ corresponds to the geometric vector $\overrightarrow{OC}$. (See Figure 8.6.) The addition operation on $V$ thus appears to conform to the parallelogram law characteristic of forces, velocities, and other vector quantities in physics.

We may therefore view Example 3 as follows:

1. We have a set $V$ consisting, in this example, of ordered pairs of rational numbers. These ordered pairs may be conveniently referred to as vectors. There is a natural binary operation $(+)$ on $V$ called *addition of vectors* such that the system $\{V, +\}$ is an abelian group.
2. Next, we also have a set $S$ which in this example is the set $Q$ of rational numbers. We call the elements of $S$ **scalars** to distinguish them from the elements of $V$ which we call vectors. Because the scalars in $S$ are all rational numbers, the system $\{S, +, \cdot\}$ is a field. The symbols $+$ and $\cdot$ refer, respectively, to addition and multiplication of rational numbers.
3. Finally, we have still another multiplication operation (which may be designated by $\circ$ but more usually by juxtaposition) whereby any scalar, that is, any element $a \in S$, may operate on any vector, that is, on any element $\vec{v} \in V$, to produce a product $a\vec{v}$. This product is always an element of $V$, that is, a vector.

It is quite straightforward to verify that the various operations described here have the following properties: If $a$ and $b$ are scalars (elements of $S$) and if $\vec{u}$ and $\vec{v}$ are vectors (elements of $V$), then

$(v_1)$  $a(\vec{u} + \vec{v}) = (a\vec{u}) + (a\vec{v})$
$(v_2)$  $(a + b)\vec{u} = (a\vec{u}) + (b\vec{u})$
$(v_3)$  $(a \cdot b)\vec{u} = a(b\vec{u})$
$(v_4)$  $1\vec{u} = \vec{u}$

where 1 is the unit element of the field $\{S + \cdot\}$ and where each of the additions and multiplications indicated here has an obvious, unambiguous interpretation as defined in the discussion above.

We find a similar situation in Example 4.

1. The set of vectors $V$ is the set of complex numbers. Each element $\vec{z} \in V$ may be expressed in the form

$$\vec{z} = x + iy$$

where $x$ and $y$ are real numbers and $i = \sqrt{-1}$. Alternatively, the complex number

$$\vec{z} = x + iy$$

may be expressed as an ordered pair of real numbers $(x, y)$ and is usually represented, using a rectangular coordinate system, by a geometric vector joining the origin to the point whose coordinates are $(x, y)$.

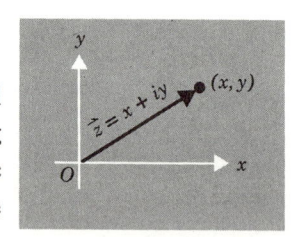

The sum of a pair of complex numbers is obtained by addition of corresponding coordinates. This defines a binary operation $(+)$ on $V$ which obeys the parallelogram law, and the system $\{V, +\}$ is easily seen to be an abelian group.

**Figure 8.7**

2. The set of scalars $S$ is the set of real numbers. We already know that the system $\{S, +, \cdot\}$ is a field if $+$ and $\cdot$ refer, respectively, to addition and multiplication of real numbers.

3. The operation of multiplying a real number $a$ (scalar) by a complex number $\vec{z} = x + iy$ (vector) is defined by

$$a\vec{z} = a(x + iy) = (a \cdot x) + i(a \cdot y)$$

The product is clearly a complex number, that is, a vector in set $V$. Moreover, if $a$ and $b$ are real numbers (scalars) and if $\vec{u}$ and $\vec{v}$ are complex numbers (vectors), we can readily verify that

$(v_1)$    $a(\vec{u} + \vec{v}) = (a\vec{u}) + (a\vec{v})$
$(v_2)$    $(a + b)\vec{u} = (a\vec{u}) + (b\vec{u})$
$(v_3)$    $(a \cdot b)\vec{u} = a(b\vec{u})$
$(v_4)$    $1\vec{u} = \vec{u}$

It is now apparent that there is a common algebraic structure underlying Examples 3 and 4. This algebraic structure, known as a vector space, can be defined by generalizing slightly the properties we have observed.

## 8.2 Definition of a Vector Space

**Definition.** Let $\{V, +\}$ be an abelian group and let $\{S, +, \cdot\}$ be a field.* Let the members of $V$ be called **vectors** and the members of $S$ be called **scalars**. For each vector $\vec{v} \in V$ and each scalar $a \in S$ let the product $a\vec{v}$ be defined as a vector (member of $V$). Then $\{V, +\}$ is called a **vector space** over the field $\{S, +, \cdot\}$ if and only if, for all scalars $a, b \in S$, and for all vectors $\vec{u}$ and $\vec{v} \in V$, the following properties hold (Axioms for a vector space).

*Axioms for a Vector Space.*

$(v_1)$    $a(\vec{u} + \vec{v}) = (a\vec{u}) + (a\vec{v})$
$(v_2)$    $(a + b)\vec{u} = (a\vec{u}) + (b\vec{u})$
$(v_3)$    $(a \cdot b)\vec{u} = a(b\vec{u})$
$(v_4)$    $1\vec{u} = \vec{u}$

It is now evident that the additive group of complex numbers is a vector space over the field of real numbers (Example 4) and the additive group of ordered pairs of rational numbers is a vector space over the field of rational numbers (Example 3).

Let us examine some additional examples of vector spaces.

**Example 1.** Let $S = R$ (the set of real numbers) and let $V$ be the set of all ordered triples of real numbers, that is,

$$V = \{(x, y, z) \mid x \in R, y \in R, \text{ and } z \in R\}$$

If we define addition of ordered triples by

$$(x_1, y_1, z_1) + (x_2, y_2, z_2) = (x_1 + x_2, y_1 + y_2, z_1 + z_2)$$
$$\text{for all } x_1, y_1, z_1, x_2, y_2, z_2 \text{ in } R$$

then $\{V, +\}$ is easily seen to be an abelian group (see Exercise 1b). Moreover, let us define multiplication of a real number by an ordered triple as follows:

$$a(x, y, z) = (ax, ay, az) \quad \text{for all } a \in S \text{ and all } (x, y, z) \in V$$

We can now readily verify that $\{V, +\}$ is a vector space over the field $\{R, +, \cdot\}$.

---

* The addition operation in the group $\{V, +\}$ need not be the same as the addition operation in the field $\{S, +, \cdot\}$. It is not necessary, however, to use different addition symbols because the context will always indicate the type of addition involved.

***Example 2.*** Let $S$ be the set of real numbers $R$, and $V$ the set of all $2 \times 2$ matrices whose elements are real numbers. We have already observed that if $+$ refers to addition of matrices, then $\{V, +\}$ is an abelian group. If we define multiplication of any real number $a$ by a $2 \times 2$ matrix $\begin{bmatrix} x & y \\ u & v \end{bmatrix}$ as follows:

$$a \begin{bmatrix} x & y \\ u & v \end{bmatrix} = \begin{bmatrix} ax & ay \\ au & av \end{bmatrix}$$

then we can readily verify that $\{V, +\}$ is a vector space over the field $\{R, +, \cdot\}$. We leave this verification as an exercise. (See Exercise 2 below.)

Additional examples of vector spaces appear in the exercises.

## EXERCISES 8.2

1. (a) Let $V$ be the set of all ordered pairs of rational numbers and let addition $+$ of ordered pairs be defined as in Section 8.2. If $\vec{u} = (x_1, y_1)$ and $\vec{v} = (x_2, y_2)$ then $\vec{u} + \vec{v} = (x_1 + x_2, y_1 + y_2)$
Verify that the system $\{V, +\}$ is a commutative group.
   (b) Let $V$ be the set of all ordered triples of real numbers and define addition of ordered triples as in Example 2, namely
   $(x_1, y_1, z_1) + (x_2, y_2, z_2) = (x_1 + x_2, y_1 + y_2, z_1 + z_2)$
   Verify that $\{V, +\}$ is an abelian group.
2. Verify in detail that the following systems are vector spaces:
   (a) Example 3, Section 8.1.
   (b) Example 4, Section 8.1.
   (c) Example 2, Section 8.2.
3. Establish that each of the following systems is not a vector space:
   (a) The system in Example 1, Section 8.1.
   (b) The system in Example 2, Section 8.1.
   (c) The system in Exercises 1, 2, 3, and 4, Exercises 8.1.
4. Decide whether or not the systems in Exercises 5 and 6 of Exercises 8.1 are vector spaces. Try to modify these systems to obtain vector spaces (if any of them are not).
5. Decide whether or not each of the following is a vector space.
   (a) Let $S = \{\{0, 1\}, +(\bmod 2), \cdot\}$
   $V = \{\text{All polynomials in } x \text{ with coefficients 0 or 1}, +(\bmod 2)\}$

## 8.2 Definition of a Vector Space

Take for scalars the elements of $S$. Take for vectors the polynomials of $V$.

(b) Let the scalars come from the field

$$\{\{0, 1, 2\}, +(\text{mod } 3), \cdot (\text{mod } 3)\}$$

and the vectors from the group

$$\{\{0, 1, 2\}, +(\text{mod } 3)\}$$

Let multiplication of a scalar by a vector be defined by

$$a(\vec{v}) = a \cdot \vec{v} \;(\text{mod } 3) \quad \text{for all } a, \vec{v} \in \{0, 1, 2\}$$

(c) Let the scalars come from the field $\{Q, +, \cdot\}$ where $Q$ is the set of rational numbers, and let the vectors come from the group $\{\sqrt{2}Q, +\}$ where $\sqrt{2}Q = \{\sqrt{2}x \mid x \in Q\}$. Let $a(\vec{v}) = a \cdot \vec{v}$ where $a \in Q$, $\vec{v} \in \sqrt{2}Q$, and "$\cdot$" denotes ordinary multiplication.

(d) Suppose that in (c) we define $a(\vec{v}) = a + \vec{v}$ (ordinary addition). Will $\{\sqrt{2}Q, +\}$ be a vector space over $\{Q, +, \cdot\}$ now? Explain.

6. Let $V$ be the set of all polynomials of the form $ax + b$ where $a$ and $b$ are rational numbers. Determine a field $\{S, +, \cdot\}$ of scalars $S$ and define a suitable addition operation on $V$ as well as a suitable multiplication of scalars by vectors so that $\{V, +\}$ becomes a vector space over $\{S, +, \cdot\}$.

7. In Exercise 5b
   (a) identify the zero scalar; the zero vector.
   (b) prove that $0(\vec{v}) = a(\vec{0}) = \vec{0}$ for every vector $\vec{v}$ and scalar $a$.
   (c) conversely, show that if $a(\vec{v}) = \vec{0}$ with $a \neq 0$, then $\vec{v} = \vec{0}$.
   (d) show that $(-\vec{v}) = (-1)(\vec{v})$ for every vector $\vec{v}$.

8. In this exercise, define all operations "naturally." Consider the polynomials in a single variable $x$. Let $V$ be the set of all polynomials of degree $n$ with real coefficients. Determine whether $\{V, +\}$ is a vector space over the field of real numbers $\{R, +, \cdot\}$ for the following values of the degree $n$.
   (a) $n \leq 2$ (b) $n = 2$ (c) $n > 2$

9. Let $V$ be the set of all ordered pairs of real numbers $(x, y)$ that satisfy the equation

$$x + 2y = 0$$

Let $\{R, +, \cdot\}$ be the field of real numbers. If $a \in R$, define
$$a(x, y) = (ax, ay),$$
$$(x, y) + (r, s) = (x + r,\ y + s)$$

In relation to these definitions decide whether or not $\{V, +\}$ is a vector space over $\{R, +, \cdot\}$.

10. Let $V$ be the set of all ordered triples of real numbers $(x, y, z)$ that satisfy the equation
$$x + 2y + 3z = 0$$

Let $\{R, +, \cdot\}$ be the field of real numbers. If $a \in R$, define
$$a(x, y, z) = (ax, ay, az)$$
$$(x, y, z) + (r, s, t) = (x + r,\ y + s,\ z + t)$$

(a) In relation to these definitions decide whether or not $\{V, +\}$ is a vector space over $\{R, +, \cdot\}$.

(b) If $z$ is held fixed with the value $z = 0$, do we still have a vector space? How is this new system related to the one described in Exercise 9?

(c) If $y$ is kept fixed with the value $y = 0$, do we still have a vector space?

## 8.3 ELEMENTARY VECTOR SPACE THEOREMS

There are three essential ingredients in any vector space:

1. An additive abelian group $\{V, +\}$ whose elements (members of $V$) are called vectors.
2. A field $\{F, +, \cdot\}$ whose elements (members of $F$) are called scalars.
3. A multiplication operation whereby any scalar $a$ may serve as an operator on any vector $\vec{v}$ to produce a vector $a\vec{v}$.

Within the group $\{V, +\}$ the operation $+$ (addition of vectors) is governed by the axioms and theorems for abelian groups. Within the field $\{F, +, \cdot\}$ the operations $+$ and $\cdot$ (addition and multiplication of scalars) are governed by the axioms and theorems for fields. These aspects of a vector space are therefore already somewhat familiar to us, and we may use these group and field properties freely.

A vector space, however, is also governed by the additional axioms $v_1$ to $v_4$. These axioms relate the new vector space operation (multiplication of a

## 8.3 Elementary Vector Space Theorems

vector by a scalar) to the group operation (addition of vectors) and the field operations (addition and multiplication of scalars). The field of scalars $\{F, +, \cdot\}$ and the group of vectors $\{V, +\}$ are therefore not completely independent systems. Using the vector space axioms in conjunction with group and field properties, we can derive interrelationships among vectors and scalars that must hold true in any vector space.

Consider, for example, the additive identity for the system of vectors and the additive identity for the system of scalars. To distinguish these two additive identities, we denote the former by "$\vec{0}$" and the latter by the usual symbol 0. Because these are additive identities, we have

$$0 + 0 = 0 \quad \text{and} \quad \vec{0} + \vec{0} = \vec{0}$$

Now let **v** be any vector and let $a$ be any scalar. Then

$$(0 + 0)\vec{v} = 0\vec{v} \quad \text{and} \quad a(\vec{0} + \vec{0}) = a\vec{0}$$

But by Axioms $v_1$ and $v_2$ (distributive laws) we know that

$$(0 + 0)\vec{v} = 0\vec{v} + 0\vec{v} \quad \text{and} \quad a(\vec{0} + \vec{0}) = a\vec{0} + a\vec{0}$$

Consequently,

$$0\vec{v} + 0\vec{v} = 0\vec{v} \quad \text{and} \quad a\vec{0} + a\vec{0} = a\vec{0}$$

In each of these equations the operation $+$ signifies vector addition. Since $\{V, +\}$ is a group whose identity element is $\vec{0}$, we may apply the Cancellation Theorem (Theorem 2.5) to each of these equations to obtain

$$0\vec{v} = \vec{0} \quad \text{and} \quad a\vec{0} = \vec{0}$$

These equalities, which hold for any vector $\vec{v}$ and any scalar $a$, establish a strong connection between the zero vector $\vec{0}$ and the zero scalar 0. We express this relationship as follows.

**THEOREM 8.1** Let $\{V, +\}$ be a vector space over the field $\{F, +, \cdot\}$ and let $\vec{0}$ and 0 be their respective additive identities. Then for any scalar $a \in F$ and any vector $\vec{v} \in V$

$$0\vec{v} = \vec{0} \quad \text{and} \quad a\vec{0} = \vec{0}$$

Next, let $a$ be a scalar, $v$ a vector, and suppose that

$$a\vec{v} = \vec{0}$$

What can we conclude from this equality?

CASE 1. If $a \neq 0$, then $a$ has a multiplicative inverse $\frac{1}{a}$ in the field of scalars. In this case

$$\frac{1}{a}(a\vec{v}) = \frac{1}{a}\vec{0}$$

Applying Axiom $v_3$ to the left member of this equation and applying Theorem 6.36 to the right member, we obtain

$$\left(\frac{1}{a} \cdot a\right)\vec{v} = \vec{0}$$

that is,

$$1\vec{v} = \vec{0}$$

Finally, applying Axiom $v_4$ to the left member, we obtain

$$\vec{v} = \vec{0}$$

CASE 2. If $\vec{v} \neq \vec{0}$, it now follows that $a = 0$ because by Case 1 if $a \neq 0$, then $\vec{v} = \vec{0}$, contradicting $\vec{v} \neq \vec{0}$. In short, we have now proved the following theorem.

**THEOREM 8.2** Let $\{V, +\}$ be a vector space over a field $\{F, +, \cdot\}$. Let $a$ be a scalar and $\vec{v}$ a vector.
(a) If $a\vec{v} = \vec{0}$ and $a \neq 0$, then $\vec{v} = \vec{0}$.
(b) If $a\vec{v} = \vec{0}$ and $\vec{v} \neq \vec{0}$, then $a = 0$.

Theorems 8.1 and 8.2 may conveniently be summarized by saying that in a vector space a product is zero if and only if at least one of the factors is zero. (Here the word "zero" may refer either to the zero vector or the zero scalar and the word "product" to either of the two multiplication operations of the vector space.)

Our next theorem establishes a connection between additive inverses of vectors and additive inverses of scalars.

**THEOREM 8.3** Let $\{V, +\}$ be a vector space over a field $\{F, +, \cdot\}$. Let $a$ be any scalar and $\vec{v}$ any vector. Then

$$-(a\vec{v}) = (-a)\vec{v} = a(-\vec{v})$$

### 8.3 Elementary Vector Space Theorems

*Proof:* (Supply reasons.)

$$(a+(-a))\vec{v} = (a+(-a))\vec{v}$$

But

$$a+(-a) = 0$$

Therefore

$$(a+(-a))\vec{v} = 0\vec{v}$$

and

$$a\vec{v}+(-a)\vec{v} = \vec{0}$$

Hence the additive inverse* of the vector $a\vec{v}$ is $(-a)\vec{v}$, that is,

$$-(a\vec{v}) = (-a)\vec{v}$$

This completes the first part of the proof. For the second part note that

$$\vec{v}+(-\vec{v}) = \vec{0}$$
$$a(\vec{v}+(-\vec{v})) = a\vec{0}$$
$$a\vec{v}+a(-\vec{v}) = \vec{0}$$

Hence $a(-\vec{v})$ is also the additive inverse of $a\vec{v}$.

If we choose $a = 1$ in Theorem 7.2, we obtain

$$-(1\vec{v}) = (-1)\vec{v} = 1(-\vec{v})$$

When we apply Axiom $v_2$, this equality reduces to

$$-\vec{v} = (-1)\vec{v}$$

We express this as follows.

**THEOREM 8.4** Let $\{V, +\}$ be a vector space over a field $\{F, +, \cdot\}$ and let $\vec{v}$ be any vector in $V$. Then $(-1)\vec{v}$ is the additive inverse of $\vec{v}$:

$$-\vec{v} = (-1)\vec{v}$$

To illustrate the significance of Theorems 8.1 to 8.4 suppose that $\{F, +, \cdot\}$ is the field of real numbers and $\{V, +\}$ is the group of ordered

---

* Recall that in any group inverses are unique. (See Theorem 2.3.)

pairs of real numbers under the operation of addition of corresponding coordinates. This means that if $(x_1, y_1)$ and $(x_2, y_2)$ are ordered pairs of real numbers, their sum is defined by

$$(x_1, y_1) + (x_2, y_2) = (x_1 + x_2, \; y_1 + y_2)$$

The abelian group $\{V, +\}$ will be a vector space over $\{F, +, \cdot\}$ if we define the product of any real number $r$ by the ordered pair of real numbers $(x, y)$ as

$$r(x, y) = (rx, ry)$$

In the vector space thus defined the scalar additive identity is simply the real number 0, whereas the vector additive identity, $\vec{0}$, is the ordered pair $(0, 0)$:

$$\vec{0} = (0, 0)$$

If $r$ is any scalar (real number) and $\vec{v}$ is any vector, that is, $\vec{v} = (x, y)$ where $x, y$ are real numbers, then Theorem 8.1 asserts that

$$0(x, y) = (0, 0) \quad \text{and} \quad a(0, 0) = (0, 0)$$

Of course, these results could have been obtained directly by applying the definition for multiplying a real number by an ordered pair of real numbers, namely

$$0(x, y) = (0 \cdot x, \; 0 \cdot y) = (0, 0) = \vec{0}$$

and

$$a(0, 0) = (a \cdot 0, \; a \cdot 0) = (0, 0) = \vec{0}$$

However, Theorem 8.1 shows that this sort of result holds in every vector space not merely in this one.

Similarly, in this particular example, Theorem 8.2 asserts the following:

(a) If $a(x, y) = (0, 0)$ and $a \neq 0$, then $(x, y) = (0, 0)$.
(b) If $a(x, y) = (0, 0)$ and $(x, y) \neq (0, 0)$, then $a = 0$.

The equality $(x, y) = (0, 0)$ implies that $x = 0$ and $y = 0$.*

---

* *Note:* It is a general property of ordered pairs that $(x, y) = (u, v)$ if and only if $x = u$ and $y = v$. This property can be deduced from the following definition of an ordered pair attributed to Norbert Wiener: $(x, y) = \{\{x\}, \{x, y\}\}$. (See P. R. Halmos, *Naive Set Theory*, page 23.) A similar property holds for ordered triples and, in general, for ordered $n$-tuples.

## 8.3 Elementary Vector Space Theorems

If we choose $a = 1$ and interpret the ordered pair $(x, y)$ as the complex number $x + iy$, Theorem 8.2 implies that a complex number is zero only if its "real part" $x$ and its "imaginary part" $y$ are both zero.

It is convenient to adopt a uniform notation for the type of vector space we have been considering. Let $V_2$ be the set of all ordered pairs $(x_1, x_2)$ of real numbers, $V_3$ the set of all ordered triples $(x_1, x_2, x_3)$ of real numbers, and, in general, let $V_n$ be the set of all ordered $n$-tuples of real numbers $(x_1, x_2, \ldots, x_n)$. The systems

$$\{V_2, +\}, \{V_3, +\}, \ldots, \{V_n, +\}$$

will be abelian groups, provided that we define addition of $n$-tuples as

$$(x_1, x_2, \ldots, x_n) + (y_1, y_2, \ldots, y_n) = (x_1 + y_1, x_2 + y_2, \ldots, x_n + y_n)$$

For $n = 1$, we interpret $V_1$ to be simply the set $R$ of real numbers. For each $n$ we call the elements of the set $V_n$ **vectors**.

If we now select the field of real numbers $\{R, +, \cdot\}$ as our field of scalars and if we define the product of any scalar $r$ by any vector $(x_1, x_2, \ldots, x_n)$ as

$$r(x_1, x_2, \ldots, x_n) = (rx_1, rx_2, \ldots, rx_n)$$

then each abelian group $\{V_n, +\}$ becomes a vector space over the field $\{R, +, \cdot\}$.

Theorems 8.1 to 8.4 along with Axioms $v_1, v_2, v_3$, and $v_4$ also play an important role in vector algebra. Consider, for example, the vector equation

$$a\vec{x} + \vec{b} = \vec{c}$$

where $a$ is a scalar different from 0, and $\vec{b}$ and $\vec{c}$ are vectors. We can solve this equation for the vector $\vec{x}$ as follows. Add $-\vec{b}$ to each side, noting that $-\vec{b} + (-\vec{b}) = \vec{0}$.

$$a\vec{x} + \vec{0} = \vec{c} + (-\vec{b})$$
$$a\vec{x} = \vec{c} + (-1)\vec{b}$$
$$\vec{x} = \frac{1}{a}[\vec{c} + (-1)\vec{b}]$$
$$\vec{x} = \frac{1}{a}\vec{c} + \frac{-1}{a}\vec{b}$$

The reader should decide which axioms and theorems are used here.

For the special case $a = 1$ we readily see that $\vec{x} + \vec{b} = \vec{c}$ if and only if

$$\vec{x} = \vec{c} + (-1)\vec{b}$$

We may also write this as

$$\vec{x} = \vec{c} - \vec{b}$$

where the difference $\vec{c} - \vec{b}$ is defined as in any additive group to mean $\vec{c} + (-\vec{b})$.

## EXERCISES 8.3

1. In any vector space prove that if $\vec{x}$ and $\vec{y}$ are vectors, then

$$-(\vec{x} + \vec{y}) = (-\vec{x}) + (-\vec{y})$$

2. In any vector space prove that if $\vec{x}$ and $\vec{y}$ are vectors and $a$ is a scalar different from 0, then

$$a\vec{x} = a\vec{y} \quad \text{if and only if} \quad \vec{x} = \vec{y}$$

3. In any vector space prove that if $\vec{x}$ is a vector other than $\vec{0}$ and if $a$ and $b$ are scalars, then

$$a\vec{x} = b\vec{x} \quad \text{if and only if} \quad a = b$$

4. In any vector space prove that if $\vec{x}$ is a vector and $a$ is a scalar, then

$$a\vec{x} = (-a)(-\vec{x})$$

5. Recall that $\{V_2, +\}$ is a vector space over $\{R, +, \cdot\}$ when we define

$$(x_1, x_2) + (y_1, y_2) = (x_1 + x_2, y_1 + y_2)$$
$$r(x_1, x_2) = (rx_1, rx_2)$$

For this vector space, verify
(a) Theorem 8.3.
(b) Theorem 8.4.

6. As in Exercise 5, verify Theorems 8.3 and 8.4 for the vector space $\{V_3, +\}$ over $\{R, +, \cdot\}$.

7. Let $P_2$ be the set of all polynomials in $x$ with real coefficients of the form

$$ax^2 + bx + c$$

where $a$, $b$, and $c$ are real numbers. With the customary interpretation of the additions and multiplications that arise,
(a) show that $\{P_2, +\}$ is an abelian group.

(b) show that $\{P_2, +\}$ is a vector space over the field of real numbers $\{R, +, \cdot\}$.
(c) Verify (1) Theorem 8.1.
    (2) Theorem 8.2.
    (3) Theorem 8.3.
    (4) Theorem 8.4.

## 8.4 LINEAR DEPENDENCE, LINEAR INDEPENDENCE, AND BASES

Suppose that $\{V, +\}$ is a vector space over the field $\{S, +, \cdot\}$. If $\vec{v}_1$ and $\vec{v}_2$ are any vectors in $V$ and if $c_1$ and $c_2$ are any scalars in $S$, then each of the products $c_1\vec{v}_1$ and $c_2\vec{v}_2$ is a vector in $V$ and hence the sum

$$(1) \qquad c_1\vec{v}_1 + c_2\vec{v}_2$$

is also a vector in $V$ because $\{V, +\}$ is a group. Furthermore, if we add another product $c_3\vec{v}_3$ to this sum, we obtain

$$(c_1\vec{v}_1 + c_2\vec{v}_2) + c_3\vec{v}_3$$

which again represents a vector in $V$. Because addition is associative in $\{V, +\}$ we may express sums involving three or more terms without parentheses and write simply

$$(2) \qquad c_1\vec{v}_1 + c_2\vec{v}_2 + c_3\vec{v}_3$$

Expressions like (1) and (2) are called **linear combinations** of the vectors involved. More precisely, a linear combination is defined as follows.

**Definition.** If $\vec{v}_1, \vec{v}_2, \ldots, \vec{v}_n$ are vectors and if $c_1, c_2, \ldots, c_n$ are scalars, then

$$c_1\vec{v}_1 + c_2\vec{v}_2 + \cdots + c_n\vec{v}_n$$

is called a **linear combination** of the vectors $\vec{v}_1, \vec{v}_2, \ldots, \vec{v}_n$.

Clearly, any linear combination of vectors in $V$ is again a vector in $V$. If we call this new vector $\vec{v}$, that is, if

$$\vec{v} = c_1\vec{v}_1 + c_2\vec{v}_2 + \cdots + c_n\vec{v}_n$$

we say that the vector $\vec{v}$ is **linearly dependent** on the set of vectors

$$\vec{v}_1, \vec{v}_2, \ldots, \vec{v}_n$$

## 8 ∥ Vector Spaces

It follows immediately that the zero vector $\vec{0}$ is linearly dependent on any set of vectors because we can write

$$(3) \qquad \vec{0} = 0\vec{v}_1 + 0\vec{v}_2 + \cdots + 0\vec{v}_n$$

This particular linear combination (one in which the scalar coefficients $c_1, c_2, \ldots, c_n$ are all zero) is called the **trivial combination** of the vectors $\vec{v}_1, \vec{v}_2, \ldots, \vec{v}_n$. All other linear combinations are called **nontrivial**. Therefore

> $c_1\vec{v}_1 + c_2\vec{v}_2 + \cdots + c_n\vec{v}_n$
>
> is a **nontrivial linear combination** if and only if at least one of the scalars $c_1, c_2, \ldots, c_n$ is not zero.

As an example of these ideas let us consider the special vector space $\{V_2, +\}$ over the field $\{R, +, \cdot\}$. In this vector space let us choose three vectors, for example,

$$\vec{v}_1 = (-1, 2), \qquad \vec{v}_2 = (2, -4), \qquad \vec{v}_3 = (0, 3)$$

and three scalars, for example,

$$c_1 = 3, \qquad c_2 = -1, \qquad c_3 = 2$$

We form the linear combination

$$\begin{aligned} c_1\vec{v}_1 + c_2\vec{v}_2 + c_3\vec{v}_3 &= 3(-1, 2) + (-1)(2, -4) + 5(0, 3) \\ &= (-3, 6) + (-2, 4) + (0, 15) \\ &= (-5, 25) \end{aligned}$$

and observe that this is again a vector in $V_2$.

The trivial linear combination is

$$\begin{aligned} 0\vec{v}_1 + 0\vec{v}_2 + 0\vec{v}_3 &= 0(-1, 2) + 0(2, -4) + 0(0, 3) \\ &= (0, 0) + (0, 0) + (0, 0) \\ &= (0, 0) \\ &= \vec{0} \end{aligned}$$

a result that is not surprising because we expect any trivial combination of vectors to be $\vec{0}$ (in view of Theorem 8.1). The question now arises whether it is also possible for a nontrivial combination of vectors to be $\vec{0}$. In this illustration the answer is Yes. Let us choose the scalars

$$c'_1 = 2, \qquad c'_2 = 1, \qquad \text{and} \qquad c'_3 = 0$$

### 8.4 Linear Dependence, Linear Independence, and Bases

and form the linear combinations

$$\begin{aligned} c_1'\vec{v}_1 + c_2'\vec{v}_2 + c_3'\vec{v}_3 &= 2\vec{v}_1 + 1\vec{v}_2 + 0\vec{v}_3 \\ &= 2(-1, 2) + 1(2, -4) + 0(0, 3) \\ &= (-2, 4) + (2, -4) + (0, 0) \\ &= (0, 0) \\ &= \vec{0} \end{aligned}$$

Thus it is indeed possible for a nontrivial linear combination to be zero.

Whenever a nontrivial linear combination of vectors is $\vec{0}$, we can solve for at least one of the vectors. For example, in the present case, since

$$2\vec{v}_1 + 1\vec{v}_2 + 0\vec{v}_3 = \vec{0}$$

we can solve for $\vec{v}_1$

$$\vec{v}_1 = \left(-\frac{1}{2}\right)\vec{v}_2 + 0\vec{v}_3$$

or alternatively, we can solve for $\vec{v}_2$

$$\vec{v}_2 = -2\vec{v}_1 + 0\vec{v}_3$$

In each case we have expressed one of the three vectors as a linear combination of the other two, that is, we have shown that at least one of the vectors is linearly **dependent** on the remaining set of two vectors. In situations such as this we call the original set of vectors $\{\vec{v}_1, \vec{v}_2, \vec{v}_3\}$ a **linearly dependent** set.

**Definition.** A set of vectors $\{\vec{v}_1, \vec{v}_2, \ldots, \vec{v}_n\}$ is called **linearly dependent** if and only if some nontrivial linear combination of these vectors is $\vec{0}$. A set of vectors that is not linearly dependent is called **linearly independent**.

Observe from this definition that if $\{\vec{v}_1, \vec{v}_2, \ldots, \vec{v}_n\}$ is a linearly independent set and if

$$c_1\vec{v}_1 + c_2\vec{v}_2 + \cdots + c_n\vec{v}_n = \vec{0}$$

then it must necessarily follow that

$$c_1 = 0, \quad c_2 = 0, \quad \ldots, \quad c_n = 0$$

This fairly obvious observation is surprisingly useful.

For a great many vector spaces there is an important systematic technique that can be used to determine whether a given set of vectors is linearly dependent or linearly independent. We shall study this method in Chapter 9. Meanwhile the following situation merits our special attention.

We have seen that if we denote the set of all ordered pairs of real numbers by $V_2$, then $\{V_2, +\}$ is a vector space over the field of real numbers $\{R, +, \cdot\}$. Let us consider the two special vectors $(1, 0)$ and $(0, 1)$ in $V_2$. We designate them by $\mathbf{e}_1$ and $\mathbf{e}_2$, respectively.

$$\mathbf{e}_1 = (1, 0) \quad \text{and} \quad \mathbf{e}_2 = (0, 1)$$

Suppose $\vec{x}$ is any vector in $V_2$, say

$$\vec{x} = (x_1, x_2)$$

where $x_1$ and $x_2$ are real numbers. Let us form the linear combination $x_1 \mathbf{e}_1 + x_2 \mathbf{e}_2$. We obtain

$$\begin{aligned} x_1 \mathbf{e}_1 + x_2 \mathbf{e}_2 &= x_1 (1, 0) + x_2 (0, 1) \\ &= (x_1, 0) + (0, x_2) \\ &= (x_1, x_2) \\ &= \vec{x}. \end{aligned}$$

This shows that every vector $\vec{x}$ in $V_2$ can be expressed as a linear combination of the two special vectors $\mathbf{e}_1$ and $\mathbf{e}_2$. (In fact the coefficients of $\mathbf{e}_1$ and $\mathbf{e}_2$ in this linear combination are, respectively, the two coordinates $x_1$ and $x_2$ which appear in the ordered pair that defines the vector $\vec{x}$.)

Now we have already observed that in any vector space, every linear combination of vectors is again a vector. In particular, *every linear combination of $\mathbf{e}_1$ and $\mathbf{e}_2$ expresses some vector of $V_2$. It follows therefore that $V_2$ is identical with the set of all linear combinations of $\mathbf{e}_1$ and $\mathbf{e}_2$.* We can think of all vectors in $V_2$ as "built up" from or "generated by" the two special vectors $\mathbf{e}_1$, $\mathbf{e}_2$. This idea is also expressed by saying that the set $\{\mathbf{e}_1, \mathbf{e}_2\}$ "spans" $V_2$.

Since any vector $\vec{x}$ in $V_2$ is linearly dependent on $\{\mathbf{e}_1, \mathbf{e}_2\}$, it is clear that the set $\{\mathbf{e}_1, \mathbf{e}_2, \vec{x}\}$ is a dependent set. In fact, from

$$\vec{x} = x_1 \mathbf{e}_1 + x_2 \mathbf{e}_2$$

we obtain

$$\vec{0} = x_1 \mathbf{e}_1 + x_2 \mathbf{e}_2 + (-1)\vec{x}$$

thus exhibiting $\vec{0}$ as a nontrivial linear combination of $\mathbf{e}_1$, $\mathbf{e}_2$, and $\vec{x}$.

On the other hand, the set $\{\mathbf{e}_1, \mathbf{e}_2\}$ is itself a linearly independent set. To prove this, we observe that if $c_1$ and $c_2$ are any scalars (real numbers) such

that
$$c_1 \mathbf{e}_1 + c_2 \mathbf{e}_2 = \vec{0}$$
then
$$c_1(1, 0) + c_2(0, 1) = (0, 0)$$
$$(c_1, 0) + (0, c_2) = (0, 0)$$
Therefore
$$(c_1, c_2) = (0, 0)$$
$$c_1 = 0 \quad \text{and} \quad c_2 = 0$$

Thus we see that the set $\{\mathbf{e}_1, \mathbf{e}_2\}$ has the following two properties:

1. $\{\mathbf{e}_1, \mathbf{e}_2\}$ is linearly independent.
2. $\{\mathbf{e}_1, \mathbf{e}_2\}$ spans $V_2$.

Any set of vectors which has these two properties is called a **basis** for the vector space $\{V_2, +\}$. The notion of a basis applies to more general vector spaces as well. However, before we can define a basis precisely, we must first define a **spanning set** more generally than we have done previously.

**Definition.** Let $\{V, +\}$ be a vector space over some specified field and let $S$ and $T$ be subsets of $V$. We say that $S$ **spans** $T$ if and only if every vector in $T$ can be expressed as a linear combination of vectors in $S$.

For example, we saw above that the set
$$S = \{\mathbf{e}_1, \mathbf{e}_2\}$$
spans the set $T = V_2$. Note, however, that this definition can be applied more generally to any subsets $S$ and $T$ of $V$. For example, since $\{\mathbf{e}_1, \mathbf{e}_2\}$ spans all of $V_2$, then clearly $\{\mathbf{e}_1, \mathbf{e}_2\}$ will span any subset of $V_2$.

We now formulate our definition of a basis for a vector space.

**Definition.** Let $\{V, +\}$ be a vector space over a field $\{S, +, \cdot\}$ and let
$$B = \{\mathbf{b}_1, \mathbf{b}_2, \ldots, \mathbf{b}_n\}$$
be a (finite)* subset of vectors in $V$. $B$ is called a (finite)* **basis** for $V$ if and only if $B$ has the following two properties:

1. $B$ is linearly independent and
2. $B$ spans $V$.

---
* We confine our discussion to vector spaces with a finite basis. It is also possible to consider spaces with an infinite basis, but these are beyond the scope of this book.

## EXERCISES 8.4

1. Let $V_2$ be the set of all ordered pairs of real numbers so that $\{V_2, +\}$ is a vector space over the field of real numbers $\{R, +, \cdot\}$. Decide whether or not the following vectors are linearly independent or linearly dependent.
   - (a) $(2, -6), (-1, 3)$
   - (b) $(2, -6), (1, 3)$
   - (c) $(0, 1), (1, 1)$
   - (d) $(4, -6), (-6, 9)$
   - (e) $(4, -6), (6, 9)$
   - (f) $(0, 0), (1, 1)$
   - (g) $(2, 2), (3, 3)$

2. For the vector space described in Exercise 1, find $c_1$ and $c_2$ such that
   - (a) $c_1(1, 2) + c_2(2, 3) = (2, 5)$
   - (b) $c_1(1, 2) + c_2(2, 3) = (1, 4)$
   - (c) $c_1(1, 2) + c_2(2, 3) = (0, 0)$
   - (d) $c_1(1, 2) + c_2(2, 3) = (a, b)$
   
   Is $\{(1, 2), (2, 3)\}$ linearly independent? Does $\{(1, 2), (2, 3)\}$ span $\{V_2, +\}$? Is $\{(1, 2), (2, 3)\}$ a basis for $\{V_2, +\}$?

3. Let $V_2^0$ be the set of all ordered triples in $V_3$ whose third coordinate is zero.
   - (a) Show that $\{V_2^0, +\}$ is a vector space over the field of real numbers.
   - (b) Does the set $\{(1, 1, 0), (1, 0, 1), (0, 1, 1)\}$ span $V_2^0$?
   - (c) Is this set a basis for $V_2^0$? Explain.

4. Let $\{V_2, +\}$ be the vector space described in Exercise 1. Prove that if $(a, b), (c, d)$ are linearly independent vectors in $V_2$ and $(e, f)$ any vectors in $V_2$, then for some real numbers $r, s$
$$r(a, b) + s(c, d) = (e, f)$$

5. Let $\{V, +\}$ be any vector space over some field $\{F, +, \cdot\}$. Prove that if $\vec{v}_1$ and $\vec{v}_2$ are linearly independent vectors in $V$ and if
$$c_1\vec{v}_1 + c_2\vec{v}_2 = d_1\vec{v}_1 + d_2\vec{v}_2$$
then $c_1 = d_1$ and $c_2 = d_2$. Extend this result to any linearly independent set of vectors $\{\vec{v}_1, \vec{v}_2, \ldots, \vec{v}_n\}$ in $V$.

6. Let $\{\vec{v}_1, \vec{v}_2, \ldots, \vec{v}_n\}$ be a linearly independent set of vectors that span a vector space. Prove that every vector $\vec{v}$ of this space has a unique representation as a linear combination of the vectors $\vec{v}_1, \vec{v}_2, \ldots, \vec{v}_n$.

### 8.4 Linear Dependence, Linear Independence, and Bases 303

7. Let $\{V_2, +\}$ be the vector space over $\{R, +, \cdot\}$ as defined in Exercise 1. Decide whether or not each of the following sets
   (a) spans the vector space.
   (b) forms a basis for the vector space.
   
   (1) $\{(1, 2), (2, 1)\}$
   (2) $\{(1, 2), (2, 1), (1, 1)\}$
   (3) $\{(1, -2), (-2, 4), (1, 1)\}$
   (4) $\{(2, -6), (-1, 3)\}$
   (5) $\{(2, -6), (2, 0), (0, -6)\}$
   (6) $\{(1, 0), (2, 1), (3, 2)\}$
   (7) $\{(a, b), (a + 1, b + 1)\}$
   (8) $\{(a, b), (a + 1, b - 1)\}$

# REVIEW EXERCISES

1. Let $\{V, +\}$ be a vector space over the field $\{F, +, \cdot\}$. Let $V_1$ and $V_2$ each be subsets of $V$ such that $\{V_1, +\}$ and $\{V_2, +\}$ are each vector spaces over $\{F, +, \cdot\}$. Prove that $\{V_1 \cap V_2, +\}$ is a vector space over $\{F, +, \cdot\}$.
   
   (*Note:* This result may be stated briefly as follows:
   
   The intersection of two vector subspaces is again a vector subspace.)

2. Decide whether or not each of the following systems is a vector space.
   (a) Take for vectors the set of all polynomials in $x$ with coefficients 0, 1, 2 with addition of polynomials computed by adding corresponding coefficients modulo 3.
   Take for scalars the set of numbers $\{0, 1, 2\}$ with both addition and multiplication of scalars computed modulo 3.
   Let the product of a polynomial (vector) by a scalar be computed by multiplying each coefficient by that scalar (modulo 3).
   (b) Take for vectors the set of all polynomials in $x$ with coefficients 0, 1, 2, 3 with addition of polynomials computed by adding corresponding coefficients modulo 4. Take for scalars the set of numbers $\{0, 1, 2, 3\}$ with both addition and multiplication of scalars computed modulo 4, and the product of a scalar by a polynomial computed by multiplying each coefficient by that scalar.

(c) Let $S = \{0, 1, 2\}$
$$V = \{(x, y) \mid x \in \{0, 1, 2\}, y \in \{0, 1, 2, 3\}\}$$
Take for scalars the numbers in set $S$ where addition and multiplication of scalars are computed modulo 3.
Take for vectors the ordered pairs in set $V$ with addition of vectors defined by
$$(x_1, y_1) + (x_2, y_2) = (x_1 + x_2 (\text{mod } 3), y_1 + y_2 (\text{mod } 4))$$
and multiplication of a vector by a scalar defined by
$$r(x, y) = (rx(\text{mod } 3), y)$$

3. Given a vector space $\{V, +\}$ over the field $\{F, +, \cdot\}$. Let
$$W = \{(x, y) \mid x \text{ and } y \text{ are elements of } V\}$$
Let addition of ordered pairs in $W$ be defined by
$$(x_1, y_1) + (x_2, y_2) = (x_1 + x_2, y_1 + y_2)$$
for all $x_1, x_2, y_1, y_2$ in $V$. Let multiplication of ordered pairs in $W$ by scalars in $F$ be defined by
$$r(x, y) = (rx, ry)$$
for all $r \in F$ and all $x, y$ in $V$.
   (a) Show that $\{W, +\}$ is a vector space over $\{F, +, \cdot\}$.
   (b) Find two subsystems of $W$ which are vector spaces and are each isomorphic to the original vector space $\{V, +\}$ over $\{F, +, \cdot\}$.

4. Determine a vector space having exactly
   (a) four elements.
   (b) five elements.

★5. (*Problem for investigation*) Does there exist a vector space $\{V, +\}$ over some field $\{F, +, \cdot\}$ such that $V$ has exactly six elements?

*John von Neumann (1903-1957)*

# 9 | Gauss-Jordan Method

The beginner should not expect that we will be able to develop miracle methods for solving equations. An actual solution is still best found by the elementary method of successive elimination.

EMIL ARTIN

## 9.1 DETERMINATION OF LINEAR DEPENDENCE OR INDEPENDENCE BY GAUSSIAN ELIMINATION

The definition of linear dependence and linear independence has the following two important consequences.

1. If a set of vectors is **linearly dependent**, then at least one of these vectors can be expressed as a linear combination of the remaining vectors.
2. If a set of vectors $\{\vec{v}_1, \vec{v}_2, \ldots, \vec{v}_n\}$ is **linearly independent** and if $x_1, x_2, \ldots, x_n$ are scalars, then
$$x_1\vec{v}_1 + x_2\vec{v}_2 + \cdots + x_n\vec{v}_n = \vec{0}$$
if and only if $x_1 = 0, x_2 = 0, \ldots, x_n = 0$.

Let us see how these observations can be used to determine the linear dependence or independence of sets of vectors in specific situations. The examples are based on the vector space $\{V_2, +\}$ over the field $\{R, +, \cdot\}$, but the ideas apply to far more general vector spaces.

***Example 1.*** Consider the set of vectors $\{\vec{v}_1, \vec{v}_2\}$ where $\vec{v}_1 = (3, 2)$ and $\vec{v}_2 = (4, 3)$. We wish to determine whether the set $\{\vec{v}_1, \vec{v}_2\}$ is linearly dependent or linearly independent. We therefore examine an arbitrary linear combination $x_1\vec{v}_1 + x_2\vec{v}_2$, where $x_1$ and $x_2$ are scalars, and observe that this combination can be $\vec{0}$ if and only if

$$x_1(3, 2) + x_2(4, 3) = (0, 0)$$

This equation is equivalent to each of the following:

$$(3x_1, 2x_1) + (4x_2, 3x_2) = (0, 0)$$
$$(3x_1 + 4x_2, \; 2x_1 + 3x_2) = (0, 0)$$
$$3x_1 + 4x_2 = 0 \quad \text{and} \quad 2x_1 + 3x_2 = 0$$

Our problem is therefore equivalent to that of determining all possible solutions of a pair of simultaneous linear equations

(1) $$\begin{cases} 3x_1 + 4x_2 = 0 \\ 2x_1 + 3x_2 = 0 \end{cases}$$

Although there are many methods for solving such systems of equations, we focus attention here on a particularly elegant and widely applicable technique known as the **Gauss-Jordan Complete Elimination Method**. At this stage the method will not appear to be the simplest way of solving this particular problem, but, as we shall see presently, the procedure can be streamlined. Its complete generality and basic simplicity make it a powerful tool for handling problems far more general than the particular example confronting us here. Furthermore, the method is so systematic that it can be readily programmed for a computer.

Returning to our problem, we divide both members of the first equation by the coefficient of $x_1$, namely by 3. This coefficient 3 is called the **pivot** for this stage of our calculation. We obtain in this manner the equivalent system of equations

(1') $$\begin{cases} 1x_1 + \dfrac{4}{3}x_2 = 0 \\ 2x_1 + 3x_2 = 0 \end{cases}$$

### 9.1 Determination of Linear Dependence

Any pair of values of $x_1$, $x_2$ that satisfies the system (1) must satisfy the system (1′) and conversely. (See Exercise 2a.)
We now retain the first equation, but replace the second equation by a new one formed by adding to this second equation $-2$ times the first equation.*

$$\begin{cases} 1x_1 + \frac{4}{3}x_2 = 0 \\ 0x_1 + \frac{1}{3}x_2 = 0 \end{cases}$$

This new system of equations is still equivalent to the original system (1). (See Exercise 2b.) Thus far we have eliminated $x_1$ from the second equation. This completes the first stage of the process.

We proceed next to eliminate $x_2$ from the first equation using a completely similar procedure. We divide both members of the second equation by the coefficient of $x_2$, namely by $\frac{1}{3}$. This coefficient $\frac{1}{3}$ is called the **pivot** for this new stage of our calculation. We obtain in this manner another system (still equivalent to the original system (1)).

$$\begin{cases} 1x_1 + \frac{4}{3}x_2 = 0 \\ 0x_1 + 1x_2 = 0 \end{cases}$$

We now retain the new second equation, but replace the first equation by another one obtained by adding to the first equation $-\frac{4}{3}$ times the second equation. This yields the new system

(2)
$$\begin{cases} 1x_1 + 0x_2 = 0 \\ 0x_1 + 1x_2 = 0 \end{cases}$$

But system (2) obviously has a unique solution for each variable, namely

$$x_1 = 0 \quad \text{and} \quad x_2 = 0$$

---

* This is a brief way of describing the following operation: We multiply each term of the first equation by $-2$ and add the resulting products to corresponding terms of the second equation. The new equation thus obtained replaces the original second equation.

Moreover, this solution is also the unique solution to the original system (1) because (2) is equivalent to (1).

Returning to the vectors $\vec{v}_1$ and $\vec{v}_2$, we now see that the only linear combination $x_1\vec{v}_1 + x_2\vec{v}_2$ which can be $\vec{0}$ is the trivial combination $0\vec{v}_1 + 0\vec{v}_2$. The set of vectors $\{\vec{v}_1, \vec{v}_2\}$ is therefore by definition a linearly independent set.

**Example 2.** Suppose that we had started with the vectors

$$\vec{v}_1 = (-1, 2) \quad \text{and} \quad \vec{v}_2 = (2, -4)$$

If we now form the linear combination $x_1\vec{v}_1 + x_2\vec{v}_2$, we observe that this linear combination can be $\vec{0}$ if and only if

$$x_1(-1, 2) + x_2(2, -4) = (0, 0)$$

This equation is equivalent to each of the following:

$$(-x_1, 2x_1) + (2x_2, -4x_2) = (0, 0)$$
$$(-x_1 + 2x_2, 2x_1 - 4x_2) = (0, 0)$$
$$-x_1 + 2x_2 = 0 \quad \text{and} \quad 2x_1 - 4x_2 = 0$$

This time our problem is equivalent to that of determining all possible solutions of the system

(3) $\quad \begin{cases} -1x_1 + 2x_2 = 0 \\ 2x_1 - 4x_2 = 0 \end{cases}$

Applying the Gauss-Jordan technique, we divide the first equation by the pivot $-1$, that is, by the coefficient of $x_1$, to obtain

$$\begin{cases} 1x_1 - 2x_2 = 0 \\ 2x_1 - 4x_2 = 0 \end{cases}$$

Next we replace the second equation by adding to it $-2$ times the new first equation

(4) $\quad \begin{cases} 1x_1 - 2x_2 = 0 \\ 0x_1 + 0x_2 = 0 \end{cases}$

This completes the first stage of the process. However, this time the elimination of $x_1$ from the second equation has also resulted in eliminating $x_2$ from this equation, so we cannot begin the second stage. Nevertheless the new system (4) is equivalent to the original system (3). But system (4) is clearly also equivalent to the single equation

(5) $$1x_1 - 2x_2 = 0$$

because the second equation in (4), namely $0x_1 + 0x_2 = 0$, is trivially true for all values of $x_1$ and $x_2$.

The single equation (5) has infinitely many solutions other than the trivial one ($x_1 = 0$ and $x_2 = 0$). For example, one solution of (5) might be expressed as

$$x_1 = 2 \qquad x_2 = 1$$

Another solution of (5) might be

$$x_1 = 10 \qquad x_2 = 5$$

and clearly we can obtain infinitely many more solutions by choosing $x_2$ half of $x_1$. Returning now to the vectors

$$\vec{v}_1 = (-1, 2) \quad \text{and} \quad \vec{v}_2 = (2, -4)$$

we see that these vectors are linearly dependent because we have, for example,

$$\begin{aligned} 2\vec{v}_1 + 1\vec{v}_2 &= 2(-1, 2) + 1(2, -4) \\ &= (-2, 4) + (2, -4) \\ &= (-2+2, 4-4) \\ &= (0, 0) \\ 2\vec{v}_1 + 1\vec{v}_2 &= \vec{0} \end{aligned}$$

Note that we can solve this last equation for either vector in terms of the other. For example, we have

$$\vec{v}_1 = -\frac{1}{2}\vec{v}_2$$

or, equivalently,

$$\vec{v}_2 = -2\vec{v}_1$$

Either vector in Example 2 is therefore linearly dependent on the other.

These two examples indicate that the Gauss-Jordan method is applicable whether the vectors are dependent or independent. In Section 9.2, we shall streamline the method and in subsequent sections we shall see that the method is applicable to even more general problems than those considered here.

## EXERCISES 9.1

1. Prove that the following systems of equations are equivalent by showing that a solution of one system is also a solution of the other system.

   (a) $\begin{cases} 3x_1 + 4x_2 = 1 \\ 2x_1 + 3x_2 = 1 \end{cases}$ and $\begin{cases} 1x_1 + \frac{4}{3}x_2 = \frac{1}{3} \\ 0x_1 + 1x_2 = 1 \end{cases}$

   (b) $\begin{cases} 3x_1 + 4x_2 = 1 \\ 2x_1 + 3x_2 = 1 \end{cases}$ and $\begin{cases} 1x_1 + 0x_2 = -1 \\ 0x_1 + 1x_2 = 1 \end{cases}$

   (c) $\begin{cases} 6x_1 + 4x_2 = 2 \\ 9x_1 + 6x_2 = 3 \end{cases}$ and $\begin{cases} 4x_1 + \frac{8}{3}x_2 = \frac{4}{3} \\ 12x_1 + 8x_2 = 4 \end{cases}$

   (d) $\begin{cases} 6x_1 + 4x_2 = 2 \\ 9x_1 + 6x_2 = 4 \end{cases}$ and $\begin{cases} 4x_1 + \frac{8}{3}x_2 = \frac{4}{3} \\ 12x_1 + 8x_2 = 5 \end{cases}$

   How does the solution set for (d) differ from that for (c)?

2. In this exercise all letters stand for real numbers. Prove that the system $\begin{cases} ax_1 + bx_2 = p \\ cx_1 + dx_1 = q \end{cases}$ is equivalent to each of the following systems. (*Hint:* Prove that every pair of numbers $(x_1, x_2)$ that satisfies the original system must satisfy each of the systems (a), (b), and (c). Also prove that every pair of numbers $(x_1, x_2)$ that satisfies any of the systems (a), (b), or (c) must satisfy the original system.)

   (a) $\begin{cases} rax_1 + rbx_2 = rp \\ cx_1 + dx_2 = q \end{cases}$ provided $r \neq 0$

   (b) $\begin{cases} ax_1 + bx_2 = p \\ (ra + c)x_1 + (rb + d)x_2 = rp + q \end{cases}$

   (c) $\begin{cases} ax_1 + bx_2 = p \\ (ra + sc)x_1 + (rb + sd)x_2 = rp + sq \end{cases}$ provided $s \neq 0$

3. Show that if two systems of equations are each equivalent to a third system, then the first two systems are equivalent.

4. Prove that if $\{\vec{v}_1, \vec{v}_2, \vec{v}_3\}$ is linearly dependent, then at least one of the vectors can be expressed in terms of the others.

5. Using the Gauss-Jordan method, find the solution sets for each of the following systems.

   (a) $\begin{cases} 2x_1 - 3x_2 = 0 \\ 3x_1 + 4x_2 = 0 \end{cases}$

(b) $\begin{cases} 3x_1 - x_2 = 0 \\ 2x_1 + 3x_2 = 0 \end{cases}$

(c) $\begin{cases} 3x_1 - x_2 = 0 \\ -9x_1 + 3x_2 = 0 \end{cases}$

6. Use the Gauss-Jordan method to decide whether the following sets of vectors are linearly dependent or linearly independent in the vector space $\{V_2, +\}$ over $\{R, +, \cdot\}$.

    (a) $\{(2, 3), (3, 4)\}$           (b) $\{(2, 3), (3, -4)\}$

    (c) $\{(2, 3), (4, 6)\}$           (d) $\{(2, 3), (4, -6)\}$

    (e) $\{(2, 3), (6, 4)\}$

7. Use the Gauss-Jordan method to find the solution sets for each of the following systems.

    (a) $\begin{cases} 2x_1 - 3x_1 = 1 \\ 3x_1 + 4x_2 = 2 \end{cases}$

    (b) $\begin{cases} 3x_1 - x_2 = 1 \\ 2x_1 + 3x_2 = 0 \end{cases}$

    (c) $\begin{cases} 3x_1 - x_2 = 1 \\ -9x_1 + 3x_2 = 2 \end{cases}$

## 9.2 AN ABBREVIATED ALGORITHM FOR DETERMINING LINEAR DEPENDENCE OR INDEPENDENCE

The computations in the Gauss-Jordan procedure are more easily visualized if the variables ($x_1$, $x_2$, etc.) are omitted and only the coefficients of these variables are used. The Gauss-Jordan method then takes the form of a repetitive sequence of operations performed on the *coefficient matrix* of each system of equations that arises in the process.

As a simple illustration consider Example 1 of Section 9.1. There we sought to determine all possible solutions of the system

(1) $\begin{cases} 3x_1 + 4x_2 = 0 \\ 2x_1 + 3x_2 = 0 \end{cases}$

If we omit the variables $x_1$ and $x_2$, we obtain the **matrix**

(A) $\begin{bmatrix} ③ & 4 & | & 0 \\ 2 & 3 & | & 0 \end{bmatrix}$

This matrix has two rows and three columns. Each row of matrix (A) represents one of the equations (1). The vertical dashed line indicates the position of the equality sign in each equation. The encircled numeral ③ indicates the first pivot. Dividing the first row of matrix (A) by the pivot 3, we obtain a new first row

$$\begin{bmatrix} 1 & \frac{4}{3} & \vdots & 0 \end{bmatrix}$$

We now multiply this new first row by $-2$ and add it to the second row of matrix (A), thereby producing a new second row

(B) $$\begin{bmatrix} 1 & \frac{4}{3} & \vdots & 0 \\ 0 & \left(\frac{1}{3}\right) & \vdots & 0 \end{bmatrix}$$

This matrix corresponds to a new system of equations equivalent to the original system (1). In the new matrix (B) we choose $\left(\frac{1}{3}\right)$ as a pivot. We divide the second row by the new pivot $\frac{1}{3}$, obtaining a new second row

$$\begin{bmatrix} 0 & 1 & \vdots & 0 \end{bmatrix}$$

We now multiply this new second row by $-\frac{4}{3}$ and add it to the first row in matrix (B), thereby obtaining a new matrix

(C) $$\begin{bmatrix} 1 & 0 & \vdots & 0 \\ 0 & 1 & \vdots & 0 \end{bmatrix}$$

This matrix (C) corresponds to the following system of equations:

(2) $$\begin{cases} 1x_1 + 0x_2 = 0 \\ 0x_1 + 1x_2 = 0 \end{cases}$$

In this system, the coefficient of the variables $x_1$ and $x_2$ form a $2 \times 2$ **unit** matrix

$$\begin{bmatrix} 1 & 0 \\ 0 & 1 \end{bmatrix}$$

The appearance of this unit matrix to the left of the dashed vertical line clearly signifies that the original set of vectors $\vec{v}_1, \vec{v}_2$ is linearly independent because, with these coefficients, the value of each of the variables must be zero.

Applying this abbreviated technique to Example 2 of Section 9.1, we start with the matrix

$$\begin{bmatrix} \boxed{-1} & 2 & | & 0 \\ 2 & -4 & | & 0 \end{bmatrix}$$

Dividing the first row by the pivot $\boxed{-1}$, we obtain a new first row

$$\begin{bmatrix} 1 & -2 & | & 0 \end{bmatrix}$$

Multiplying this row by $-2$ and adding it to the second row of the preceding matrix, we obtain

$$\begin{bmatrix} 1 & -2 & | & 0 \\ 0 & 0 & | & 0 \end{bmatrix}$$

At this point it is no longer possible to choose a new pivot in the second row because all the entries in this row are zeros. The matrix to the left of the dashed vertical line is now a **singular*** $2 \times 2$ matrix

$$\begin{bmatrix} 1 & -2 \\ 0 & 0 \end{bmatrix}$$

and since the elements of this matrix are coefficients for the variables $x_1, x_2$, the values of the variables need not be zero. (We have already observed this fact in Section 9.1.) The appearance of such a singular matrix apparently signifies that the original set of vectors in this example is linearly dependent.

Taking our cue from these two examples, we see that the problem of determining the linear dependence or independence of a pair of vectors in $V_2$ can be reduced to a simple algorithmic procedure applied to a matrix formed from the coordinates of the vectors. In both examples the column at the extreme right of each matrix consisted solely of zeros and this column remained the same throughout the process. We may therefore drop this column altogether and proceed as follows:

---

* For the present it will suffice to reserve the term "singular matrix" to mean a matrix in which a row of zeros appears.

## 9 ‖ Gauss-Jordan Method

*Example 1.*                       *Example 2.*

$\vec{v}_1 = (3, 2), \quad \vec{v}_2 = (4, 3)$         $\vec{v}_1 = (-1, 2), \quad \vec{v}_2 = (2, -4)$

First we set up a matrix whose columns are the coordinates of $\vec{v}_1$ and $\vec{v}_2$.

$$\begin{bmatrix} ③ & 4 \\ 2 & 3 \end{bmatrix} \qquad\qquad \begin{bmatrix} ⊖1 & 2 \\ 2 & -4 \end{bmatrix}$$

Choose a nonzero element (encircled above) and use it as the pivot of a Gauss-Jordan elimination step to transform that column into a "basic unit column" as follows:

$$\begin{bmatrix} 1 & \frac{4}{3} \\ 0 & ①/③ \end{bmatrix} \qquad\qquad \begin{bmatrix} 1 & -2 \\ 0 & 0 \end{bmatrix}$$

In Example 1 a new pivot $①/③$ is available in the second row, but in Example 2 no such pivot is available. The essential difference between the two examples is now evident: a second elimination step can be carried out in Example 1 but not in Example 2. The final matrices are therefore

$$\begin{pmatrix}\text{Unit}\\ \text{matrix}\end{pmatrix} \begin{bmatrix} 1 & 0 \\ 0 & 1 \end{bmatrix} \qquad \begin{pmatrix}\text{Singular}\\ \text{matrix}\end{pmatrix} \begin{bmatrix} 1 & -2 \\ 0 & 0 \end{bmatrix}$$

The unit matrix indicates that the vectors $\vec{v}_1$ and $\vec{v}_2$ are linearly independent in Example 1 and the singular matrix indicates that the vectors are linearly dependent in Example 2.

    The algorithmic procedure we have described applies equally well to vector spaces of more than two dimensions. For example, let $V_3$ be the set of all ordered triples $(x_1, x_2, x_3)$ with real coordinates. We saw in Section 8.3 that $\{V_3, +\}$ is a vector space over the field $\{R, +, \cdot\}$. If we select three vectors in $V_3$, say

$$\vec{v}_1 = (2, 3, -1), \qquad \vec{v}_2 = (4, 2, 0), \qquad \vec{v}_3 = (1, 0, -1)$$

we can decide whether or not this set of vectors is linearly independent as follows: We set up a matrix whose *columns* are the coordinates of $\vec{v}_1$, $\vec{v}_2$, and $\vec{v}_3$

$$\begin{bmatrix} ② & 4 & 1 \\ 3 & 2 & 0 \\ -1 & 0 & -1 \end{bmatrix}$$

## 9.2 An Abbreviated Algorithm

Choosing the encircled element ②as a pivot we perform a Gauss-Jordan elimination step, reducing the first column to a unit vector

$$\begin{bmatrix} 1 & 2 & \frac{1}{2} \\ 0 & -4 & \frac{-3}{2} \\ 0 & 2 & \frac{-1}{2} \end{bmatrix}$$

Next, choosing the encircled element -4 as a pivot we perform another Gauss-Jordan elimination step, reducing the second column to a unit vector.

$$\begin{bmatrix} 1 & 0 & -\frac{1}{4} \\ 0 & 1 & \frac{3}{8} \\ 0 & 0 & -\frac{5}{4} \end{bmatrix}$$

Finally, choosing the encircled element $-\frac{5}{4}$ as a pivot we perform a third Gauss-Jordan elimination step, reducing the third column to a unit vector.

$$\begin{bmatrix} 1 & 0 & 0 \\ 0 & 1 & 0 \\ 0 & 0 & 1 \end{bmatrix}$$

Since this is a unit matrix, we conclude that the vectors $\vec{v}_1$, $\vec{v}_2$, and $\vec{v}_3$ are linearly independent. If we had selected the vectors

$$\vec{v}_1 = (2, 3, -1), \quad \vec{v}_2 = (4, 2, 0), \quad \vec{v}_3 = (0, -4, 2)$$

then our starting matrix would have looked like this

$$\begin{bmatrix} 2 & 4 & 0 \\ 3 & 2 & -4 \\ -1 & 0 & 2 \end{bmatrix}$$

Performing a Gauss-Jordan elimination step to reduce the first column, we obtain

$$\begin{bmatrix} 1 & 2 & 0 \\ 0 & -4 & -4 \\ 0 & 2 & 2 \end{bmatrix}$$

However, this time a second Gauss-Jordan elimination step produces a singular matrix
$$\begin{bmatrix} 1 & 0 & -2 \\ 0 & 1 & 1 \\ 0 & 0 & 0 \end{bmatrix}$$
showing that the vectors $\vec{v}_1$, $\vec{v}_2$, and $\vec{v}_3$ are linearly dependent. The linear dependence of this particular set of three vectors can be verified by observing that
$$2\vec{v}_1 - 1\vec{v}_2 + 1\vec{v}_3 = (4, 6, -2) - (4, 2, 0) + (0, -4, 2)$$
$$= (0, 0, 0) = \vec{0}$$

There is a systematic way of determining the coefficients of $\vec{v}_1$, $\vec{v}_2$, and $\vec{v}_3$ by keeping track of the row operations performed. Although we shall not pursue this matter here, we shall study some further ramifications of the Gauss-Jordan method in Section 9.2 and also in Chapter 10.

## EXERCISES 9.2

*Use the abbreviated technique shown in Section 9.2 to find the solution set for each of the following systems.*

1. $\begin{cases} x_1 - 3x_2 = 0 \\ 2x_1 + 5x_2 = 0 \end{cases}$

2. $\begin{cases} 2x_1 + 5x_2 = 0 \\ x_1 - 3x_2 = 0 \end{cases}$

3. $\begin{cases} 3x_1 - 2x_2 = 0 \\ 2x_1 + 5x_2 = 0 \end{cases}$

4. $\begin{cases} 6x_1 - 2x_2 = 0 \\ -9x_1 + 3x_2 = 0 \end{cases}$

5. $\begin{cases} x_1 - 3x_2 = 0 \\ 2x_1 + 5x_2 = 11 \end{cases}$

6. $\begin{cases} 2x_1 + 5x_2 = 13 \\ x_1 - 3x_2 = 1 \end{cases}$

7. $\begin{cases} 6x_1 - 2x_2 = 1 \\ -9x_1 + 3x_2 = 2 \end{cases}$

8. $\begin{cases} x_1 + x_2 - x_3 = 0 \\ x_1 + 4x_2 - 3x_3 = 0 \\ 4x_1 + x_2 - 2x_3 = 0 \end{cases}$

9. $\begin{cases} x_1 + x_2 + x_3 = 6 \\ x_1 + 2x_2 + 3x_3 = 14 \\ 2x_1 + x_2 - x_3 = 1 \end{cases}$

10. For each pair of vectors $\vec{v}_1$, $\vec{v}_2$ listed, form a matrix whose columns are the coordinates of $\vec{v}_1$ and $\vec{v}_2$. Then use the Gauss-Jordan elimination procedure to reduce the matrix either to a unit matrix or a singular matrix and indicate whether $\vec{v}_1$ and $\vec{v}_2$ are linearly independent or linearly dependent.
    (a) $\vec{v}_1 = (4, 6)$, $\vec{v}_2 = (6, 8)$
    (b) $\vec{v}_1 = (4, 6)$, $\vec{v}_2 = (6, 9)$

(c) $\vec{v}_1 = (4, 6)$, $\vec{v}_2 = (6, -9)$
(d) $\vec{v}_1 = (4, 6)$, $\vec{v}_2 = (-6, -9)$
(e) $\vec{v}_1 = (4, 6)$, $\vec{v}_2 = (-9, -6)$

11. For what value or values of $a$ will each of the following pairs of vectors be linearly dependent?

(a) $\vec{v}_1 = (2, 3)$, $\vec{v}_2 = (8, a)$
(b) $\vec{v}_1 = (2, 3)$, $\vec{v}_2 = (4 + 2a, 5 + 2a)$
(c) $\vec{v}_1 = (2, a)$, $\vec{v}_2 = (5, 3a)$
(d) $\vec{v}_1 = (2, a)$, $\vec{v}_2 = (6, 3a)$
(e) $\vec{v}_1 = (2, a)$, $\vec{v}_2 = (0, 3)$

12. Let $V_3$ be the set of all ordered triples of real numbers determining a vector space as in Exercise 3, Exercises 8.4. Decide whether or not each of the following sets of vectors are linearly independent or dependent.

(a) $\{(2, 4, 6), (-1, 2, 4), (1, -2, 3)\}$
(b) $\{(2, 4, 6), (-1, 2, 4), (4, 0, -2)\}$
(c) $\{(2, 4, 6), (-1, 2, 4), (0, 0, 1)\}$

13. Let $V_3$ be the set of all ordered triples of real numbers such that $\{V_3, +\}$ is the abelian group in which addition is defined as

$$(a_1, a_2, a_3) + (b_1, b_2, b_3) = (a_1 + b_1, a_2 + b_2, a_3 + b_3)$$

Let $\{R, +, \cdot\}$ be the field of real numbers and $\{V_3, +\}$ the vector space over $R$. Decide whether or not each of the following sets of vectors are linearly independent or linearly dependent.

(a) $\{(2, -6, 4), (-1, 3, -2), (0, 1, 0)\}$
(b) $\{(2, -6, 4), (1, 3, 2), (0, 1, 0)\}$
(c) $\{(1, 0, 0), (0, 1, 0), (0, 0, 1)\}$
(d) $\{(1, 1, 0), (1, 0, 1), (1, 1, 0)\}$
(e) $\{(1, 1, -1), (1, -1, 1), (-1, 1, 1)\}$
(f) $\{(1, 1, 1), (2, 2, 2), (1, 2, 3)\}$

14. For the vector space $\{V_3, +\}$ over $\{R, +, \cdot\}$ of Exercise 13, determine $c_1, c_2, c_3$ such that

(a) $c_1(1, 1, 0) + c_2(1, 0, 1) + c_3(0, 1, 1) = (1, 0, 0)$
(b) $c_1(1, 1, 0) + c_2(1, 0, 1) + c_3(0, 1, 1) = (0, 0, 0)$
(c) $c_1(1, 1, 0) + c_2(1, 0, 1) + c_3(0, 1, 1) = (a, b, c)$.

Is $\{(1, 1, 0), (1, 0, 1), (0, 1, 1)\}$ linearly independent? Does $\{(1, 1, 0), (1, 0, 1), (0, 1, 1)\}$ span $V_3$? Is $\{(1, 1, 0), (1, 0, 1), (0, 1, 1)\}$ a basis for $V_3$?

## 9.3 THE GAUSS-JORDAN COMPLETE ELIMINATION PROCEDURE

Although we have used the Gauss-Jordan method merely to ascertain linear dependence or linear independence of a set of vectors, the method also enables one to solve any system of linear equations. For example, suppose we want to solve the system

$$\begin{cases} 2x_1 - x_2 = 6 \\ 3x_1 - 2x_2 = 11 \end{cases}$$

We represent this system by the matrix

$$\begin{bmatrix} ② & -1 & 6 \\ 3 & -2 & 11 \end{bmatrix}$$

Using ② as the pivot we obtain

$$\begin{bmatrix} 1 & -\frac{1}{2} & 3 \\ & & \end{bmatrix}$$

Multiplying this row by $-3$ and adding it to the second row, we get

$$\begin{bmatrix} 1 & -\frac{1}{2} & 3 \\ 0 & \left(-\frac{1}{2}\right) & 2 \end{bmatrix}$$

Using $\left(-\frac{1}{2}\right)$ as the new pivot in the second row, we obtain

$$\begin{bmatrix} 0 & 1 & -4 \end{bmatrix}$$

When we multiply this last row by $\frac{1}{2}$ and add it to the first row of the preceding matrix, we have

$$\begin{bmatrix} 1 & 0 & 1 \\ 0 & 1 & -4 \end{bmatrix}$$

This matrix represents the system

$$\begin{cases} 1x_1 + 0x_2 = 1 \\ 0x_1 + 1x_2 = -4 \end{cases}$$

### 9.3 The Gauss-Jordan Complete Elimination Procedure

or, more simply,

$$\begin{cases} x_1 = 1 \\ x_2 = -4 \end{cases}$$

This final system is equivalent to the original system and exhibits the (unique) solution of that system.

The Gauss-Jordan procedure is equally applicable to systems of linear equations with more than two variables. As an illustration let us solve the following system of three linear equations in three variables:

(1) $\quad 2x - 4y + z = 0$
(2) $\quad x - 3y - 2z = -4$
(3) $\quad 3x + y + 6z = 2$

We begin by eliminating the variable $x$ from each equation except the first. To do this we select the coefficient of $x$ in Equation (1) as the first pivot. Dividing both members of Equation (1) by the pivot 2, we obtain the equivalent equation

(1') $\quad x - 2y + \frac{1}{2}z = 0$

Next we multiply each side of this new Equation (1') by $-1$ and add the resulting terms to the corresponding terms of Equation (2), obtaining

(2') $\quad 0x - 1y - \frac{5}{2}z = -4$

Finally, we multiply each side of Equation (1') by $-3$ and add the resulting terms to the corresponding terms of Equation (3).

(3') $\quad 0x + 7y + \frac{9}{2}z = 2$

The system (1'), (2'), (3') is equivalent to the system (1), (2), (3), which means that both systems have the same solutions for $x, y, z$.

Each of the Equations (2') and (3') now contains only *two* variables, that is, one variable less than Equation (1'). Let us therefore repeat the process on these two equations using the coefficient of $y$, namely $-1$, as a pivot. Dividing both sides of Equation (2') by the pivot $-1$ yields

(2'') $\quad 0x + 1y + \frac{5}{2}z = 4$

Next we multiply both sides of Equation (2″) by −7 and add the resulting terms to the corresponding terms of Equation (3′):

$$(3″) \qquad 0x + 0y - 13z = -26$$

Finally, we divide both sides of this equation by −13 and obtain

$$(3‴) \qquad 0x + 0y + 1z = 2$$

Equation (3‴) now contains only one variable, that is, one less than Equation (2″). The original system (1), (2), (3) has therefore been transformed into the equivalent **triangular** system

$$(1') \qquad 1x - 2y + \frac{1}{2}z = 0$$

$$(2″) \qquad 1y + \frac{5}{2}z = 4$$

$$(3‴) \qquad 1z = 2$$

The value $z = 2$ from Equation (3‴) can now be "substituted back" into the preceding equation to obtain the value of $y$, namely $y = -1$. These values for $y$ and $z$ are then substituted into Equation (1′) to obtain the value for $x$, namely $x = -3$.

The triangular procedure described above was developed by Gauss before 1848.* About 1870 it occurred to Wilhelm Jordan, Professor of Geodesy at the Technical Hochschule in Germany, that the reverse substitution could be conveniently treated by continuing the elimination procedure as follows:

In the triangular system (1′), (2″), (3‴) multiply each side of Equation (2″) by 2 and add the resulting terms to the corresponding terms of Equation (1′):

$$(1″) \qquad 1x + 0y + \frac{11}{2}z = 8$$

Next multiply each side of Equation (3‴) by $-\frac{11}{2}$ and add the resulting terms to the corresponding terms of Equation (1″) to get

$$(1‴) \qquad 1x + 0y + 0z = -3$$

---

* We are indebted to Professor Albert Tucker of Princeton for this interesting historical reference.

## 9.3 The Gauss-Jordan Complete Elimination Procedure

Finally, multiply each side of Equation (3''') by $-\frac{5}{2}$ and add the resulting terms to the corresponding terms of Equation (2'') to obtain

$$(2''') \qquad 0x + 1y + 0z = -1$$

The original system (1), (2), (3) is thereby transformed by the Gauss-Jordan procedure into the equivalent **diagonal** system

$$\begin{aligned}(1''') & \qquad 1x \qquad\qquad\qquad = -3 \\ (2''') & \qquad\qquad 1y \qquad\quad = -1 \\ (3''') & \qquad\qquad\qquad 1z = \phantom{-}2\end{aligned}$$

This is by far the simplest system equivalent to the original, for it displays the complete solution at once.

As already seen for the case of two variables, the Gauss-Jordan procedure can be streamlined considerably. We observe that the operations affect only the coefficients of the variables and the numbers on the right side of each equation. We therefore detach these coefficients and numbers so that they form a rectangular array of 12 numbers, in this case a 3 × 4 matrix. We now have a representation of the system (1), (2), (3) in *tableau* form.

(4)

| x | y | z | |
|---|---|---|---|
| ② | −4 | 1 | 0 |
| 1 | −3 | −2 | −4 |
| 3 | 1 | 6 | 2 |

(Initial system)

The encircled entry ② is the coefficient of $x$ which we used previously as our initial pivot. Dividing the first row by the pivot 2 yields a new first row with the $x$-entry equal to 1.

| x | y | z | |
|---|---|---|---|
| 1 | −2 | $\frac{1}{2}$ | 0 |

(New first row)

## 9 ‖ Gauss-Jordan Method

This, of course, merely represents Equation (1'). Now multiply each entry in this new first row by $-1$ and add to the corresponding entry of the previous second row, that is, the second row of matrix (4). This yields a new second row in which the coefficient of $x$ is zero.

$$\begin{array}{ccc} x & y & z \end{array}$$

$$\left[\begin{array}{ccc|c} 1 & -2 & \frac{1}{2} & 0 \\ 0 & -1 & -\frac{5}{2} & -4 \end{array}\right] \quad \text{(New second row)}$$

Next return to the new first row, multiply each entry by $-3$, and add to the corresponding entry of the previous third row, that is, the third row of matrix (4). This yields a new third row in which the $x$-entry is again zero. The coefficient matrix now looks like this.

$$\begin{array}{ccc} x & y & z \end{array}$$

(5) $\quad \left[\begin{array}{ccc|c} 1 & -2 & \frac{1}{2} & 0 \\ 0 & -1 & -\frac{5}{2} & -4 \\ 0 & 7 & \frac{9}{2} & 2 \end{array}\right] \quad \text{(New third row)}$

The new matrix (5) clearly represents the system of Equations (1'), (2'), (3'), with the variable $x$ eliminated from each equation except the first. We call this procedure for eliminating a variable from all equations but one an **iteration**. An iteration is begun by selecting a nonzero coefficient of the variable to be eliminated and designating this entry as a pivot. (Such a pivot must exist for any variable which is actually present in at least one of the equations.) Dividing its row by the pivot produces a new row with the entry 1 in the pivotal position. Then adding appropriate multiples of this new row to each of the other rows replaces them by new rows with the entry 0 in the pivotal column. This completes the iteration.

Returning now to our system of equations, we begin a second iteration. Our purpose this time is to eliminate the variable $y$ from each equation except the second. [Note that this combines the Gauss "partial" elimination of $y$ (from the third equation only) with Jordan's "complete" elimination of $y$ (from the first equation also).] We therefore select the entry $-1$ from the second row of matrix (5) and use this entry as a pivot. We divide each number in the second row by $-1$, thus producing a new second row with 1 as the $y$-entry.

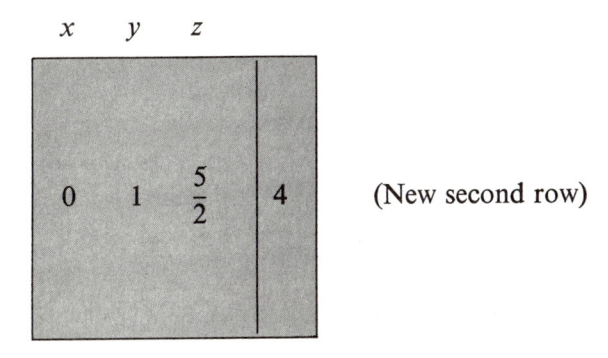

$$\begin{array}{cccc} x & y & z & \\ 0 & 1 & \frac{5}{2} & 4 \end{array}$$ (New second row)

This, of course, merely represents the Equation (2″) above. Multiply this new second row by 2 and add to the previous first row, that is, the first row of matrix (5). This yields a new first row with 0 as the $y$-entry.

$$\begin{array}{cccc} x & y & z & \\ 1 & 0 & \frac{11}{2} & 8 \\ 0 & 1 & \frac{5}{2} & 4 \end{array}$$ (New first row)

The new first row represents Equation (1″). Return to the new second row, multiply each entry by $-7$, and add to the previous third row, that is, the third row of matrix (5) to obtain a new third row in which the $y$-entry is again 0. This completes our second iteration and our coefficient matrix now looks like this

(6)

|   | x | y | z |   |
|---|---|---|---|---|
|   | 1 | 0 | $\frac{11}{2}$ | 8 |
|   | 0 | 1 | $\frac{5}{2}$ | 4 |
|   | 0 | 0 | $-13$ | $-26$ |

(New third row)

The new third row represents Equation (3").

The next step should be fairly evident by now. Perform a third iteration whose purpose is to eliminate $z$ from each equation except the last. This is accomplished by selecting the third row entry $-13$ as the obvious pivot. Divide each entry of this row by $-13$, thus making the $z$-entry unity. Then add appropriate multiples of this new third row to each of the other rows so as to reduce their $z$-entries to 0. The final result of this third iteration is the array

(7)

|   | x | y | z |   |
|---|---|---|---|---|
|   | 1 | 0 | 0 | $-3$ |
|   | 0 | 1 | 0 | $-1$ |
|   | 0 | 0 | 1 | 2 |

(Final system)

This final array represents the diagonal system (1'''), (2'''), (3''') which is, of course, equivalent to the original system of Equations (1), (2), (3). The solution of the system of equations is neatly displayed by the last column. Each entry in this column denotes the value of one of the variables, the variable corresponding to an entry 1 in the same row.

The complete process which we have described in minute detail is summarized in the following sequence of "tableaux."* Each tableau represents a system of equations equivalent to the given system.

* The same type of tableau plays an important role in linear programming.

### 9.3 The Gauss-Jordan Complete Elimination Procedure

| x | y | z | |
|---|---|---|---|
| ②  | −4 | 1 | 0 |
| 1 | −3 | −2 | −4 |
| 3 | 1 | 6 | 2 |

Initial Tableau
(It represents the given system of equations)
② is selected as the pivot for a first iteration.

| x | y | z | |
|---|---|---|---|
| 1 | −2 | $\frac{1}{2}$ | 0 |
| 0 | (−1) | $-\frac{5}{2}$ | −4 |
| 0 | 7 | $\frac{9}{2}$ | 2 |

Tableau after First Iteration
−1 is selected as the pivot for another iteration.

| x | y | z | |
|---|---|---|---|
| 1 | 0 | $\frac{11}{2}$ | 8 |
| 0 | 1 | $\frac{5}{2}$ | 4 |
| 0 | 0 | (−13) | −26 |

Tableau after Second Iteration
−13 is selected as the pivot for a third iteration.

| x | y | z | |
|---|---|---|---|
| 1 | 0 | 0 | −3 |
| 0 | 1 | 0 | −1 |
| 0 | 0 | 1 | 2 |

Final Tableau
(After third iteration)

An alert reader will no doubt observe ways of simplifying certain parts of the computation by judicious manipulation of the equations of the system. For example, in the original system (1), (2), (3), if the second equation is subtracted from the first and also from the third, the result, after dividing the new third equation by 2, is

| x | y | z | |
|---|---|---|---|
| 1 | −1 | 3 | 4 |
| ① | −3 | −2 | −4 |
| 1 | 2 | 4 | 3 |

Now iterate using ①as the pivot.

|   | x | y | z |   |
|---|---|---|---|---|
| 1 | −1 | 3 | 4 |
| 0 | −2 | −5 | −8 |
| 0 | 3 | 1 | −1 |

Next, add the third row to the second,

| 1 | −1 | 3 | 4 |
| 0 | ① | −4 | −9 |
| 0 | 3 | 1 | −1 |

Iterate again using ①as the pivot.

| 1 | 0 | −1 | −5 |
| 0 | 1 | −4 | −9 |
| 0 | 0 | 13 | 26 |

Divide the last row by 13 and use the resulting 1 as the pivot for the final iteration.

| 1 | 0 | 0 | −3 |
| 0 | 1 | 0 | −1 |
| 0 | 0 | 1 | 2 |

Although such manipulation of the equations between successive iterations circumvents the appearance of fractions in the tableaux, it is less systematic than the straightforward use of the Gauss-Jordan procedure. The reader is urged to practice the method until he is quite proficient in its use.

Each of the systems studied here had a unique solution. It is, of course, possible that a system of equations may not have a unique solution. In fact, there may be more than one solution or there may be no solution. (Some possibilities are illustrated in Exercises 8, 9, and 10.) Nevertheless, the Gauss-Jordan method is applicable in these cases as well.

## 9.3 The Gauss-Jordan Complete Elimination Procedure

## EXERCISES 9.3

In Exercises 1–7, solve each system of linear equations in two ways: (a) by Gauss's method of reducing them to triangular form and then substituting back and (b) by the Gauss-Jordan Complete Elimination Procedure.

1. $\begin{cases} x - 3y = 5 \\ 3x + 2y = 4 \end{cases}$

2. $\begin{cases} 3x + 4y = 17 \\ 6x - 10y = 5 \end{cases}$

3. $\begin{cases} x - y + z = -2 \\ x - 2y - 2z = -1 \\ 2x + y + 3z = 1 \end{cases}$

4. $\begin{cases} x + y - z = 3 \\ 2y + z = 10 \\ 5x - y - 2z = -3 \end{cases}$

5. $\begin{cases} -3x - 5y + 2z = 4 \\ 2y + z = -1 \\ x + y - z = 3 \end{cases}$

(*Hint:* Interchange the first and third equations. This will simplify subsequent computation.)

6. $\begin{cases} x + y + z - u = 1 \\ x - y + 3z + 2u = 2 \\ 2x + y + 3z + u = -2 \\ x - 2y + z + 3u = 10 \end{cases}$

7. $\begin{cases} 2x + y - z - 3u = 11 \\ -4x - 3y + 6z - u = -5 \\ x + y + u = 0 \\ 5x + 4z - 5u = 10 \end{cases}$

(*Hint:* Write the third equation first.)

8. Use the Gauss-Jordan method to solve the following system of equations for $x$, $y$, and $z$ in terms of $u$ and $v$.

$\begin{cases} 2x - y + 3z + 2u - v = 14 \\ 3x + y + u + 2v = 5 \\ -2x - 2z + u = -10 \end{cases}$

9. Try to solve the following system of equations by the Gauss-Jordan method.

$\begin{cases} x + 2y - z = 2 \\ 3x - y = 5 \\ x + 9y - 4z = 3 \end{cases}$

Does the method yield a unique solution for $x$, $y$, and $z$? Explain.

10. Try to solve the following system of equations by the Gauss-Jordan method.
$$\begin{cases} x - y + 2z = 3 \\ 2x \quad - z = -1 \\ \quad -2y + 5z = 8 \end{cases}$$
Does the method yield a unique solution for $x$, $y$, and $z$? Explain.

11. If each equation in a system of linear equations has zero for its right member (that is, if the number on the right of the equality sign is zero), the system is called *homogeneous*. Show that a homogeneous system will always have at least one solution. What is the solution?

## REVIEW EXERCISES

1. Try to solve each of the following linear systems
    (1) by Gauss's method of reducing them to triangular form and then substituting back and
    (2) by the Gauss-Jordan Complete Elimination Procedure.
    
    (a) $\begin{cases} 2x + 3y = -1 \\ 3x + 5y = 4 \end{cases}$
    
    (b) $\begin{cases} 2x + 3y - z = 9 \\ x - y + z = 1 \\ 4x + y = 10 \end{cases}$
    
    (c) $\begin{cases} 2x + 3y - z = 9 \\ x - y + z = 1 \\ 4x + y + z = 10 \end{cases}$
    
    (d) $\begin{cases} 2x + 3y - z = 9 \\ x - y + z = 1 \\ 4x + y + z = 11 \end{cases}$

2. Decide whether or not $\vec{v}_1$, $\vec{v}_2$, $\vec{v}_3$ are linearly independent.
    (a) $\vec{v}_1 = (1, 1, 1)$, $\vec{v}_2 = (1, 2, 3)$, $\vec{v}_3 = (-1, 0, 1)$
    (b) $\vec{v}_1 = (1, -1, 1)$, $\vec{v}_2 = (1, 2, 3)$, $\vec{v}_3 = (-1, 0, 1)$
    (c) $\vec{v}_1 = (1, -1, -1)$, $\vec{v}_2 = (1, 2, 3)$, $\vec{v}_3 = (1, 0, 1)$

3. For what value or values of $a$ will each of the following vectors be linearly dependent?
    (a) $\vec{v}_1 = (4, -3)$, $\vec{v}_2 = (6, a)$
    (b) $\vec{v}_1 = (4, -3)$, $\vec{v}_2 = (-6, a)$
    (c) $\vec{v}_1 = (1, 1, 1)$, $\vec{v}_2 = (1, 2, 3)$, $\vec{v}_3 = (-2, 0, a)$
    (d) $\vec{v}_1 = (1, -1, 1)$, $\vec{v}_2 = (1, 2, 3)$, $\vec{v}_3 = (-2, 0, a)$
    (e) $\vec{v}_1 = (1, -1, -1)$, $\vec{v}_2 = (1, 2, 3)$, $\vec{v}_3 = (a, 0, a)$
    (f) $\vec{v}_1 = (a, 1, 1)$, $\vec{v}_2 = (a, 2, 3)$, $\vec{v}_3 = (a, 5, -2)$
    (g) $\vec{v}_1 = (a, 2, 3)$, $\vec{v}_2 = (1, a, 3)$, $\vec{v}_3 = (1, 2, 3)$

*Arthur Cayley (1821-1895)*

# 10 | Matrix Algebra

> The invention of matrices illustrates once more the power and suggestiveness of a well-devised notation; it also exemplifies the fact, which some mathematicians are reluctant to admit, that a trivial notational device may be the germ of a vast theory having innumerable applications.
>
> E. T. BELL

## 10.1 MATRICES AND LINEAR SYSTEMS

In Section 9.3 we solved the system of linear equations

(1) $$\begin{cases} 2x - 4y + z = 0 \\ x - 3y - 2z = -4 \\ 3x + y + 6z = 2 \end{cases}$$

to obtain a unique solution for $x, y, z$

(2) $$\begin{cases} x = -3 \\ y = -1 \\ z = 2 \end{cases}$$

It is convenient to represent the three separate equations which make up this solution (2) by the single matrix equation

(3) $$\begin{bmatrix} x \\ y \\ z \end{bmatrix} = \begin{bmatrix} -3 \\ -1 \\ 2 \end{bmatrix}$$

The expressions appearing on each side of the equality sign are called **one-column matrices**. They are also often called **column vectors**. An equation such as (3) is interpreted as stating that corresponding entries of the column vectors are equal.

In the actual computation of solution (2), we found it convenient to represent the system (1) by the (initial) tableau

(4)
$$\begin{array}{ccc|c} x & y & z & \\ \hline 2 & -4 & 1 & 0 \\ 1 & -3 & -2 & -4 \\ 3 & 1 & 6 & 2 \end{array}$$

In this tableau the coefficients of the variables $x$, $y$, $z$ form a rectangular array of numbers

$$\begin{bmatrix} 2 & -4 & 1 \\ 1 & -3 & -2 \\ 3 & 1 & 6 \end{bmatrix}$$

This array is a **three-column matrix**, also called the **coefficient matrix** of the system of Equations (1). Let us designate this coefficient matrix by $A$. The numbers 0, −4, and 2 in the right-hand column of the tableau (4) form a one-column matrix or **column vector**

$$\begin{bmatrix} 0 \\ -4 \\ 2 \end{bmatrix}$$

Let us designate this column vector by $B$.

Tableau (4) now strongly suggests the following "operational" interpretation of the system of Equations (1).

> When the coefficient matrix $A$ operates upon the values of the variables $x$, $y$, $z$, it produces the column vector $B$.

If we regard the (unknown) values of the variables $x$, $y$, $z$ as forming a one-column matrix or column vector as in Equation (3) and if we denote this column vector

$$\begin{bmatrix} x \\ y \\ z \end{bmatrix}$$

by $X$, then we may express our operational interpretation of the system of Equations (1) as follows:

(5)
$$\begin{bmatrix} 2 & -4 & 1 \\ 1 & -3 & -2 \\ 3 & 1 & 6 \end{bmatrix} \begin{bmatrix} x \\ y \\ z \end{bmatrix} = \begin{bmatrix} 0 \\ -4 \\ 2 \end{bmatrix}$$

We can write this even more briefly as

(5') $$AX = B$$

However, in order that (5) or (5') truly represents the system of Equations (1), it is necessary that the coefficient matrix $A$ operate on the column vector $X$ in the following manner: The entries of each *row* of $A$ must be multiplied, respectively, by the corresponding entries of the *column* vector $X$ and the sum of these products must be equated to the entry in the corresponding row in $B$. This *row by column* rule yields the following matrix equation analogous to (3):

$$\begin{bmatrix} 2x - 4y + 1z \\ 1x - 3y - 2z \\ 3x + 1y + 6z \end{bmatrix} = \begin{bmatrix} 0 \\ -4 \\ 2 \end{bmatrix}$$

This single matrix equation clearly represents the original system (1) consisting of three separate linear equations.

The operation introduced here, whereby *each row* of matrix $A$ is multiplied by the column vector $X$ and the results assembled to form a new column vector, is fundamental to our discussion. This operation is called **multiplication** of the matrix $A$ by the one-column matrix $X$. The resulting one-column matrix is called the **product** of $A$ by $X$ and is denoted by $AX$.

> The product $AX$ is defined only when the number of **columns** in matrix $A$ is the same as the number of **entries** in the column vector $X$.

(This requirement is automatically fulfilled in the illustration.) If $A$ and $X$ have this property, then the product $AX$ can be defined regardless of how many rows there are in matrix $A$. For example, if $A$ represents a matrix having two columns such as

$$\begin{bmatrix} 2 & -4 \\ 1 & 3 \\ -2 & 5 \\ 3 & -1 \end{bmatrix}$$

and if $X$ is a column vector with 2 entries,

$$\begin{bmatrix} x \\ y \end{bmatrix}$$

then

$$AX = \begin{bmatrix} 2 & -4 \\ 1 & 3 \\ -2 & 5 \\ 3 & -1 \end{bmatrix} \begin{bmatrix} x \\ y \end{bmatrix} = \begin{bmatrix} 2x - 4y \\ 1x + 3y \\ -2x + 5y \\ 3x - 1y \end{bmatrix}$$

Thus $AX$ is a column vector having the same number of rows as the matrix $A$ (in this case, four rows).

We shall presently develop a further generalization of the matrix multiplication operation, but before we do let us return to the system of Equations (1) and make another important observation. By the Gauss-Jordan procedure we transformed the (initial) tableau (4) into the (final) tableau

(6)

| $x$ | $y$ | $z$ |     |
|---|---|---|-----|
| 1 | 0 | 0 | $-3$ |
| 0 | 1 | 0 | $-1$ |
| 0 | 0 | 1 | 2   |

This represents a system of equations equivalent to (1), and in matrix notation this system can be written

(7)
$$\begin{bmatrix} 1 & 0 & 0 \\ 0 & 1 & 0 \\ 0 & 0 & 1 \end{bmatrix} \begin{bmatrix} x \\ y \\ z \end{bmatrix} = \begin{bmatrix} -3 \\ -1 \\ 2 \end{bmatrix}$$

or even more briefly

(7′)
$$IX = C$$

where we have defined

(8) $$I = \begin{bmatrix} 1 & 0 & 0 \\ 0 & 1 & 0 \\ 0 & 0 & 1 \end{bmatrix} \quad \text{and} \quad C = \begin{bmatrix} -3 \\ -1 \\ 2 \end{bmatrix}$$

In matrix Equation (7) let us compute the product of the vector $I$ by the column vector $X$, obtaining

$$\begin{bmatrix} 1x + 0y + 0z \\ 0x + 1y + 0z \\ 0x + 0y + 1z \end{bmatrix} = \begin{bmatrix} -3 \\ -1 \\ 2 \end{bmatrix}$$

that is,

$$\begin{bmatrix} x \\ y \\ z \end{bmatrix} = \begin{bmatrix} -3 \\ -1 \\ 2 \end{bmatrix}$$

Equation (7) therefore merely reduces to Equation (3). Equation (3) is succinctly expressed by

(7″) $$X = C$$

A comparison of (7′) with (7″) also shows that the matrix $I$ defined in (8) behaves like an *identity element* when it operates on a column vector. This simply means that it leaves $X$ unaltered when it operates on $X$. Later we will see that this type of square matrix behaves like an identity element for matrix multiplication in general.

Let us return to the system (1) and its solution (2). If we alter the entries on the right side of the equations in (1), we may expect to obtain a solution different from (2). For example, suppose we replace the original right-hand entries 0, $-4$, and 2 by the new entries $-1$, $-3$, and 1. Since we expect a different solution, we rewrite the new system using new variables $x', y', z'$.

(9) $$\begin{cases} 2x' - 4y' + z' = -1 \\ x' - 3y' - 2z' = -3 \\ 3x' + y' + 6z' = 1 \end{cases}$$

The new initial tableau is very much like (4) except for the right-hand column.

(10)

| $x'$ | $y'$ | $z'$ |    |
|------|------|------|----|
| 2    | $-4$ | 1    | $-1$ |
| 1    | $-3$ | $-2$ | $-3$ |
| 3    | 1    | 6    | 1  |

## 10 ‖ Matrix Algebra

If we perform the same iterations as before, we find that the new final tableau is very similar to (6) except for the right-hand column

(11)
$$\begin{array}{ccc|c} x' & y' & z' & \\ \hline 1 & 0 & 0 & 2 \\ 0 & 1 & 0 & 1 \\ 0 & 0 & 1 & -1 \end{array}$$

The matrix equation corresponding to (10) is

(12)
$$\begin{bmatrix} 2 & -4 & 1 \\ 1 & -3 & -2 \\ 3 & 1 & 6 \end{bmatrix} \begin{bmatrix} x' \\ y' \\ z' \end{bmatrix} = \begin{bmatrix} -1 \\ -3 \\ 1 \end{bmatrix}$$

and the equivalent matrix equation corresponding to (11) is

(13)
$$\begin{bmatrix} 1 & 0 & 0 \\ 0 & 1 & 0 \\ 0 & 0 & 1 \end{bmatrix} \begin{bmatrix} x' \\ y' \\ z' \end{bmatrix} = \begin{bmatrix} 2 \\ 1 \\ -1 \end{bmatrix}$$

Let us now designate by $X'$, $B'$, $C'$ as the new column vectors:

$$X' = \begin{bmatrix} x' \\ y' \\ z' \end{bmatrix} \qquad B' = \begin{bmatrix} -1 \\ -3 \\ 1 \end{bmatrix} \qquad C' = \begin{bmatrix} 2 \\ 1 \\ -1 \end{bmatrix}$$

Then the two matrix Equations (12) and (13) are abbreviated very nicely by writing

(12')  $\qquad\qquad\qquad AX' = B'$

(13')  $\qquad\qquad\qquad IX' = C'$

Both of these matrix equations are clearly equivalent to

(13")  $\qquad\qquad\qquad X' = C'$

which is merely shorthand for the solution

$$\begin{cases} x' = 2 \\ y' = 1 \\ z' = -1 \end{cases}$$

We readily verify that these values satisfy the system of Equations (9).

## 10.1 Matrices and Linear Systems 339

Now, because the two systems (1) and (9) have the same coefficient matrix $A$, both systems can be represented by a single matrix equation rather than by the two separate matrix Equations (5) and (12). We simply combine these two matrix equations into one.

(14)
$$\begin{bmatrix} 2 & -4 & 1 \\ 1 & -3 & -2 \\ 3 & 1 & 6 \end{bmatrix} \begin{bmatrix} x & x' \\ y & y' \\ z & z' \end{bmatrix} = \begin{bmatrix} 0 & -1 \\ -4 & -3 \\ 2 & 1 \end{bmatrix}$$

In (14) we merely juxtaposed the two single columns which make up the vectors $X$ and $X'$ to form a two-column matrix

$$W = \begin{bmatrix} x & x' \\ y & y' \\ z & z' \end{bmatrix} = [X \quad X']$$

Similarly, we juxtaposed the columns in $B$ and $B'$ to form a two-column matrix

$$D = \begin{bmatrix} 0 & -1 \\ -4 & -3 \\ 2 & 1 \end{bmatrix} = [B \quad B']$$

With this notation, (14) may now be abbreviated simply as

$$A[X \quad X'] = [B \quad B']$$

or even more elegantly,

(14') $$AW = D$$

However, in order that (14) or (14') truly represents the *two* systems of Equations (1) and (9), it is again necessary to interpret properly the operation of multiplying the three-column matrix $A$ by the two-column matrix $W$. We have no choice but to use the row by column rule to multiply the matrix $A$ by each of the column vectors $X$ and $X'$ and to equate each of these matrix products to $B$ and $B'$, respectively. This requires that we define the product $AW$ as

(15) $$AW = A[X \quad X'] = [AX \quad AX']$$

that is,

$$\begin{bmatrix} 2 & -4 & 1 \\ 1 & -3 & -2 \\ 3 & 1 & 6 \end{bmatrix} \begin{bmatrix} x & x' \\ y & y' \\ z & z' \end{bmatrix} = \begin{bmatrix} 2x - 4y + z & 2x' - 4y' + z' \\ x - 3y - 2z & x' - 3y' - 2z' \\ 3x + y + 6z & 3x' + y' + 6z' \end{bmatrix}$$

Furthermore, when we equate the two-column matrix $[AX \ AX']$ to the two-column matrix $[B \ B']$, we are required to equate corresponding column vectors so as to yield $AX = B$, $AX' = B'$. But this, in turn, simply means that we must equate entries occupying corresponding positions whenever we equate two such matrices.

Notice that in order to define the product of the matrix $A$ by the matrix $W$, we had to multiply $A$ by each of the column vectors in $W$. We have already pointed out that the number of columns in matrix $A$ must be the same as the number of entries in each column of matrix $W$. This requirement can also be described by stating that

> The number of **columns** in matrix $A$ must be the same as the number of **rows** in matrix $W$. Two matrices $A$ and $W$ that have this property are called **conformable** (for multiplication).

Whenever this requirement is fulfilled, that is, whenever $A$ and $W$ are conformable, then the matrix product $AW$ can be defined no matter how many rows there are in matrix $A$ or how many columns there are in matrix $W$. We merely multiply the matrix $A$, in turn, by each column vector in $W$ using the row by column rule. The resulting product matrix will have as many rows as matrix $A$ and as many columns as matrix $W$. For example, suppose that

$$A = \begin{bmatrix} 3 & 5 \\ 2 & -1 \\ -4 & 0 \\ 1 & 3 \end{bmatrix} \quad \text{and} \quad W = \begin{bmatrix} x & u & s \\ y & v & t \end{bmatrix}$$

Then

$$AW = \begin{bmatrix} 3x + 5y & 3u + 5v & 3s + 5t \\ 2x - 1y & 2u - 1v & 2s - 1t \\ -4x + 0y & -4u + 0v & -4s + 0t \\ 1x + 3y & 1u + 3v & 1s + 3t \end{bmatrix}$$

Once again we may describe the matrix multiplication rule very neatly as follows: We "partition" the matrix $W$ into its three-column vectors

$$W = [W_1 \ W_2 \ W_3], \quad \text{where} \quad W_1 = \begin{bmatrix} x \\ y \end{bmatrix}, \quad W_2 = \begin{bmatrix} u \\ v \end{bmatrix}, \quad \text{and} \quad W_3 = \begin{bmatrix} s \\ t \end{bmatrix}$$

Then

(16) $$AW = A[W_1 \ W_2 \ W_3] = [AW_1 \ AW_2 \ AW_3]$$

In the special case where $A$ is the matrix

$$I = \begin{bmatrix} 1 & 0 \\ 0 & 1 \end{bmatrix}$$

Equation (16) becomes

$$IW = [IW_1 \ IW_2 \ IW_3] = [W_1 \ W_2 \ W_3] = W$$

indicating that $I$ behaves like an identity element for matrix multiplication. We shall generalize these ideas in Section 10.2.

## EXERCISES 10.1

1. Express each of the following systems of linear equations as a single matrix equation in the form $AX = B$.

   (a) $\begin{cases} x + y = 5 \\ x - y = 1 \end{cases}$

   (b) $\begin{cases} 3x - 2y + z = 3 \\ x \phantom{-2y} - 3z = 5 \\ 2x - 5y \phantom{- 3z} = -1 \end{cases}$

   (c) $\begin{cases} x + y + z + w = 9 \\ 2x + y - 3z - w = -2 \end{cases}$

   (d) $\begin{cases} 2x + 3y = 4 \\ x - 2y = 9 \\ -x + 4y = -3 \end{cases}$

   (e) $x + y = 10$

   (f) $\begin{cases} x \phantom{+y} = 3 \\ \phantom{x+} y = 2 \end{cases}$

2. Compute the following matrix products:

   (a) $\begin{bmatrix} 2 & 3 \\ 4 & 1 \end{bmatrix} \begin{bmatrix} 1 & 2 & 1 \\ 3 & 1 & 2 \end{bmatrix}$

   (b) $\begin{bmatrix} 1 & 2 \\ 3 & 1 \\ -1 & 2 \end{bmatrix} \begin{bmatrix} 4 & 0 & 1 \\ 1 & 2 & 3 \end{bmatrix}$

   (c) $\begin{bmatrix} 4 & 0 & 1 \\ 1 & 2 & 3 \end{bmatrix} \begin{bmatrix} 1 & 2 \\ 3 & 1 \\ -1 & 2 \end{bmatrix}$ (Compare this with (b))

(d) $\begin{bmatrix} 3 & 1 \\ 6 & 2 \end{bmatrix} \begin{bmatrix} -1 & 3 \\ 3 & -9 \end{bmatrix}$ (Comment)

(e) $\begin{bmatrix} 1 & 2 & 3 \\ 4 & 5 & 6 \\ 7 & 8 & 9 \end{bmatrix} \begin{bmatrix} 1 & 4 & 7 \\ 2 & 5 & 8 \\ 3 & 6 & 9 \end{bmatrix}$

(f) $\begin{bmatrix} 1 & -2 & 3 \\ -4 & 3 & -2 \\ 0 & 1 & 2 \end{bmatrix} \begin{bmatrix} 1 & 0 & 3 & -1 \\ 2 & 1 & 0 & -5 \\ 0 & 1 & 4 & 0 \end{bmatrix}$

(g) $\begin{bmatrix} 2 & 0 & 0 \\ 0 & 3 & 0 \\ 0 & 0 & 5 \end{bmatrix} \begin{bmatrix} \frac{1}{2} & 0 & 0 \\ 0 & \frac{1}{3} & 0 \\ 0 & 0 & \frac{1}{5} \end{bmatrix}$

(h) $\begin{bmatrix} 7 & 0 & -3 \\ 0 & -4 & 0 \\ -5 & 0 & 1 \end{bmatrix} \begin{bmatrix} 1 & 0 & 3 \\ 0 & 2 & 0 \\ 5 & 0 & 7 \end{bmatrix}$

(i) $\begin{bmatrix} 1 & 0 & 0 \\ 0 & 1 & 0 \\ 0 & 0 & 1 \end{bmatrix} \cdot \begin{bmatrix} a & b & c \\ d & e & f \\ g & h & i \end{bmatrix}$ (j) $\begin{bmatrix} a & b & c \\ d & e & f \\ g & h & i \end{bmatrix} \cdot \begin{bmatrix} 1 & 0 & 0 \\ 0 & 1 & 0 \\ 0 & 0 & 1 \end{bmatrix}$

(k) What generalization is suggested by (i) and (j)?

3. Compute the matrix products.

(a) $\left( \begin{bmatrix} 2 & 3 \\ 4 & 5 \end{bmatrix} \cdot \begin{bmatrix} 1 & 2 \\ 3 & 4 \end{bmatrix} \right) \cdot \begin{bmatrix} 1 & 3 \\ 2 & 5 \end{bmatrix}$

(b) $\begin{bmatrix} 2 & 3 \\ 4 & 5 \end{bmatrix} \cdot \left( \begin{bmatrix} 1 & 2 \\ 3 & 4 \end{bmatrix} \cdot \begin{bmatrix} 1 & 3 \\ 2 & 5 \end{bmatrix} \right)$

(c) $\left( \begin{bmatrix} 2 & 3 \\ 4 & 5 \end{bmatrix} \cdot \begin{bmatrix} 1 & 2 & 3 \\ 4 & 5 & 7 \end{bmatrix} \right) \begin{bmatrix} -1 \\ 2 \\ 3 \end{bmatrix}$

(d) $\begin{bmatrix} 2 & 3 \\ 4 & 5 \end{bmatrix} \cdot \left( \begin{bmatrix} 1 & 2 & 3 \\ 4 & 5 & 7 \end{bmatrix} \begin{bmatrix} -1 \\ 2 \\ 2 \end{bmatrix} \right)$

(e) What generalization is suggested by (a), (b), (c), and (d)?

(f) $\begin{bmatrix} 5 & 7 \\ 2 & 3 \end{bmatrix} \cdot \begin{bmatrix} 3 & -7 \\ -2 & 5 \end{bmatrix}$

(g) $\begin{bmatrix} 3 & -7 \\ -2 & 5 \end{bmatrix} \cdot \begin{bmatrix} 5 & 7 \\ 2 & 3 \end{bmatrix}$

(h) $\begin{bmatrix} 4 & 3 \\ 9 & 7 \end{bmatrix} \cdot \begin{bmatrix} 7 & -3 \\ -9 & 4 \end{bmatrix}$

(i) $\begin{bmatrix} 7 & -3 \\ -9 & 4 \end{bmatrix} \cdot \begin{bmatrix} 4 & 3 \\ 9 & 7 \end{bmatrix}$

(j) $\begin{bmatrix} 1 & 0 & 0 \\ 2 & 1 & 1 \\ 3 & 1 & 2 \end{bmatrix} \cdot \begin{bmatrix} 1 & 0 & 0 \\ -1 & 2 & -1 \\ -1 & -1 & 1 \end{bmatrix}$

(k) $\begin{bmatrix} 1 & 0 & 0 \\ -1 & 2 & -1 \\ -1 & -1 & 1 \end{bmatrix} \cdot \begin{bmatrix} 1 & 0 & 0 \\ 2 & 1 & 1 \\ 3 & 1 & 2 \end{bmatrix}$

(l) What generalization is suggested by (f), (g), (h), (i), (j), and (k)?

## 10.2 GENERAL DEFINITIONS: MATRICES AND MATRIX PRODUCTS

Let us pause to reformulate the ideas of Section 10.1 in a more precise and more general form. We begin with two sets of natural numbers

$$I = \{1, 2, \ldots, m\} \quad \text{and} \quad J = \{1, 2, \ldots, n\}$$

where $m$ and $n$ may, but need not be distinct. Let $i$ be a variable whose domain is the set $I$* and let $j$ be a variable whose domain is the set $J$. The variables $i$ and $j$ are called **indices** and are used in the following way. If with each $i \in I$ there is associated a unique real number $a(i)$, we designate this number by $a_i$. The set of numbers $a_1, a_2, \ldots, a_m$ is also designated by

$$a_i \quad (i = 1, 2, \ldots, m)$$

Similarly,

$$b_j \quad (j = 1, 2, \ldots, n)$$

designates the numbers $b_1, b_2, \ldots, b_n$. Finally if $(i, j)$ is an ordered pair of natural numbers, where $i \in I$ and $j \in J$† and if with each such ordered pair $(i, j)$ there is associated a number $a(i, j)$, then we shall designate this number by $a_{ij}$. (In most of our work we shall assume that $a_{ij}$ is a real number, although many of our results apply equally well if $a_{ij}$ is a complex number or, even more generally, a member of any specified ring.) The set of all values

---

* This merely means that the symbol "$i$" may denote any member of set $I$ in the present context.

† The set of all such ordered pairs is called the **cartesian product** $I \times J$.

of $a_{ij}$ as $i$ varies over $I$, and $j$ varies over $J$, consists, of $m \times n$ numbers (not necessarily distinct) which we can conveniently display as

$$\begin{bmatrix} a_{11} & a_{12} & \cdots & a_{1n} \\ a_{21} & a_{22} & \cdots & a_{2n} \\ \vdots & \vdots & & \vdots \\ a_{m1} & a_{m2} & \cdots & a_{mn} \end{bmatrix}$$

This rectangular array consists of $m$ rows and $n$ columns and can be abbreviated by writing

$$[a_{ij}] \quad (i = 1, 2, \ldots, m \\ j = 1, 2, \ldots, n)$$

Moreover, when the values of $m$ and $n$ are clear from the context, we simply write

$$[a_{ij}]$$

Any such rectangular array is called an **m × n matrix** or, when the values of $m$ and $n$ are clear, simply a **matrix**.*

If the index $i$ is held fixed, then the ordered set of elements

$$\{a_{i1}, a_{i2}, \ldots, a_{in}\}$$

is called the $i$th row of the matrix $[a_{ij}]$. The $i$th row may be displayed as a $1 \times n$ matrix

$$[a_{i1} \quad a_{i2} \quad \cdots \quad a_{in}]$$

A $1 \times n$ matrix is also called a **row vector**.

If the index $j$ is held fixed, then the ordered set of elements

$$\{a_{1j}, a_{2j}, \ldots, a_{mj}\}$$

is called the $j$th *column* of the matrix $[a_{ij}]$. The $j$th column may be displayed as an $m \times 1$ matrix

$$\begin{bmatrix} a_{1j} \\ a_{2j} \\ \vdots \\ a_{mj} \end{bmatrix}$$

Such an $m \times 1$ matrix is also called a **column vector**.

---

* From this point of view, a matrix is a **function** whose domain is $I \times J$ and whose range is a subset of some specified ring, usually the real numbers. (See Kelley, *Introduction to Modern Algebra*, page 294.)

## 10.2 General Definitions: Matrices and Matrix Products

Because it is awkward as well as space consuming to display a column vector in this manner, we shall also represent a column vector by an ordered $n$-tuple enclosed in parentheses. For example, the preceding column vector can be displayed using parentheses instead of square brackets and with the entries separated by commas as

$$(a_{1j}, a_{2j}, \ldots, a_{mj})$$

Do not confuse this with the analogous expression in square brackets

$$[a_{1j} \quad a_{2j} \quad \cdots \quad a_{mj}]$$

which represents a row vector ($1 \times m$ matrix).*

Now suppose we are given two matrices $A$ and $B$. If the number of columns in $A$ is the same as the number of rows in $B$, we call $A$ and $B$ *conformable* for multiplication (in that order). This means that we can represent $A$ and $B$ as

(1) $$A = [a_{ij}] \quad B = [b_{jk}] \quad \begin{array}{l}(i = 1, 2, \ldots, m \\ j = 1, 2, \ldots, n \\ k = 1, 2, \ldots, p)\end{array}$$

where $m$, $n$, and $p$ are natural numbers. The important fact here is that for each $i$, the $i$th row vector of matrix $A$

(2) $$[a_{i1} \quad a_{i2} \quad \cdots \quad a_{in}]$$

has $n$ entries and for each $k$, the $k$th column vector of matrix $B$

(3) $$(b_{1k}, b_{2k}, \ldots, b_{nk})$$

has also $n$ entries. If we multiply the values of corresponding entries in these two vectors and add these products, we obtain

(4) $$a_{i1}b_{1k} + a_{i2}b_{2k} + \cdots + a_{in}b_{nk}$$

This sum of products is called the **inner product** of the $i$th row vector of $A$ by the $k$th column vector of $B$. It is often denoted by

$$\sum_{j=1}^{} a_{ij} b_{jk}$$

---

* This convention is used, for example, by E. Nering, *Linear Algebra and Matrix Theory*, John Wiley, New York.

(We also call this sum of products simply the *product* of the $i$th row of $A$ by the $k$th column of $B$.) Its value is a number $c_{ik}$, where

(5) $$c_{ik} = \sum_{j=1}^{n} a_{ij} b_{jk}$$

Notice that in the summation the indices $i$ and $k$ are held fixed, whereas the repeated index $j$ varies from 1 to $n$.

Now if we continue to hold $k$ fixed but allow $i$ to vary (from 1 to $m$), then we obtain $m$ values of $c_{ik}$ which we can assemble as a column vector

(6) $$C_k = (c_{1k}, c_{2k}, \ldots, c_{mk})$$

This column vector has the same number $m$ of entries as each column vector of matrix $A$. Now let $B_k$ denote the $k$th column vector of matrix $B$, that is,

(7) $$B_k = (b_{1k}, b_{2k}, \ldots, b_{mk})$$

Since the entries of $C_k$ are obtained by multiplying in turn each row of matrix $A$ by the column vector $B_k$, we define the product of $A$ by $B_k$ to be the column vector $C_k$

(8) $$AB_k = C_k$$

Having defined the product of matrix $A$ by a column vector of matrix $B$, it is a natural step to a definition of the product of matrix $A$ by matrix $B$. Although we continue to use the row by column rule, we no longer hold the remaining index $k$ fixed, but allow it to vary (from 1 to $p$). In this way we obtain $p$ products, each of which is a column vector

$$AB_1 = C_1, \quad AB_2 = C_2, \quad \ldots, \quad AB_p = C_p$$

We assemble these column vectors $C_k$ side by side to form an $m \times p$ matrix

$$C = [C_1 \quad C_2 \quad \cdots \quad C_p]$$

and we call this new matrix the **product** of the (conformable) matrices $A$ and $B$. In other words, the product

(9) $$AB = C$$

is defined by

(10) $$A[B_1 \quad B_2 \quad \cdots \quad B_p] = [AB_1 \quad AB_2 \quad \cdots \quad AB_p]$$

where $AB_k = C_k$.

## 10.2 General Definitions: Matrices and Matrix Products

From (6) we observe that the $i$th entry of the column vector $C_k$ has the value $c_{ik}$ defined in (5). Therefore the product matrix $C$ may also be expressed as

(11) $\qquad C = [c_{ik}] \qquad \text{where} \qquad c_{ik} = \sum_{j=1}^{n} a_{ij} b_{jk}$

In many books (11) is used as the definition of $AB$. We prefer (10) as more natural and easier to remember because it exhibits more clearly the row by column rule for multiplying matrices.

It should now be clear that a single matrix equation such as

(12) $\qquad\qquad\qquad AX = B$

can represent one or more systems of simultaneous linear equations. Thus if

$A = [a_{ij}]$ is an $m \times n$ matrix,
$X = (x_1, x_2, \ldots, x_n)$ is an $n \times 1$ column vector, and
$B = (b_1, b_2, \ldots, b_m)$ is an $m \times 1$ column vector,

then (12) represents the system

(13) $\qquad \begin{cases} a_{11}x_1 + a_{12}x_2 + \cdots + a_{1n}x_n = b_1 \\ a_{21}x_1 + a_{22}x_2 + \cdots + a_{2n}x_n = b_2 \\ \vdots \qquad \vdots \qquad \vdots \qquad \vdots \qquad \vdots \\ a_{m1}x_1 + a_{m2}x_2 + \cdots + a_{mn}x_n = b_m \end{cases}$

This system consists of $m$ equations in $n$ variables.

If $X$ is an $n \times p$ matrix and $B$ is an $m \times p$ matrix instead of one-column matrices, then (12) represents $p$ such systems of equations. Equation (14) in Section 10.1 is an example of this.

## EXERCISES 10.2

Let $A$ be any $m \times n$ matrix. Denote the $i$th row of $A$ by $A^i$ and denote the $j$th column of $A$ by $A_j$. Then $A$ may be partitioned either by rows

$$A = \begin{bmatrix} A^1 \\ A^2 \\ \vdots \\ A^i \\ \vdots \\ A^m \end{bmatrix} \qquad \text{(Note: Here } A^2 \text{ means the second row of matrix } A. \text{ The 2 is an index not an exponent.)}$$

or by columns

$$A = [A_1 \quad A_2 \quad \ldots \quad A_j \quad \ldots \quad A_n]$$

## 10 ∥ Matrix Algebra

**1.** If $A = \begin{bmatrix} 1 & 4 & 7 \\ 2 & 5 & 8 \\ 3 & 6 & 9 \end{bmatrix}$ and $I = \begin{bmatrix} 1 & 0 & 0 \\ 0 & 1 & 0 \\ 0 & 0 & 1 \end{bmatrix}$, determine each of the following products:

(a) $AI_1$  (b) $AI_2$  (c) $AI_3$
(d) $A^1 I_1$  (e) $A^1 I_2$  (f) $A^1 I_3$
(g) $A^2 I_1$  (h) $A^2 I_2$  (i) $A^3 I_2$
(j) $A_1 I^1$  (k) $A_2 I^3$  (l) $AI$
(m) $AA_1$  (n) $AA_2$  (o) $A^1 A$
(p) $A^2 A$  (q) $AA$  (r) $A^3 A_3$
(s) $A^1 A_1$  (t) $A^1 A_2$  (u) $A_1 A^1$
(v) $A_1 A^2$

**2.** In Exercise 1, let

$$A = \begin{bmatrix} a_{11} & a_{12} & a_{13} \\ a_{21} & a_{22} & a_{23} \\ a_{31} & a_{32} & a_{33} \end{bmatrix}$$

Answer (a) to (v) and deduce some appropriate generalizations using the notation above.

**3.** If an $m \times n$ matrix $A$ is partitioned by rows, that is,

$$A = \begin{bmatrix} A^1 \\ A^2 \\ \vdots \\ A^m \end{bmatrix}$$

and an $n \times p$ matrix $B$ is partitioned by columns, that is,

$$B = [B_1 \ B_2 \ \cdots \ B_n]$$

express the product $AB$ in terms of products of rows $A^i$ by columns $B_k$. Do you notice anything interesting about the result?

**4.** Let $A = [a_1 \ a_2 \ a_3]$ be a $1 \times 3$ row matrix,
$B = (b_1, \ b_2, \ b_3)$ be a $3 \times 1$ column matrix,
$C = [c_1 \ c_2 \ c_3 \ c_4]$ be a $1 \times 4$ row matrix.
Prove $(AB)C = A(BC)$. Try to generalize. (For further discussion of this point, see Section 10.3.)

**5.** Compute each of the following:

(a) $\sum_{i=1}^{4} i$  (b) $\sum_{i=1}^{4} 2i$

(c) $\sum_{i=1}^{4} (i+1)$  (d) $\sum_{i=1}^{4} i^2$

(e) $\sum_{i=1}^{4} 2^i$ (f) $\sum_{i=1}^{4} \frac{1}{i}$

(g) $\sum_{i=1}^{4} i(i+1)$ (h) $\sum_{j=1}^{2} \sum_{i=1}^{3} (i+j)$

(i) $\sum_{i=1}^{3} \sum_{j=1}^{2} (i+j)$ (j) $\sum_{j=1}^{2} \sum_{i=1}^{3} (i \cdot j)$

(k) $\sum_{i=1}^{3} \sum_{j=1}^{2} (ij)$ (l) $\sum_{j=1}^{2} \sum_{i=1}^{3} (2i+j)$

(m) $\sum_{i=1}^{3} \sum_{j=1}^{2} (2i+j)$ (n) $\sum_{j=1}^{2} \sum_{i=1}^{3} (2i-j)$

(o) $\sum_{i=1}^{3} \sum_{j=1}^{2} (2i-j)$ (p) $\sum_{j=1}^{2} \sum_{i=1}^{3} (ij^2)$

(q) $\sum_{j=1}^{2} \left[ j^2 \sum_{i=1}^{3} i \right]$ (r) $\sum_{i=1}^{3} \sum_{j=1}^{2} (ij^2)$

(s) $\sum_{i=1}^{3} \left[ i \sum_{j=1}^{2} j^2 \right]$ (t) $\sum_{j=1}^{2} \sum_{i=1}^{3} (a_i b_j)$

(u) $\sum_{j=1}^{3} \left( b_j \sum_{i=1}^{2} a_i \right)$ (v) $\sum_{i=1}^{3} \sum_{j=1}^{2} (a_i b_j)$

(w) $\sum_{i=1}^{3} \left( a_i \sum_{j=1}^{2} b_j \right)$

**6.** Write out the expanded expression for each of the following:

(a) $\sum_{i=1}^{4} x^i$ (b) $\sum_{i=1}^{4} x_i$

(c) $\sum_{i=1}^{4} ix_i$ (d) $\sum_{i=1}^{4} a_i$

(e) $\sum_{i=1}^{4} a_i b_i$ (f) $\sum_{j=1}^{4} a_{1j} b_{2j}$

(g) $\sum_{j=1}^{4} a_{ij} b_{jk}$ (h) $\sum_{j=1}^{4} (a_{ij} + b_{jk})$

(i) $\sum_{j=1}^{4} (a_{ij} b_{jk})^2$ (j) $\sum_{j=1}^{4} (a_{ij} b_{jk})^j$

**7.** Write the matrix $[c_{ij}]$ if $\begin{cases} i = 1, 2, 3 \\ j = 1, 2, 3, 4 \end{cases}$ and $c_{ij}$ is

(a) $i + j$ (b) $i - j$

(c) $ij$ (d) $i^2 + j^2$

(e) $\sum_{k=1}^{2} (i^k + k^j)$  (f) $\sum_{k=1}^{2} a_{ik} b_{kj}$

(g) $\sum_{k=1}^{2} (i^k \div k^j)$

8. Let

$$A = \begin{bmatrix} 5 & -2 & 3 \\ 0 & 2 & 4 \\ 1 & 6 & -1 \end{bmatrix} \quad \text{and} \quad B = \begin{bmatrix} 5 & 0 & 1 \\ -2 & 2 & 6 \\ 3 & 4 & -1 \end{bmatrix}$$

Compute:
(a) $AB$ (b) $BA$ (c) $AA$
(d) $BB$ (e) $A^1 B_1$ (f) $A^1 B_2$
(g) $A_1 B^1$ (h) $A_3 B^3$ (i) $(AB)A$
(j) $A(BA)$ (k) $(BA)B$ (l) $B(AB)$

## 10.3 OTHER OPERATIONS WITH MATRICES

Let $V_m$ be the set of all ordered $m$-tuples

$$(x_1, x_2, \ldots, x_m)$$

whose coordinates $x_1, x_2, \ldots, x_m$ are real numbers. We have already agreed to add $n$-tuples by adding corresponding coordinates.

(1) $(x_1, x_2, \ldots, x_m) + (y_1, y_2, \ldots, y_m) = (x_1 + y_1, x_2 + y_2, \ldots, x_m + y_m)$

With this addition operation, the system $\{V_m, +\}$ is an abelian group. Moreover, if $r$ is any real number and if we define the product $r(x_1, x_2, \ldots, x_m)$ as

(2) $\qquad r(x_1, x_2, \ldots, x_m) = (rx_1, rx_2, \ldots, rx_m)$

then the system $\{V_m, +\}$ becomes a vector space over the field of real numbers $\{R, +, \cdot\}$.

The above operations with $m$-tuples lead quite naturally to analogous operations with matrices. Recall that we have already agreed to identify an ordered $m$-tuple with an $m \times 1$ matrix (column vector). Consequently, Equation (1) may be interpreted as

(3) $\qquad \begin{bmatrix} x_1 \\ x_2 \\ \vdots \\ x_m \end{bmatrix} + \begin{bmatrix} y_1 \\ y_2 \\ \vdots \\ y_m \end{bmatrix} = \begin{bmatrix} x_1 + y_1 \\ x_2 + y_2 \\ \vdots \\ x_m + y_m \end{bmatrix}$

## 10.3 Other Operations with Matrices

and Equation (2) may be interpreted as

(4)
$$r \begin{bmatrix} x_1 \\ x_2 \\ \vdots \\ x_m \end{bmatrix} = \begin{bmatrix} rx_1 \\ rx_2 \\ \vdots \\ rx_m \end{bmatrix}$$

Equations (3) and (4) merely state that column vectors are added or multiplied by scalars in exactly the same manner as the $m$-tuples that represent these column vectors. The calculations are performed in each case in *coordinate-wise* fashion, which is certainly the most natural way to define these operations. We shall therefore adopt (3) and (4) as the definitions for addition of column vectors and multiplication of a column vector by a scalar.

Suppose next that $A$ and $B$ are each $m \times n$ matrices (that is, each matrix has $m$ rows and $n$ columns). Let us denote the $n$ columns of matrix $A$ by $A_1, A_2, \ldots, A_n$ and denote the $n$ columns of matrix $B$ by $B_1, B_2, \ldots, B_n$. Each of these columns is an $m \times 1$ matrix (column vector) and may therefore be considered an element (vector) in $V_m$.

As in Sections 10.1 and 10.2, let us view each of the matrices $A$ and $B$ as assembled from its column vectors by juxtaposition of these vectors to form the matrix. We therefore express these matrices as

$$A = [A_1 \ A_2 \ \cdots \ A_n] \qquad B = [B_1 \ B_2 \ \cdots \ B_n]$$

Now just as we added vectors by adding corresponding coordinates, it appears very natural to "add" matrices $A$ and $B$ by adding their corresponding column vectors. We therefore define

(5)
$$A + B = [A_1 + B_1 \ A_2 + B_2 \ \cdots \ A_n + B_n]$$

The vector sums $A_1 + B_1, A_2 + B_2, \ldots$ are calculated coordinate-wise in accordance with the definition in Equation (3). Each coordinate of one of the column vectors $A_i$ is added to the corresponding coordinate of the corresponding column vector $B_i$.

> It follows that each element of matrix $A$ is added to the corresponding element of matrix $B$.

## 10 ‖ Matrix Algebra

This is precisely the rule we used in Chapter 4 for addition of $2 \times 2$ matrices. The same rule applies to $m \times n$ matrices. Notice that the sum $A + B$ is not defined unless

1. $A$ and $B$ have the same number of column vectors and
2. the column vectors of $A$ have the same number of coordinates as the column vectors of $B$.

These two requirements imply that $A$ and $B$ must have the same number of rows as well as the same number of columns in order to "conform" for addition.

Similarly, we define multiplication of a matrix $A$ by a real number (scalar) $r$ in a manner analogous to the definition of multiplication of a vector by a scalar $r$; we multiply each column vector of $A$ by $r$.

(6) $\qquad rA = r[A_1 \; A_2 \; \cdots \; A_n] = [rA_1 \; rA_2 \; \cdots \; rA_n]$

The products $rA_1, rA_2, \ldots$ are calculated coordinate-wise as defined in Equation (4).

> It follows that each element of matrix $A$ is multiplied by the scalar $r$.

As an example of these operations suppose that $A$ and $B$ are the following $3 \times 4$ matrices:

$$A = \begin{bmatrix} 1 & 2 & 3 & 4 \\ 5 & 6 & 7 & 8 \\ 6 & 0 & -1 & -2 \end{bmatrix} \qquad B = \begin{bmatrix} 4 & 5 & 6 & 7 \\ 8 & 9 & 0 & -1 \\ -2 & -3 & 4 & 5 \end{bmatrix}$$

Then

$$A + B = \begin{bmatrix} 1+4 & 2+5 & 3+6 & 4+7 \\ 5+8 & 6+9 & 7+0 & 8+(-1) \\ 9+(-2) & 0+(-3) & -1+4 & -2+5 \end{bmatrix}$$

$$= \begin{bmatrix} 5 & 7 & 9 & 11 \\ 13 & 15 & 7 & 7 \\ 7 & -3 & 3 & 3 \end{bmatrix}$$

$$3A = \begin{bmatrix} 3(1) & 3(2) & 3(3) & 3(4) \\ 3(5) & 3(6) & 3(7) & 3(8) \\ 3(9) & 3(0) & 3(-1) & 3(-2) \end{bmatrix} = \begin{bmatrix} 3 & 6 & 9 & 12 \\ 15 & 18 & 21 & 24 \\ 27 & 0 & -3 & -6 \end{bmatrix}$$

## 10.3 Other Operations with Matrices

Notice that the resulting matrices $A + B$ and $3A$ are also $3 \times 4$ matrices.

In general, if $A$ and $B$ are $m \times n$ matrices and $r$ is any real number, then $A + B$ and $rA$ are again $m \times n$ matrices. This observation suggests that we study more closely several mathematical systems whose elements are matrices.

First let us denote the set of all $m \times n$ matrices with real elements by $V_{m,n}$ and let us consider the system $\{V_{m,n}, +\}$ where $+$ signifies addition of matrices as defined above. Clearly, this operation is closed (that is, $+$ is a binary operation on $V_{m,n}$). Is the operation $+$ associative? To verify this let $A, B, C \in V_{m,n}$ and let

$$A = [A_1 \quad A_2 \quad \cdots \quad A_n], \quad B = [B_1 \quad B_2 \quad \cdots \quad B_n], \quad C = [C_1 \quad C_2 \quad \cdots \quad C_n]$$

The sum $(A + B) + C$ is expressed by

$$(A + B) + C = [(A_1 + B_1) + C_1 \quad (A_2 + B_2) + C_2 \quad \cdots \quad (A_n + B_n) + C_n]$$

We see that if we can prove that addition of vectors is associative, it will follow at once that addition of matrices is associative. But addition of vectors is defined by Equation (3) from which it is immediately apparent that the associativity of addition of vectors is a direct consequence of associativity of addition of real numbers. The key step here is the equality

$$\begin{bmatrix} (x_1 + y_1) + z_1 \\ (x_2 + y_2) + z_2 \\ \cdots \cdots \cdots \\ (x_n + y_n) + z_n \end{bmatrix} = \begin{bmatrix} x_1 + (y_1 + z_1) \\ x_2 + (y_2 + z_2) \\ \cdots \cdots \cdots \\ x_n + (y_n + z_n) \end{bmatrix}$$

Therefore addition of matrices in $V_{m,n}$ is indeed associative.

We leave it to the reader (see Exercise 1) to verify that there is an identity element for addition of matrices in $V_{m,n}$, that each matrix in $V_{m,n}$ has an additive inverse, and finally that addition of $m \times n$ matrices is commutative. This completes the proof of the following theorem.

**THEOREM 10.1** If $V_{m,n}$ is the set of all $m \times n$ matrices whose elements are real numbers, then the system $\{V_{m,n}, +\}$ is an abelian group.

Now we know that the elements of this abelian group, the $m \times n$ matrices, can be multiplied by real numbers (scalars) in the manner defined by Equation (6). It is practically a routine exercise to verify that multiplication by scalars satisfies the vector space axioms $(v_1)$, $(v_2)$, $(v_3)$, and $(v_4)$ (see Exercise 3). This verification proves the following theorem.

**THEOREM 10.2**  If $V_{m,n}$ is the set of all $m \times n$ matrices with real elements, then the abelian group $\{V_{m,n}, +\}$ is a vector space over the field of real numbers $\{R, +, \cdot\}$.

Thus having viewed an $m \times n$ matrix as an object assembled from $n$ column vectors chosen from a vector space, namely $\{V_m, +\}$, we now find that every $m \times n$ matrix may in turn be considered as a new kind of vector. The new vector is an element of $V_{m,n}$ where $\{V_{m,n}, +\}$ is itself a vector space over the field of real numbers. This observation provides a glimpse into the great generality and utility of the notion of a vector space.

One more mathematical system commands our attention very naturally at this point. We can not only add matrices but we can also multiply them, provided they are conformable for multiplication. To avoid problems of conformability let us confine our attention to *square* matrices, those having the same number of rows as columns. In short, let us consider only $n \times n$ matrices where $n$ is a fixed positive integer.

Suppose that we denote the set of all $n \times n$ matrices with real elements by $M_n$. If $A$ and $B$ are any pair of such matrices, then the sum $A + B$ and the product $A \cdot B$ are uniquely defined. The sum $A + B$ is the $n \times n$ matrix obtained by adding corresponding elements of matrices $A$ and $B$. The product $A \cdot B$ is also an $n \times n$ matrix obtained by using the row by column multiplication rule defined in Sections 10.1 and 10.2. Clearly, $+$ and $\cdot$ are binary operations on the set $M_n$ of all $n \times n$ matrices. What can one then say of the system $\{M_n, +, \cdot\}$? If you have not already guessed the answer, it is embodied in the following theorem.

**THEOREM 10.3**  If $M_n$ is the set of all $n \times n$ matrices with real elements, then the system $\{M_n, +, \cdot\}$ is a ring.

Theorem 10.3 is the natural generalization of Theorem 4.1 which we proved in detail in Section 4.2. Most of the proof involving the $2 \times 2$ matrices applies with only minor modifications to the general case of $n \times n$ matrices. Two items merit special attention, however, in this more general case.

First, let us prove that multiplication of $n \times n$ matrices distributes over addition of $n \times n$ matrices, that is, let us prove that if $A$, $B$, and $C$ are $n \times n$ matrices, then

$$A(B + C) = AB + AC$$

## 10.3 Other Operations with Matrices

Using the notation introduced in Sections 10.1 and 10.2, let us first express matrices $B$ and $C$ in terms of their column vectors

$$B = [B_1 \quad B_2 \quad \cdots \quad B_n] \qquad C = [C_1 \quad C_2 \quad \cdots \quad C_n]$$

Then by definition (5)

$$B + C = [B_1 + C_1 \quad B_2 + C_2 \quad \cdots \quad B_n + C_n]$$

Hence

$$A(B + C) = A[B_1 + C_1 \quad B_2 + C_2 \quad \cdots \quad B_n + C_n]$$

Using our definition of matrix multiplication (see Equation (10) in Section 10.2) this becomes

(7) $\qquad A(B + C) = [A(B_1 + C_1) \quad A(B_2 + C_2) \quad \cdots \quad A(B_n + C_n)]$

On the other hand,

$$AB = A[B_1 \quad B_2 \quad \cdots \quad B_n] = [AB_1 \quad AB_2 \quad \cdots \quad AB_n]$$
$$AC = A[C_1 \quad C_2 \quad \cdots \quad C_n] = [AC_1 \quad AC_2 \quad \cdots \quad AC_n]$$

Hence, by our definition of addition,

(8) $\qquad AB + AC = [AB_1 + AC_1 \quad AB_2 + AC_2 \quad \cdots \quad AB_n + AC_n]$

Comparing Equation (7) and (8) we see that our task reduces to proving that corresponding column vectors in these two equations are the same, that is, we must prove that

$$A(B_1 + C_1) = AB_1 + AC_1$$
$$A(B_2 + C_2) = AB_2 + AC_2$$
$$\cdots\cdots\cdots\cdots$$
$$A(B_n + C_n) = AB_n + AC_n$$

We accomplish this task by proving the following lemma.

LEMMA. If $A$ is an $n \times n$ matrix and if $X$ and $Y$ are $n \times 1$ column vectors, then

$$A(X + Y) = AX + AY$$

*Proof:* Let

$$X = (x_1, x_2, \ldots, x_n)$$
$$Y = (y_1, y_2, \ldots, y_n)$$

and let $A = [a_{ij}]$. The product of the matrix $A$ by the column vector $X$ is obtained by multiplying in turn each row of matrix $A$

by the column vector $X$ and assembling these $n$ inner products to form the column vector $AX$. (See Equations (6), (7), and (8) of Section 10.2.) The $i$th row of $A$ is the $1 \times n$ matrix

$$[a_{i1} \ a_{i2} \ \cdots \ a_{in}]$$

Hence the $i$th coordinate of $AX$ is

$$a_{i1} x_1 + a_{i2} x_2 + \cdots + a_{in} x_n = \sum_{j=1}^{n} a_{ij} x_j$$

Similarly, the $i$th coordinate of $AY$ is

$$a_{i1} y_1 + a_{i2} y_2 + \cdots + a_{in} y_n = \sum_{j=1}^{n} a_{ij} y_j$$

Hence the $i$th coordinate of $AX + AY$ is

$$\sum_{j=1}^{n} a_{ij} x_j + \sum_{j=1}^{n} a_{ij} y_j = \sum_{j=1}^{n} (a_{ij} x_j + a_{ij} y_j) = \sum_{j=1}^{n} a_{ij}(x_j + y_j)$$

where we have used the distributivity property for real numbers to obtain the last sum. But this sum is precisely the $i$th coordinate of the column vector $A(X + Y)$. Since this is true for each coordinate (for each $i = 1, 2, \ldots, n$), it follows that

$$A(X + Y) = AX + AY$$

This completes the proof of our lemma.

Returning now to Equations (7) and (8), it follows that for matrices $A$, $B$, and $C$

$$A(B + C) = AB + AC$$

In a very similar manner we can prove that

$$(B + C)A = BA + CA$$

(See Exercise 5.)

Thus we see that this ring property for real matrices is essentially a consequence of ring properties for real numbers. The proof that multiplication of $n \times n$ matrices is associative also requires some special attention because the method used for $2 \times 2$ matrices was already somewhat cumbersome. Let us see what modifications are needed in order to prove the associativity of multiplication in the present more general case.

## 10.3 Other Operations with Matrices 357

Our problem is to show that if $A$, $B$, and $C$ are $n \times n$ matrices, then

$$A(BC) = (AB)C$$

Using the notation introduced in Sections 10.1 and 10.2, let us first express the matrix $C$ in terms of its column vectors $C_1, C_2, \ldots, C_n$. Thus

$$C = [C_1 \quad C_2 \quad \cdots \quad C_n]$$

Then

$$BC = B[C_1 \quad C_2 \quad \cdots \quad C_n] = [BC_1 \quad BC_2 \quad \cdots \quad BC_n]$$
$$\therefore A(BC) = A[BC_1 \quad BC_2 \quad \cdots \quad BC_n]$$
(9) $\quad \therefore A(BC) = [A(BC_1) \quad A(BC_2) \quad \cdots \quad A(BC_n)]$

On the other hand,

$$(AB)C = (AB)[C_1 \quad C_2 \quad \cdots \quad C_n]$$
(10) $\quad (AB)C = [(AB)C_1 \quad (AB)C_2 \quad \cdots \quad (AB)C_n]$

Comparing (9) and (10) we see that our task reduces to proving that corresponding column vectors in these two equations are the same, that is, we must prove that

$$A(BC_1) = (AB)C_1$$
$$A(BC_2) = (AB)C_2$$
$$\cdots$$
$$A(BC_n) = (AB)C_n$$

We accomplish this by proving the following lemma.

LEMMA. If $A$ and $B$ are $n \times n$ matrices and if $X$ is an $n \times 1$ column vector, then

$$A(BX) = (AB)X$$

*Proof:* Let $X = (x_1, x_2, \ldots, x_n)$ and let $A = [a_{ij}] =$ and $B = [b_{jk}]$ where each index $i, j, k$ varies independently from 1 to $n$ inclusive. Using the row by column rule, $BX$ is the column vector whose $j$th coordinate is

$$\sum_{k=1}^{n} b_{jk} x_k$$

Hence, using the row by column rule again, $A(BX)$ is the column vector whose $i$th coordinate is

$$\sum_{j=1}^{n} \left[ a_{ij} \sum_{k=1}^{n} b_{jk} x_k \right]$$

The distributivity property for real numbers permits us to express the $i$th coordinate as

(11) $\qquad i$th coordinate of $A(BX) = \sum_{j=1}^{n} \left[ \sum_{k=1}^{n} a_{ij} b_{jk} x_k \right]$

Next, recall that $AB$ is the matrix $[c_{ik}]$ where

$$c_{ik} = \sum_{j=1}^{n} a_{ij} b_{jk}$$

Hence, using the row by column rule, $(AB)X$ is the column vector whose $i$th coordinate is

$$\sum_{k=1}^{n} c_{ik} x_k$$

Substituting for $c_{ik}$, we obtain

(12) $\qquad i$th coordinate of $(AB)X = \sum_{k=1}^{n} \left[ \sum_{j=1}^{n} a_{ij} b_{jk} \right] x_k$

Using the distributive property for real numbers, Equation (12) can be expressed as

(13) $\qquad i$th coordinate of $(AB)X = \sum_{k=1}^{n} \left[ \sum_{j=1}^{n} a_{ij} b_{jk} x_k \right]$

Comparing (11) and (13) we see that for each value of $i$, the very same products, $a_{ij} b_{jk} x_k$ are added except that the *order of addition* is different in the two cases. But a sum of real numbers does not depend on the order of addition. Consequently, the sums in (11) and (13) are the same. Therefore, the $i$th coordinate of $A(BX)$ is the same as the $i$th coordinate of $(AB)X$ and since this is true for each $i = 1, 2, \ldots, n$, the proof of our lemma is complete.

Returning now to (7) and (8) it follows that for matrices $A$, $B$, and $C$,

$$A(BC) = (AB)C$$

Once again we see that a ring property for real matrices stems essentially from ring properties of real numbers. The fact that $n \times n$ matrices form a ring under the operations of matrix addition and matrix multiplication strongly suggests that we regard a matrix as a new kind of number. This point of view has turned out to be remarkably fruitful and has yielded deeper insights into the concept of numbers.

The study of abstract mathematical systems has, in fact, revealed an immense variety of number systems. The properties of these systems have been systematically organized under headings such as groups, rings, integral domains, fields, ordered fields, complete ordered fields, and vector spaces. This list is by no means exhaustive, but it constitutes a firm foundation on which to build still deeper insights and a broader understanding of mathematics.

## EXERCISES 10.3

1. Let $V_{m,n}$ be the set of all $m \times n$ matrices whose elements are real numbers.
    (a) Specify an $m \times n$ matrix that serves as an identity element for addition in $V_{m,n}$.
    (b) If $[a_{ij}]$ is an $m \times n$ matrix, specify an additive inverse for $[a_{ij}]$.
    (c) Show that addition of $m \times n$ matrices is commutative.
2. Let $A$, $B$, and $C$ be $n \times n$ matrices and let $I$ be the $n \times n$ identity matrix, that is, the identity element for multiplication of $n \times n$ matrices.
    If $AB = I$ and $BC = I$, prove that $A = C$.
    (*Note:* This proves that a left inverse is the same as a right inverse.)
3. Prove that $\{V_{m,n}, +\}$ is a vector space over the field of real numbers by verifying Axioms $v_1$, $v_2$, $v_3$, and $v_4$ for the operations involved.
4. Prove that for each $n$ there is only one identity element for multiplication of $n \times n$ matrices.
    (*Hint:* Let $I$ and $I'$ both be identity matrices for multiplication. Consider the product $I \cdot I'$.)
5. If $A$, $B$, and $C$ are $n \times n$ matrices, prove that

$$(A + B)C = AB + AC$$

## 10.4 MATRIX INVERSES AND LINEAR INDEPENDENCE

We have already noted that each of the special matrices

$$\begin{bmatrix} 1 & 0 \\ 0 & 1 \end{bmatrix} \quad \text{and} \quad \begin{bmatrix} 1 & 0 & 0 \\ 0 & 1 & 0 \\ 0 & 0 & 1 \end{bmatrix}$$

behaves like an identity element for matrix multiplication. This property is, in fact, characteristic of each of the special matrices $I_n$ defined (for each $n = 1, 2, 3, \ldots$) by

$$I_n = \begin{bmatrix} 1 & 0 & \cdots & 0 \\ 0 & 1 & \cdots & 0 \\ \vdots & \vdots & \vdots & \vdots \\ 0 & 0 & \cdots & 1 \end{bmatrix} \quad \begin{pmatrix} n \text{ rows} \\ n \text{ columns} \end{pmatrix}$$

Each matrix $I_n$ is formed by assembling into a single matrix the $n$ "unit" column vectors

$$\vec{e}_1 = \begin{bmatrix} 1 \\ 0 \\ \vdots \\ 0 \end{bmatrix}, \quad \vec{e}_2 = \begin{bmatrix} 0 \\ 1 \\ \vdots \\ 0 \end{bmatrix}, \quad \ldots, \quad \vec{e}_n = \begin{bmatrix} 0 \\ 0 \\ \vdots \\ 1 \end{bmatrix}$$

where the $k$th coordinate of $\vec{e}_k$ is 1 and its remaining coordinates are 0. We usually designate the matrix $I_n$ simply by $I$ whenever the value of $n$ is fixed and clear from the context. If $A$ is an $n \times n$ matrix, then we always have

$$AI = IA = A$$

To prove this, we observe first that

$$A\vec{e}_1 = \begin{bmatrix} a_{11} & a_{12} & \cdots & a_{1n} \\ a_{21} & a_{22} & \cdots & a_{2n} \\ \vdots & \vdots & \vdots & \vdots \\ a_{n1} & a_{n2} & \cdots & a_{nn} \end{bmatrix} \begin{bmatrix} 1 \\ 0 \\ \vdots \\ 0 \end{bmatrix} = \begin{bmatrix} a_{11} \\ a_{21} \\ \vdots \\ a_{n1} \end{bmatrix} = A_1$$

and similarly we show that

$$A\vec{e}_2 = A_2, \quad A\vec{e}_3 = A_3, \quad \ldots, \quad A\vec{e}_n = A_n$$

It then follows that

$$\begin{aligned} AI &= A[\vec{e}_1 \quad \vec{e}_2 \quad \cdots \quad \vec{e}_n] \\ &= [A\vec{e}_1 \quad A\vec{e}_2 \quad \cdots \quad A\vec{e}_n] \\ &= [A_1 \quad A_2 \quad \cdots \quad A_n] \\ AI &= A \end{aligned}$$

## 10.4 Matrix Inverses and Linear Independence

Next we observe that

$$IA_1 = \begin{bmatrix} 1 & 0 & 0 & \cdots & 0 \\ 0 & 1 & 0 & \cdots & 0 \\ 0 & 0 & 1 & \cdots & 0 \\ \vdots & \vdots & \vdots & \vdots & \vdots \\ 0 & 0 & 0 & \cdots & 1 \end{bmatrix} \begin{bmatrix} a_{11} \\ a_{12} \\ a_{13} \\ \vdots \\ a_{1n} \end{bmatrix}$$

$$= \begin{bmatrix} a_{11} \\ a_{12} \\ a_{13} \\ \vdots \\ a_{1n} \end{bmatrix} = A_1$$

Similarly, we show that

$$IA_2 = A_2, \quad IA_3 = A_3, \quad \ldots, \quad IA_n = A_n$$

It then follows that

$$IA = I[A_1 \; A_2 \; \cdots \; A_n]$$
$$= [IA_1 \; IA_2 \; \cdots \; IA_n]$$
$$= [A_1 \; A_2 \; \cdots \; A_n]$$
$$IA = A$$

By Theorem 1.1 it follows that the identity matrix $I$ is unique, that is, for each $n$ there is only one identity element for multiplication of $n \times n$ matrices.

An important problem arises naturally at this point. If $A$ is an $n \times n$ matrix, does there exist an inverse matrix for $A$ (with respect to multiplication)? If such an inverse matrix exists, it is usually denoted by $A^{-1}$, so that

$$AA^{-1} = A^{-1}A = I$$

When such an inverse matrix exists it can be used, for example, to produce immediately the solution of a system of equations of the form

$$AX = B$$

where $A$ is the given $n \times n$ matrix, $B$ is the given $n \times 1$ column vector, and the coordinates of the unknown $n \times 1$ vector $X$ are to be determined. The solution vector for $X$ may, in fact, be determined as follows:

$$AX = B$$
$$A^{-1}(AX) = A^{-1}B$$
$$(A^{-1}A)X = A^{-1}B$$
$$IX = A^{-1}B$$
$$X = A^{-1}B$$

(That $A^{-1}B$ is indeed a solution may be checked by substituting $A^{-1}B$ for $X$ in the equation $AX = B$.) This means that if the matrix $A$ has an inverse $A^{-1}$, then the system of equations $AX = B$ always has a unique solution that may be found simply by multiplying $A^{-1}$ by $B$ (in that order).

We have already seen examples of systems of equations of the form $AX = B$ that do not have a unique solution. (See Example 2, Section 9.1.) It follows that not every matrix $A$ has an inverse. A twofold problem therefore presents itself. Given an arbitrary $n \times n$ matrix $A$, how can we tell whether $A$ has an inverse, and if an inverse does indeed exist, how do we determine it? Both questions can be answered by a judicious application of the Gauss-Jordan Algorithm.

Let us observe first that if an inverse for matrix $A$ exists, then we can determine it by finding a matrix $Z$ such that

(1) $$AZ = I$$

In fact, if $A^{-1}$ exists and if $AZ = I$, then

$$A^{-1}(AZ) = A^{-1}I$$
$$(A^{-1}A)Z = A^{-1}$$
$$IZ = A^{-1}$$
$$Z = A^{-1}$$

In other words, if $A^{-1}$ exists, it must be the unique solution of $AZ = I$. (*Note:* It is actually possible to prove a somewhat stronger result than this. It can be shown that if $AZ = I$, then $ZA = I$. This means that if there is a matrix $Z$ such that $AZ = I$, then the matrix $A$ must have an inverse and this inverse is the matrix $Z$. A proof of this stronger theorem requires a somewhat deeper study of matrices which we cannot undertake here.*)

We have already seen (Section 10.1) that a matrix equation such as $AZ = I$ represents, in general, several systems of simultaneous equations. For example, suppose that $A$ is the $3 \times 3$ matrix

$$A = \begin{bmatrix} -1 & 2 & 1 \\ 0 & 1 & -2 \\ 1 & 3 & 1 \end{bmatrix}$$

---

\* See, for example, E. D. Nering, *Linear Algebra and Matrix Theory*, John Wiley, 1963 (Chapter 2, Section 3, Theorem 3.4).

## 10.4 Matrix Inverses and Linear Independence

In order to determine $A^{-1}$ we must determine a $3 \times 3$ matrix

$$Z = \begin{bmatrix} x_1 & x_2 & x_3 \\ y_1 & y_2 & y_3 \\ z_1 & z_2 & z_3 \end{bmatrix}$$

such that $AZ = I$, that is, such that

$$\begin{bmatrix} -1 & 2 & 1 \\ 0 & 1 & -2 \\ 1 & 3 & 1 \end{bmatrix} \begin{bmatrix} x_1 & x_2 & x_3 \\ y_1 & y_2 & y_3 \\ z_1 & z_2 & z_3 \end{bmatrix} = \begin{bmatrix} 1 & 0 & 0 \\ 0 & 1 & 0 \\ 0 & 0 & 1 \end{bmatrix}$$

If we apply the (row by column) rule for multiplication of the two matrices on the left and equate the product matrix to the matrix on the right, we obtain the following three systems of linear equations:

$$\begin{cases} -1x_1 + 2y_1 + 1z_1 = 1 \\ 0x_1 + 1y_1 - 2z_1 = 0 \\ 1x_1 + 3y_1 + 1z_1 = 0 \end{cases}, \quad \begin{cases} -1x_2 + 2y_2 + 1z_2 = 0 \\ 0x_2 + 1y_2 - 2z_2 = 1 \\ 1x_2 + 3y_2 + 1z_2 = 0 \end{cases},$$

$$\begin{cases} -1x_3 + 2y_3 + 1z_3 = 0 \\ 0x_3 + 1y_3 - 2z_3 = 0 \\ 1x_3 + 3y_3 + 1z_3 = 1 \end{cases}$$

But we have already used the Gauss-Jordan technique to solve such systems of equations quite conveniently. In fact, since the same coefficient matrix appears in all three systems (to the left of the equality signs), we can solve all three systems at once using the following single tableau:

| -1 | 2 | 1 | 1 | 0 | 0 |
|---|---|---|---|---|---|
| 0 | 1 | -2 | 0 | 1 | 0 |
| 1 | 3 | 1 | 0 | 0 | 1 |

(The pivot is the encircled $-1$.)

Applying the Gauss-Jordan procedure to this tableau we obtain, in turn, each of the following. (The pivot is encircled.)

| 1 | -2 | -1 | -1 | 0 | 0 |
|---|---|---|---|---|---|
| 0 | ① | -2 | 0 | 1 | 0 |
| 0 | 5 | 2 | 1 | 0 | 1 |

First iteration

| 1 | 0 | -5 | -1 | 2 | 0 |
|---|---|---|---|---|---|
| 0 | 1 | -2 | 0 | 1 | 0 |
| 0 | 0 | ⑫ | 1 | -5 | 1 |

Second iteration

$$\begin{bmatrix} 1 & 0 & 0 & \vdots & -\dfrac{7}{12} & -\dfrac{1}{12} & \dfrac{5}{12} \\ 0 & 1 & 0 & \vdots & \dfrac{2}{12} & \dfrac{2}{12} & \dfrac{2}{12} \\ 0 & 0 & 1 & \vdots & \dfrac{1}{12} & -\dfrac{5}{12} & \dfrac{1}{12} \end{bmatrix} \quad \text{Third iteration}$$

The final tableau represents the following three systems of equations:

$$\begin{cases} x_1 = -\dfrac{7}{12} \\ y_1 = \dfrac{2}{12} \\ z_1 = \dfrac{1}{12} \end{cases}, \quad \begin{cases} x_2 = -\dfrac{1}{12} \\ y_2 = \dfrac{2}{12} \\ z_2 = -\dfrac{5}{12} \end{cases}, \quad \begin{cases} x_3 = \dfrac{5}{12} \\ y_3 = \dfrac{2}{12} \\ z_3 = \dfrac{1}{12} \end{cases}$$

These equations yield the nine elements that make up the desired matrix $Z$. Clearly, it is unnecessary to write these equations explicitly (as we have done here) because the desired matrix $Z$ already appears fully assembled in our final tableau to the right of the vertical dashed line.

$$Z = A^{-1} = \begin{bmatrix} -\dfrac{7}{12} & -\dfrac{1}{12} & \dfrac{5}{12} \\ \dfrac{2}{12} & \dfrac{2}{12} & \dfrac{2}{12} \\ \dfrac{1}{12} & -\dfrac{5}{12} & \dfrac{1}{12} \end{bmatrix}$$

In short, we have the following simple rule for finding the inverse of a matrix $A$.

1. Set up the tableau

$$[\,A\,|\,I\,]$$

2. Perform the Gauss-Jordan iterations necessary to transform the matrix $A$ into the identity matrix $I$. These iterations automatically convert the original matrix $I$ into $A^{-1}$.

## 10.4 Matrix Inverses and Linear Independence

Now it is clear that this method will work whenever a unique solution exists for each of the systems of simultaneous equations represented by the matrix equation

$$AZ = I$$

We have already mentioned that these systems may not have a unique solution and that when this is the case the matrix $A$ cannot have an inverse. Consider, for example, the matrix

$$A = \begin{bmatrix} -1 & 2 & 1 \\ 0 & 1 & -2 \\ -1 & 0 & 5 \end{bmatrix}$$

Applying the Gauss-Jordan procedure to the matrix equation $AZ = I$, we obtain

| -1 | 2 | 1 | 1 | 0 | 0 |
|---|---|---|---|---|---|
| 0 | 1 | -2 | 0 | 1 | 0 |
| -1 | 0 | 5 | 0 | 0 | 1 |

Initial tableau

| 1 | -2 | -1 | -1 | 0 | 0 |
|---|---|---|---|---|---|
| 0 | 1 | -2 | 0 | 1 | 0 |
| 0 | -2 | 4 | -1 | 0 | 1 |

First iteration

| 1 | 0 | -5 | -1 | 2 | 0 |
|---|---|---|---|---|---|
| 0 | 1 | -2 | 0 | 1 | 0 |
| 0 | 0 | 0 | -1 | 2 | 1 |

Second iteration

At this point we see that something is clearly wrong. There is no pivot available for the third iteration. In fact, the zeros in the third row represent the equations

$$0x_1 + 0y_1 + 0z_1 = -1, \quad 0x_2 + 0y_2 + 0z_2 = 2, \quad 0x_3 + 0y_3 + 0z_3 = 1$$

which are obviously impossible. Since these equations have no solution, the matrix $A$ clearly has no inverse in this case.

It is true, however, that whenever the matrix $A$ does have an inverse this procedure will yield an inverse which will be unique. (See Exercise 7.) In this respect the situation is quite analogous to that described in Sections

9.1 and 9.2 where we developed a procedure for determining whether vectors in $V_2$ were linearly dependent or linearly independent. Briefly stated,

> An $n \times n$ matrix $A$ will have an inverse whenever its $n$ column vectors are linearly independent. The matrix $A$ will not have an inverse whenever its columns are linearly dependent.

The Gauss-Jordan procedure can always be used to decide this question. We explore some of these assertions in the Exercises.

For further study the reader is urged to consult more advanced texts in linear algebra and abstract algebra.

## EXERCISES 10.4

1. The text asserts that

$$A^{-1} = \begin{bmatrix} -\frac{7}{12} & -\frac{1}{12} & \frac{5}{12} \\ \frac{2}{12} & \frac{2}{12} & \frac{2}{12} \\ \frac{1}{12} & -\frac{5}{12} & \frac{1}{12} \end{bmatrix} \quad \text{where} \quad A = \begin{bmatrix} -1 & 2 & 1 \\ 0 & 1 & -2 \\ 1 & 3 & 1 \end{bmatrix}$$

Verify the assertion by computing each of the following.
(a) $AA^{-1}$ 
(b) $A^{-1}A$

2. Using the Gauss-Jordan procedure, find the inverse for each of the following matrices and compute both products of each matrix and its inverse.

(a) $\begin{bmatrix} 7 & 5 \\ 4 & 3 \end{bmatrix}$ 
(b) $\begin{bmatrix} 8 & 12 \\ 3 & 5 \end{bmatrix}$

(c) $\begin{bmatrix} -7 & 2 \\ -11 & 3 \end{bmatrix}$ 
(d) $\begin{bmatrix} 1 & 0 & 2 \\ 0 & 2 & 1 \\ 2 & 1 & 0 \end{bmatrix}$

(e) $\begin{bmatrix} 1 & 0 & 2 \\ 0 & 4 & 0 \\ 5 & 0 & 6 \end{bmatrix}$ 
(f) $\begin{bmatrix} 1 & 0 & 2 \\ 2 & 1 & -1 \\ 3 & 2 & 1 \end{bmatrix}$

3. Use the Gauss-Jordan procedure to find the inverse of each matrix or show that the matrix does not have an inverse.

(a) $\begin{bmatrix} 8 & -4 \\ -2 & 1 \end{bmatrix}$ 
(b) $\begin{bmatrix} 6 & 9 \\ 4 & 6 \end{bmatrix}$

(c) $\begin{bmatrix} 12 & 6 \\ 18 & 8 \end{bmatrix}$ 
(d) $\begin{bmatrix} 8 & 2 \\ 7 & 2 \end{bmatrix}$

(e) $\begin{bmatrix} 1 & 2 & 3 \\ 2 & -1 & -2 \\ 3 & 1 & 1 \end{bmatrix}$ 
(f) $\begin{bmatrix} 1 & 2 & 3 \\ 2 & -1 & 2 \\ 1 & 1 & 1 \end{bmatrix}$

4. Using the associativity of matrix multiplication, multiply both members of each equation on the left by the given matrix. What do you observe?

(a) $\begin{bmatrix} 5 & 7 \\ 2 & 3 \end{bmatrix} \begin{bmatrix} x \\ y \end{bmatrix} = \begin{bmatrix} 2 \\ 3 \end{bmatrix}$ Multiply by $\begin{bmatrix} 3 & -7 \\ -2 & 5 \end{bmatrix}$

(b) $\begin{bmatrix} 5 & 7 \\ 2 & 3 \end{bmatrix} \begin{bmatrix} x & u \\ y & v \end{bmatrix} = \begin{bmatrix} 2 & 4 \\ 3 & 5 \end{bmatrix}$ Multiply by $\begin{bmatrix} 3 & -7 \\ -2 & 5 \end{bmatrix}$

(c) $\begin{bmatrix} 4 & 3 \\ 9 & 7 \end{bmatrix} \begin{bmatrix} x & u \\ y & v \end{bmatrix} = \begin{bmatrix} 2 & 4 \\ 3 & 5 \end{bmatrix}$ Multiply by $\begin{bmatrix} 7 & -3 \\ -9 & 4 \end{bmatrix}$

(d) $\begin{bmatrix} 1 & 0 & 0 \\ 2 & 1 & 1 \\ 3 & 1 & 2 \end{bmatrix} \begin{bmatrix} x & u \\ y & v \\ z & w \end{bmatrix} = \begin{bmatrix} 2 & 5 \\ 3 & 6 \\ 4 & 7 \end{bmatrix}$ Multiply by $\begin{bmatrix} 1 & 0 & 0 \\ -1 & 2 & -1 \\ -1 & -1 & 1 \end{bmatrix}$

5. If $A$ and $B$ are $n \times n$ matrices and $A$ and $B$ each has an inverse $A^{-1}$ and $B^{-1}$, respectively, prove that the product $AB$ also has an inverse, namely

$$(AB)^{-1} = B^{-1}A^{-1}$$

6. If $A$ is an $n \times n$ matrix and if $A$ has an inverse $A^{-1}$, then $A^{-1}$ also has an inverse, namely

$$(A^{-1})^{-1} = A$$

7. (a) Prove: If matrix $A$ has an inverse and if $AZ = AW$, then $Z = W$.
  (b) Use the result in (a) to prove that if $A$ has an inverse, this inverse is unique.
  (c) Prove the uniqueness of matrix inverses by citing an appropriate theorem from Chapter 1.

# 10 ‖ Matrix Algebra

8. Let $A = \begin{bmatrix} 2 & -1 & -1 \\ 1 & 3 & 2 \\ -1 & 2 & 3 \end{bmatrix}$, $B = \begin{bmatrix} 0 & 2 & -1 \\ 3 & 1 & 2 \\ 2 & -1 & 3 \end{bmatrix}$,

$C = \begin{bmatrix} 1 & -1 & 2 \\ 2 & 1 & -1 \\ 1 & 1 & 2 \end{bmatrix}$

(a) Compute:
- (1) $(A + B) + C$
- (2) $A + (B + C)$
- (3) $2A + (3B - C)$
- (4) $(2A + 3B) - C$
- (5) $A(B + C)$
- (6) $AB + AC$
- (7) $(AB)C$
- (8) $A(BC)$
- (9) $(A + B)C$
- (10) $AC + BC$
- (11) $AB - BA$
- (12) $AC - CA$
- (13) $(A + B)(A - B)$
- (14) $AA - BB$

(b) Solve for the matrix $X$:
- (1) $A + X = B$
- (2) $2A + X = B + C$
- (3) $A - X = B$
- (4) $2X = A + B$
- (5) $2X + A = X + C$
- (6) $2X - A = B - X$
- (7) $CX = I$
- (8) $XC = I$
- (9) $CX = B$
- (10) $XC = B$
- (11) $AX + B = C$

9. Try to solve each of the following systems of equations by first converting to a matrix equation and then using the inverse of the coefficient matrix.

(a) $\begin{cases} 5x + 2y = 1 \\ 7x + 3y = 3 \end{cases}$

(b) $\begin{cases} 5x + 2y = 1 \\ 7x + 3y = 2 \end{cases}$ and $\begin{cases} 5u + 2v = 3 \\ 7u + 3v = 4 \end{cases}$

(c) $\begin{cases} x - y + 2z = 1 \\ 2x + y - z = 2 \\ x + y + 2z = 3 \end{cases}$ and $\begin{cases} u - v + 2w = 4 \\ 2u + v - w = 5 \\ u + v + 2w = 6 \end{cases}$

10. Using basic definitions for computing products of matrices, give an independent proof that
$$(AB)C = A(BC)$$
where $A = [a_1, a_2, \ldots, a_n]$ a row vector,
$B = [b_{ij}]$ $i, j = 1, 2, \ldots, n$,
$C = (c_1, c_2, \ldots, c_n)$ a column vector.

11. Let $A = \begin{bmatrix} 5 & 7 \\ 2 & 3 \end{bmatrix}$, $I = \begin{bmatrix} 1 & 0 \\ 0 & 1 \end{bmatrix}$, $O = \begin{bmatrix} 0 & 0 \\ 0 & 0 \end{bmatrix}$.

Show that $A$ satisfies the equation $X^2 - 8X + I = O$. (Here $X^2$ means $XX$.) Use this information to find $A^{-1}$.

**12.** Let $B = \begin{bmatrix} 4 & 5 \\ 2 & 3 \end{bmatrix}$, $I = \begin{bmatrix} 1 & 0 \\ 0 & 1 \end{bmatrix}$, $O = \begin{bmatrix} 0 & 0 \\ 0 & 0 \end{bmatrix}$

Determine real numbers $r, s, t$, not all zero such that
$$rB^2 + sB + tI = 0$$
and use the information to compute $B^{-1}$. (Here $B^2$ means $BB$.)

## REVIEW EXERCISES

**1.** Let $A = \begin{bmatrix} 2 & 1 & -1 \\ 1 & 1 & 1 \\ 1 & -2 & -1 \end{bmatrix}$   $B = \begin{bmatrix} 1 & 2 & 1 \\ 1 & -1 & -1 \\ 2 & 2 & -1 \end{bmatrix}$

(a) Compute:
  (1) $A + 2B$
  (2) $AB$
  (3) $BA$
  (4) $A^{-1}$ using the Gauss-Jordan procedure. Check by showing $AA^{-1} = I$. Using your result, solve the system of equations:
$$2x + y - z = -1$$
$$x + y + z = 2$$
$$x - 2y - z = 1$$
  (5) $B^{-1}$ using the Gauss-Jordan procedure. Check by showing $BB^{-1} = I$. Using this result, solve the system of equations:
$$x + 2y + z = 1$$
$$x - y - z = 0$$
$$2x + 2y - z = -2$$
  (6) Using the result for $AB$ computed in (2) compute $(AB)B^{-1}$. How could you have predicted this result?

(b) Solve for the matrix $X$:
  (1) $A + X = 2B$
  (2) $2A + X = B$
  (3) $A + 2X = 2A + B$
  (4) $AX = A + B$

**2.** Let $e = \begin{bmatrix} 1 & 0 \\ 0 & 1 \end{bmatrix}$   $r = \begin{bmatrix} 0 & 1 \\ 1 & 0 \end{bmatrix}$

$p = \begin{bmatrix} w^2 & 0 \\ 0 & w \end{bmatrix}$   $s = \begin{bmatrix} 0 & w^2 \\ w & 0 \end{bmatrix}$

$q = \begin{bmatrix} w & 0 \\ 0 & w^2 \end{bmatrix}$   $t = \begin{bmatrix} 0 & w \\ w^2 & 0 \end{bmatrix}$

where $w \neq 1$, $w^3 = 1$.

(a) Fill in the table with the computed products.

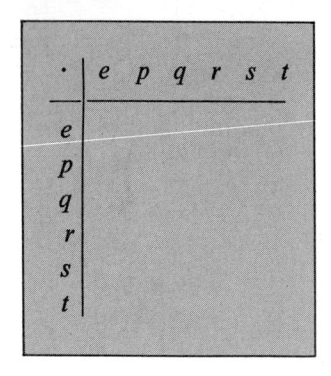

| · | e | p | q | r | s | t |
|---|---|---|---|---|---|---|
| e |   |   |   |   |   |   |
| p |   |   |   |   |   |   |
| q |   |   |   |   |   |   |
| r |   |   |   |   |   |   |
| s |   |   |   |   |   |   |
| t |   |   |   |   |   |   |

(b) Show that the mathematical system $\{\{e, p, q, r, s, t\}, \cdot\}$ is a noncommutative group.

3. Let $a = \begin{bmatrix} 0 & 0 \\ 0 & 0 \end{bmatrix}$   $c = \begin{bmatrix} 0 & 1 \\ 0 & 0 \end{bmatrix}$

$b = \begin{bmatrix} 1 & 0 \\ 0 & 0 \end{bmatrix}$   $d = \begin{bmatrix} 1 & 1 \\ 0 & 0 \end{bmatrix}$

(a) Fill in the table with the computed products.

| · | a | b | c | d |
|---|---|---|---|---|
| a |   |   |   |   |
| b |   |   |   |   |
| c |   |   |   |   |
| d |   |   |   |   |

(b) Fill in the table with the computed sums taken mod 2.

| + | a | b | c | d |
|---|---|---|---|---|
| a |   |   |   |   |
| b |   |   |   |   |
| c |   |   |   |   |
| d |   |   |   |   |

(c) Show that $\{\{a, b, c, d\}, +(\bmod 2), \cdot\}$ is a noncommutative ring.

# Answers and Solutions to Selected Exercises

## CHAPTER 1

*Exercises 1.1 (Page 3)*

1. Some interpretations for the product $a \times b$ are

| $a$ | $b$ | Interpretation of $a \times b$ |
|---|---|---|
| (i) Number of gallons of gas | Cost per gallon | Cost of $a$ gallons of gas |
| (ii) Number of coins of a certain kind | Value in cents of each coin | Value in cents of $a$ coins |
| (iii) Number of desks | Weight in pounds of each desk | Weight in pounds of $a$ desks |

Many other interpretations are possible.

2. (a) 4  (b) 3  (c) $-3$  (d) $-2\frac{2}{3}$  (e) $\frac{b}{2}$  (f) $\frac{6}{a}$  (g) $-\frac{b}{a}$
 (h) $a \times \left(-\frac{b}{a}\right) + b \stackrel{?}{=} 0$   Yes; $-b + b = 0$

3. (a) The decimal part is .25, whereas the whole number part is the product of the original whole number and a number one more.
 (b) $(n + .5)(n + .5) = n(n + 1) + .25$
   Proof: $(n + .5)(n + .5) = n^2 + n + (.5)^2$
   $= n(n + 1) + .25$

4. (a) and (b) $(n + .6)(n + .4) = n^2 + .6n + .4n + (.6)(.4)$
   $= n^2 + n + .24$
   $= n(n + 1) + .24$

371

**Answers and Solutions to Selected Exercises**

5. (a) and (b) $(n + .7)(n + .3) = n^2 + .7n + .3n + (.7)(.3)$
$= n^2 + n + .21$
$= n(n + 1) + .21$

6. Let $x + y = 1$. Then $(n + x)(n + y) = n^2 + nx + ny + xy$
$= n^2 + n(x + y) + xy$
$= n(n + 1) + xy$

7. (a) The differences are divisible by 9.
   (b) The difference between a 2-digit number and the number for its digits reversed is a multiple of 9.
   *Proof*: $(10t + u) - (10u + t) = 10t + u - 10u - t$
   $= 9t - 9u$
   $= 9(t - u)$

8. (a) The differences are divisible by 9.
   (b) The difference between any 3-digit number and the number for a shuffling or permutation of its digits is a multiple of 9.
   *Proof*: $100h + 10t + u = h + 99h + t + 9t + u$
   $= 9(11h + t) + (h + t + u)$
   Let $x, y, z$ be $h, t, u$ in some order.
   Then $100x + 10y + z = x + 99x + y + 9y + z$
   $= 9(11x + y) + (x + y + z)$
   Hence $(100h + 10t + u) - (100x + 10y + z)$
   $= 9(11h + t) + (h + t + u) - 9(11x + y) - (x + y + z)$
   $= 9(11h + t) - 9(11x + y)$ as $h + t + u = x + y + z$
   $= 9(11h + t - 11x - y)$

9. $2^2 = 4 = 3(1) + 1$
   $3^2 = 9 = 3(3) + 0$
   $4^2 = 16 = 3(5) + 1$
   $5^2 = 25 = 3(8) + 1$
   $6^2 = 36 = 3(12) + 0$
   (a) The remainder is either 0 or 1.
   (b) If the integer is $3n$, then $(3n)^2 = 9n^2 = 3(3n^2) + 0$     (Remainder is 0)
   If the integer is $3n + 1$,
   then $(3n + 1)^2 = 9n^2 + 6n + 1 = 3(3n^2 + 2n) + 1$     (Remainder is 1)
   If the integer is $3n - 1$,
   then $(3n - 1)^2 = 9n^2 - 6n + 1 = 3(3n^2 - 2n) + 1$     (Remainder is 1)

## Exercises 1.2 (Page 6)

1. (a) $n \to 5n + 1$
   $n \to 2^n$
   $n \to \dfrac{1}{1+1} + \dfrac{1}{1+2} + \dfrac{1}{1+3} + \cdots + \dfrac{1}{1+n}$
   (b) $(x, y) \to x^2 + xy + y^2$
   $(x, y) \to 2^x$
   $(x, y) \to (1 + 1) + (1 + 2) + (1 + 3) + \cdots + (1 + x + y)$

(c) $(x, y, z) \to x + y + z$
$(x, y, z) \to \dfrac{x + y + z}{3}$
$(x, y, z) \to z$
The rules in (a) define unary operations. The rules in (b) define binary operations.

2. (a) The distance of the point from the line
   (b) The line passing through the point crossing the line at right angles
   (c) The point of intersection of the lines in (b) (the foot of the perpendicular)
   These rules define binary operations.

3. (a) The point of intersection
   (b) The number of degrees in the smallest of the angles formed
   (c) The line perpendicular to both lines at their point of intersection
   These rules define binary operations.

4. (a) Its area, $a^2$; its perimeter, $4a$; the length of a diagonal, $a\sqrt{2}$
   (b) Its area, $ab$; its perimeter, $2a + 2b$; the length of a diagonal, $\sqrt{a^2 + b^2}$
   (c) Its area, $\pi r^2$; its circumference $2\pi r$; its diameter, $2r$
   The rules in (a) define unary operations. The rules in (b) define binary operations.
   The rules in (c) define unary operations.

## Exercises 1.3 (Page 9)

1. On, unary      2. In, unary       3. On, unary      4. On, unary
5. On, binary     6. On, unary       7. On, binary     8. On, unary
9. On, unary     10. On, binary     11. On, binary    12. On, neither (ternary)
13. On, binary   14. On, binary     15. In, binary ($a$ must be $\geq b$)
16. In, binary ($b \neq 0$)

## Exercises 1.4 (Page 14)

1.

| +(mod 5) | 0 | 1 | 2 | 3 | 4 |
|---|---|---|---|---|---|
| 0 | 0 | 1 | 2 | 3 | 4 |
| 1 | 1 | 2 | 3 | 4 | 0 |
| 2 | 2 | 3 | 4 | 0 | 1 |
| 3 | 3 | 4 | 0 | 1 | 2 |
| 4 | 4 | 0 | 1 | 2 | 3 |

(a) 2   (b) 2   (c) 2   (d) 4   (e) 0   (f) 3   (g) 3   (h) 2
(i) 3   (j) 4

2. (a) 0   (b) 3   (c) 0   (d) 3   (e) 1   (f) 7   (g) 6   (h) 5
   (i) 0

3. (a) 0   (b) 1   (c) 2   (d) 3   (e) 4   (f) 5   (g) 6   (h) 7
   (i) 12

# Answers and Solutions to Selected Exercises

**4.** (a) 1  (b) 0  (c) 6  (d) 5  (e) 3  (f) 6  (g) 5  (h) 4  (i) 5

**5.** (a) 1  (b) 0  (c) 7  (d) 6  (e) 4  (f) 7  (g) 6  (h) 5  (i) 6

**6.** (a) 6  (b) 6  (c) 5  (d) 5  (e) 3  (f) 3  (g) 4  (h) 4

(i)

| ×(mod 7) | 0 | 1 | 2 | 3 | 4 | 5 | 6 |
|---|---|---|---|---|---|---|---|
| 0 | 0 | 0 | 0 | 0 | 0 | 0 | 0 |
| 1 | 0 | 1 | 2 | 3 | 4 | 5 | 6 |
| 2 | 0 | 2 | 4 | 6 | 1 | 3 | 5 |
| 3 | 0 | 3 | 6 | 2 | 5 | 1 | 4 |
| 4 | 0 | 4 | 1 | 5 | 2 | 6 | 3 |
| 5 | 0 | 5 | 3 | 1 | 6 | 4 | 2 |
| 6 | 0 | 6 | 5 | 4 | 3 | 2 | 1 |

**7.** (a) 8  (b) 10  (c) 12  (d) 10  (e) 12  (f) 10  (g) 10  (h) 10

**8.**

| +(mod 10) | 0 | 1 | 2 | 3 | 4 | 5 | 6 | 7 | 8 | 9 |
|---|---|---|---|---|---|---|---|---|---|---|
| 0 | 0 | 1 | 2 | 3 | 4 | 5 | 6 | 7 | 8 | 9 |
| 1 | 1 | 2 | 3 | 4 | 5 | 6 | 7 | 8 | 9 | 0 |
| 2 | 2 | 3 | 4 | 5 | 6 | 7 | 8 | 9 | 0 | 1 |
| 3 | 3 | 4 | 5 | 6 | 7 | 8 | 9 | 0 | 1 | 2 |
| 4 | 4 | 5 | 6 | 7 | 8 | 9 | 0 | 1 | 2 | 3 |
| 5 | 5 | 6 | 7 | 8 | 9 | 0 | 1 | 2 | 3 | 4 |
| 6 | 6 | 7 | 8 | 9 | 0 | 1 | 2 | 3 | 4 | 5 |
| 7 | 7 | 8 | 9 | 0 | 1 | 2 | 3 | 4 | 5 | 6 |
| 8 | 8 | 9 | 0 | 1 | 2 | 3 | 4 | 5 | 6 | 7 |
| 9 | 9 | 0 | 1 | 2 | 3 | 4 | 5 | 6 | 7 | 8 |

(a) 0  (b) 1  (c) 1  (d) 5  (e) 5  (f) 5  (g) 5  (h) 5  (i) 5  (j) 5

**9.** (a) Sunday  (b) Monday  (c) Monday  (d) Friday  (e) Wednesday  (f) Saturday  (g) Friday  (h) Tuesday

## Exercises 1.5 (Page 23)

**1.** (a)

| +(mod 2) | 0 | 1 |
|---|---|---|
| 0 | 0 | 1 |
| 1 | 1 | 0 |

| ×(mod 2) | 0 | 1 |
|---|---|---|
| 0 | 0 | 0 |
| 1 | 0 | 1 |

(b)

| +(mod 3) | 0 | 1 | 2 |
|---|---|---|---|
| 0 | 0 | 1 | 2 |
| 1 | 1 | 2 | 0 |
| 2 | 2 | 0 | 1 |

| ×(mod 3) | 0 | 1 | 2 |
|---|---|---|---|
| 0 | 0 | 0 | 0 |
| 1 | 0 | 1 | 2 |
| 2 | 0 | 2 | 1 |

Answers and Solutions to Selected Exercises 375

(c)

| +(mod 6) | 0 | 1 | 2 | 3 | 4 | 5 |
|---|---|---|---|---|---|---|
| 0 | 0 | 1 | 2 | 3 | 4 | 5 |
| 1 | 1 | 2 | 3 | 4 | 5 | 0 |
| 2 | 2 | 3 | 4 | 5 | 0 | 1 |
| 3 | 3 | 4 | 5 | 0 | 1 | 2 |
| 4 | 4 | 5 | 0 | 1 | 2 | 3 |
| 5 | 5 | 0 | 1 | 2 | 3 | 4 |

| ×(mod 6) | 0 | 1 | 2 | 3 | 4 | 5 |
|---|---|---|---|---|---|---|
| 0 | 0 | 0 | 0 | 0 | 0 | 0 |
| 1 | 0 | 1 | 2 | 3 | 4 | 5 |
| 2 | 0 | 2 | 4 | 0 | 2 | 4 |
| 3 | 0 | 3 | 0 | 3 | 0 | 3 |
| 4 | 0 | 4 | 2 | 0 | 4 | 2 |
| 5 | 0 | 5 | 4 | 3 | 2 | 1 |

(d)

| +(mod 7) | 0 | 1 | 2 | 3 | 4 | 5 | 6 |
|---|---|---|---|---|---|---|---|
| 0 | 0 | 1 | 2 | 3 | 4 | 5 | 6 |
| 1 | 1 | 2 | 3 | 4 | 5 | 6 | 0 |
| 2 | 2 | 3 | 4 | 5 | 6 | 0 | 1 |
| 3 | 3 | 4 | 5 | 6 | 0 | 1 | 2 |
| 4 | 4 | 5 | 6 | 0 | 1 | 2 | 3 |
| 5 | 5 | 6 | 0 | 1 | 2 | 3 | 4 |
| 6 | 6 | 0 | 1 | 2 | 3 | 4 | 5 |

| ×(mod 7) | 0 | 1 | 2 | 3 | 4 | 5 | 6 |
|---|---|---|---|---|---|---|---|
| 0 | 0 | 0 | 0 | 0 | 0 | 0 | 0 |
| 1 | 0 | 1 | 2 | 3 | 4 | 5 | 6 |
| 2 | 0 | 2 | 4 | 6 | 1 | 3 | 5 |
| 3 | 0 | 3 | 6 | 2 | 5 | 1 | 4 |
| 4 | 0 | 4 | 1 | 5 | 2 | 6 | 3 |
| 5 | 0 | 5 | 3 | 1 | 6 | 4 | 2 |
| 6 | 0 | 6 | 5 | 4 | 3 | 2 | 1 |

2. (a)

| −(mod 2) | 0 | 1 |
|---|---|---|
| 0 | 0 | 1 |
| 1 | 1 | 0 |

| ÷(mod 2) | 0 | 1 |
|---|---|---|
| 0 | — | 0 |
| 1 | — | 1 |

(b)

| −(mod 3) | 0 | 1 | 2 |
|---|---|---|---|
| 0 | 0 | 2 | 1 |
| 1 | 1 | 0 | 2 |
| 2 | 2 | 1 | 0 |

| ÷(mod 3) | 0 | 1 | 2 |
|---|---|---|---|
| 0 | — | 0 | 0 |
| 1 | — | 1 | 2 |
| 2 | — | 2 | 1 |

(c)

| −(mod 6) | 0 | 1 | 2 | 3 | 4 | 5 |
|---|---|---|---|---|---|---|
| 0 | 0 | 5 | 4 | 3 | 2 | 1 |
| 1 | 1 | 0 | 5 | 4 | 3 | 2 |
| 2 | 2 | 1 | 0 | 5 | 4 | 3 |
| 3 | 3 | 2 | 1 | 0 | 5 | 4 |
| 4 | 4 | 3 | 2 | 1 | 0 | 5 |
| 5 | 5 | 4 | 3 | 2 | 1 | 0 |

| ÷(mod 6) | 0 | 1 | 2 | 3 | 4 | 5 |
|---|---|---|---|---|---|---|
| 0 | — | 0 | 0, 3 | 0, 2, 4 | 0, 3 | 0 |
| 1 | — | 1 | — | — | — | 5 |
| 2 | — | 2 | 1, 4 | — | 2, 5 | 4 |
| 3 | — | 3 | — | 1, 3, 5 | — | 3 |
| 4 | — | 4 | 2, 5 | — | 1, 4 | 2 |
| 5 | — | 5 | — | — | — | 1 |

(d)

| −(mod 7) | 0 | 1 | 2 | 3 | 4 | 5 | 6 |
|---|---|---|---|---|---|---|---|
| 0 | 0 | 6 | 5 | 4 | 3 | 2 | 1 |
| 1 | 1 | 0 | 6 | 5 | 4 | 3 | 2 |
| 2 | 2 | 1 | 0 | 6 | 5 | 4 | 3 |
| 3 | 3 | 2 | 1 | 0 | 6 | 5 | 4 |
| 4 | 4 | 3 | 2 | 1 | 0 | 6 | 5 |
| 5 | 5 | 4 | 3 | 2 | 1 | 0 | 6 |
| 6 | 6 | 5 | 4 | 3 | 2 | 1 | 0 |

| ÷(mod 7) | 0 | 1 | 2 | 3 | 4 | 5 | 6 |
|---|---|---|---|---|---|---|---|
| 0 | — | 0 | 0 | 0 | 0 | 0 | 0 |
| 1 | — | 1 | 4 | 5 | 2 | 3 | 6 |
| 2 | — | 2 | 1 | 3 | 4 | 6 | 5 |
| 3 | — | 3 | 5 | 1 | 6 | 2 | 4 |
| 4 | — | 4 | 2 | 6 | 1 | 5 | 3 |
| 5 | — | 5 | 6 | 4 | 3 | 1 | 2 |
| 6 | — | 6 | 3 | 2 | 5 | 4 | 1 |

3. (a) 5  (b) 2  (c) 4  (d) 4  (e) 3  (f) 6  (g) 2  (h) 5
   (i) 4

4. (a) 6  (b) 2  (c) 5  (d) 1  (e) 1  (f) 1  (g) 2  (h) 2

5. (a) 0  (b) 5  (c) 4  (d) 4  (e) 0  (f) 5  (g) 1  (h) 4
   (i) 3  (j) 3  (k) 2  (l) 2  (m) 1  (n) 4  (o) 6  (p) 2

6. (a) $a = a + 0n \Rightarrow a \equiv a \pmod{n}$
   (b) $a = b + kn \Rightarrow b = a + (-k)n \Rightarrow b \equiv a \pmod{n}$
   (c) $\left.\begin{array}{r} a = b + kn \\ b = c + k'n \end{array}\right\} \Rightarrow a = c + (k + k')n \Rightarrow a \equiv c \pmod{n}$
   (d) $a = b + kn \Rightarrow -a = -b + (-k)n \Rightarrow -a \equiv -b \pmod{n}$
   (e) (i) If $a = b + kn$ and $c = d + k'n$, then $a + c = b + d + (k + k')n$ and hence
       $a + c \equiv b + d \pmod{n}$
       (ii) and $a - c = b - d + (k - k')n$
       $a - c \equiv b - d \pmod{n}$
       (iii) $ac = (b + kn)(d + k'n)$
       $= bd + (kd + k'b + kk'n)n$
       $\therefore ac \equiv bd \pmod{n}$
       (iv) Counterexample suffices.
       $8 \equiv 6 \pmod{2}$
       $4 \equiv 2 \pmod{2}$
       Yet $\frac{8}{4} \not\equiv \frac{6}{2} \pmod{2}$ as $2 \not\equiv 3 \pmod{2}$

## Exercises 1.6 (Page 25)

1.

| | | + | − | × | ÷ |
|---|---|---|---|---|---|
| (a) | $\{0, \frac{1}{2}, 1, 2\}$ | No | No | No | No |
| (b) | $\{0, 1, -1\}$ | No | No | Yes | No |
| (c) | {Odd integers} | No | No | Yes | No |
| (d) | {Even integers} | Yes | Yes | Yes | No |
| (e) | {Multiples of 5} | Yes | Yes | Yes | No |
| (f) | $\{1, -1\}$ | No | No | Yes | Yes |
| (g) | {Positive rationals} | Yes | No | Yes | Yes |
| (h) | {Real numbers} | Yes | Yes | Yes | No |

2. (a) $W_5$ is closed with respect to $+\pmod 5$, $-\pmod 5$, and $\times\pmod 5$, but is not closed with respect to $\div\pmod 5$.
   (b) $W_6$ is closed with respect to $+\pmod 6$, $-\pmod 6$, and $\times\pmod 6$, but is not closed with respect to $\div\pmod 6$.
   (c) $W_7^*$ is closed with respect to $\times\pmod 7$ and $\div\pmod 7$, but is not closed with respect to $+\pmod 7$ and $-\pmod 7$.

|  | Whole Numbers | Integers | Rational Numbers | Real Numbers | Positive Real Numbers | Negative Real Numbers |
|---|---|---|---|---|---|---|
| (a) $n \to n+2$ | Yes | Yes | Yes | Yes | Yes | No |
| (b) $n \to n-2$ | No | Yes | Yes | Yes | No | Yes |
| (c) $n \to 3n$ | Yes | Yes | Yes | Yes | Yes | Yes |
| (d) $n \to \frac{n}{3}$ | No | No | Yes | Yes | Yes | Yes |
| (e) $n \to n^2$ | Yes | Yes | Yes | Yes | Yes | No |
| (f) $n \to \sqrt{n}$ | No | No | No | No | Yes | No |
| (g) $n \to 2^n$ | Yes | No | No | Yes | Yes | No |
| (h) $(a,b) \to a$ | Yes | Yes | Yes | Yes | Yes | Yes |
| (i) $(a,b) \to 2a+b$ | Yes | Yes | Yes | Yes | Yes | Yes |
| (j) $(a,b) \to 2a-b$ | No | Yes | Yes | Yes | No | No |
| (k) $(a,b) \to a^2+b^2$ | Yes | Yes | Yes | Yes | Yes | No |
| (l) $(a,b) \to a^2-b^2$ | No | Yes | Yes | Yes | No | No |
| (m) $(a,b) \to a^2+b^2-2ab$ | Yes | Yes | Yes | Yes | Yes | No |
| (n) $(a,b) \to 2^{a+b}$ | Yes | No | No | Yes | Yes | No |
| (o) $(a,b) \to |a+b|$ | Yes | Yes | Yes | Yes | Yes | No |
| (p) $(a,b) \to |a|-|b|$ | No | Yes | Yes | Yes | No | No |

## Exercises 1.7 (Page 31)

These are commutative: 1, 4, 5, 7, 8, 11, 12, 16, 17, 19, 20, 21.
These are not commutative: 2, 3, 6, 9, 10, 13, 14, 15, 18.

## Exercises 1.8 (Page 34)

These are associative triples (the others are not): 1, 2, 4, 8.
These are associative operations *in* S (the others are not): 9, 11, 12, 16, 21, 23, 25, 26, 27, 28, 30, 31.
The following are associative *on* S: 9, 11, 12, 16, 21, 23, 25, 26, 27, 28. (Observe that 30 and 31 are associative *in* S but not *on* S.)

## Exercises 1.9 (Page 39)

The following have identities as noted. The others do not.
1. 4    2. 6    5. 8    6. 1    8. 6    9. 7    11. 16    12. 5
13. 9    15. ∅    16. The given set    17. ∅    19. 1    20. $c$
21. $-2$    22. 0

## Exercises 1.10 (Page 48)

1. 6 is the identity; $2' = 8$, $4' = 4$, $6' = 6$, $8' = 2$
2. 1 is the identity; $1' = 1$, $3' = 3$
3. 0 is the identity; $0' = 0$, $1' = 3$, $2' = 2$, $3' = 1$
4. 4 is the identity; $2' = 2$, $4' = 4$
5. 0 is the identity; $0' = 0$, $1' = 4$, $2' = 3$, $3' = 2$, $4' = 1$
6. 1 is the identity; $1' = 1$, $2' = 3$, $3' = 2$, $4' = 4$
7. 0 is the identity; $a' = -a$
8. 1 is the identity; $\left(\frac{a}{b}\right)' = \frac{b}{a}$ provided $ab \neq 0$
9. 8 is the identity; $2' = 4$, $4' = 2$, $6' = 6$, $8' = 8$, $10' = 12$, $12' = 10$
10. 4 is the identity; $4' = 4$, $8' = 8$
11. No identity; no inverses
12. 10 is the identity; $5' = 5$, $10' = 10$
13. 7 is the identity; $7' = 7$
14. 1 is the identity; $\left(\frac{a}{b}\right)' = \frac{b}{a}$
15. 1 is the identity; $1' = 1$, $(-1)' = -1$
16. 16 is the identity; $4' = 4$, $8' = 12$, $12' = 8$, $16' = 16$
17. $c$ is the identity; $a' = b$, $b' = a$, $c' = c$, $d' = d$
18. $-1$ is the identity; $a' = -a - 2$
19. 0 is the identity; $b' = \frac{-b}{b+1}$ ($b \neq 1$). $-1$ has no inverse.
20. 0 is the identity; $0' = 0$, $2' = 4$, $3' = 3$, $4' = 2$

21. Theorem 1.3 (part b)
    $a \circ c = a \circ c$     $a \circ c$ names a unique element; equality is reflexive
    $a \circ c = b \circ c$     Replacement, $a = b$

22. Theorem 1.4 (part b)
    $x \circ a = b$                    Given
    $(x \circ a) \circ a' = b \circ a'$     Theorem 1.3b
    $x \circ (a \circ a') = b \circ a'$     Associativity
    $x \circ e = b \circ a'$            Replacement, $a \circ a' = e$
    $x = b \circ a'$                   Replacement, $x \circ e = x$ for all $x$
    Hence, if $x \circ a = b$ has a solution, it must be $b \circ a'$. But
    $(b \circ a') \circ a = b \circ (a' \circ a)$   Associativity
    $= b \circ e$                      Replacement, $a' \circ a = e$
    $= b$                              Replacement, $b \circ e = b$
    Hence $b \circ a'$ is a solution and the only one.

23. $a = b$ and $c = d$     Given
    $a \circ c = a \circ c$     Reflexivity
    $a \circ c = b \circ d$     Replacement

## Review Exercises (Page 49)

1.

|     | a  | b             | c             |
|-----|----|---------------|---------------|
| (a) | 11 | 60            | 61            |
| (b) | 13 | 84            | 85            |
| ★(c)| a  | $\frac{a^2-1}{2}$ | $\frac{a^2+1}{2}$ |

2. The following are operations *on* the given set: (a), (d).

3. (a) 2   (b) 8   (c) 9   (d) 3

4.

|             | (1) | (2) | (3) | (4) | (5) |
|-------------|-----|-----|-----|-----|-----|
| Closed      | Yes | Yes | Yes | Yes | Yes |
| Commutative | Yes | Yes | Yes | No  | No  |
| Associative | Yes | Yes | No* | Yes | No† |
| Identity    | No  | Yes | Yes | Yes | Yes |
| Inverse     | No  | Yes | No  | Yes | Yes |

*$(ab)c \neq a(bc)$   †$(tz)t \neq t(zt)$

# CHAPTER 2

## Exercises 2.2 (Page 60)

1. (a) Each element is its own inverse.
   (b) (1) $\{1, 3, 5, 7\}$   (2) $\{\ \}$   (3) $\{5\}$

2. (a) $\{S, A\}$   (b) $\{\ \}$   (c) $\{R, L\}$   (d) $\{\ \}$   (e) $\{R\}$   (f) $\{L\}$
   (g) $\{R, L\}$   (h) $\{\ \}$   (i) $\{A\}$   (j) $\{S, A\}$   (k) $\{L, R\}$   (l) $\{\ \}$

5. (a) $\{e, r, s, t\}$   (b) $\{q\}$   (c) $\{p\}$   (d) $\{\ \}$   (e) $\{\ \}$   (f) $\{\ \}$
   (g) $\{r' \circ t \circ s' = t\}$   (h) $\{s' \circ t \circ r' = t\}$   (i) $\{\ \}$   (j) $\{t\}$   (k) $\{s\}$
   (l) $\{\ \}$   (m) $\{p\}$   (n) $\{q\}$   (o) $\{\ \}$   (p) $\{e, r\}$   (q) $\{p\}$
   (r) $\{q\}$

6. 

| | Closed | Associative | Identity | Inverse | A Group |
|---|---|---|---|---|---|
| (a) | Yes | Yes | Yes | No | No |
| (b) | Yes | Yes | No | No | No |
| (c) | Yes | Yes | Yes | No | No |
| (d) | Yes | Yes | Yes | Yes | Yes |
| (e) | Yes | Yes | Yes | Yes | Yes |
| (f) | Yes | Yes | Yes | Yes | Yes |
| (g) | Yes | Yes | No | No | No |
| (h) | Yes | Yes | Yes | No | No |
| (i) | Yes | Yes | Yes | Yes | Yes |

## Exercises 2.3 (Page 66)

1. (a) $(p,r), (p,s), (p,t), (q,r), (q,s), (q,t), (r,s), (r,t), (s,t)$
   (e) The following are true and the rest are false: (1), (2), (4), (6), (8).
   (f) (1) $\{e\}$   (2) $\{\ \}$   (3) $\{\ \}$   (4) $\{\ \}$   (5) $\{t, q\}$
       (6) $\{r, s, t\}$   (7) $\{\ \}$   (8) $\{r, s, t\}$

2. (a) $\begin{pmatrix} 1 & 2 & 3 & 4 \\ 3 & 4 & 1 & 2 \end{pmatrix}$   (b) $\begin{pmatrix} 1 & 2 & 3 & 4 \\ 4 & 1 & 2 & 3 \end{pmatrix}$   (c) $\begin{pmatrix} 1 & 2 & 3 & 4 \\ 3 & 1 & 4 & 2 \end{pmatrix}$

   (d) $\begin{pmatrix} 1 & 2 & 3 & 4 \\ 3 & 4 & 1 & 2 \end{pmatrix}$   (e) $\begin{pmatrix} 1 & 2 & 3 & 4 \\ 3 & 4 & 1 & 2 \end{pmatrix}$

   (f) $\left[ \begin{pmatrix} 1 & 2 & 3 & 4 \\ 4 & 1 & 2 & 3 \end{pmatrix} \circ \begin{pmatrix} 1 & 2 & 3 & 4 \\ 2 & 3 & 4 & 1 \end{pmatrix} \right] \circ \begin{pmatrix} 1 & 2 & 3 & 4 \\ 3 & 4 & 1 & 2 \end{pmatrix}$

   $= \begin{pmatrix} 1 & 2 & 3 & 4 \\ 1 & 2 & 3 & 4 \end{pmatrix} \circ \begin{pmatrix} 1 & 2 & 3 & 4 \\ 3 & 4 & 1 & 2 \end{pmatrix} = \begin{pmatrix} 1 & 2 & 3 & 4 \\ 3 & 4 & 1 & 2 \end{pmatrix}$

   $= \begin{pmatrix} 1 & 2 & 3 & 4 \\ 4 & 1 & 2 & 3 \end{pmatrix} \circ \left[ \begin{pmatrix} 1 & 2 & 3 & 4 \\ 2 & 3 & 4 & 1 \end{pmatrix} \circ \begin{pmatrix} 1 & 2 & 3 & 4 \\ 3 & 4 & 1 & 2 \end{pmatrix} \right]$

   $= \begin{pmatrix} 1 & 2 & 3 & 4 \\ 4 & 1 & 2 & 3 \end{pmatrix} \circ \begin{pmatrix} 1 & 2 & 3 & 4 \\ 4 & 1 & 2 & 3 \end{pmatrix} = \begin{pmatrix} 1 & 2 & 3 & 4 \\ 3 & 4 & 1 & 2 \end{pmatrix}$

   (g) Let $U, V, W$ be any three permutations which we shall regard as the instructions. Let $D$ represent a dance group on a dance floor initially occupying positions

   | 1 | 2 | 3 | 4 |

   on the dance floor. Interpret $W(D)$ to be the dance group $D$ in positions after carrying out the instruction $W$. Let $V \circ W$ be the instruction equivalent to carrying out first instruction $W$, then instruction $V$. It now follows that

   $$(V \circ W)(D) = V(W(D))$$

   Also,

   $$[U \circ (V \circ W)](D) = U((V \circ W)(D))$$
   $$= U(V(W(D)))$$

   Similarly,

   $$[(U \circ V) \circ W](D) = (U \circ V)(W(D))$$
   $$= U(V(W(D)))$$

Hence
$$[U \circ (V \circ W)](D) = [(U \circ V) \circ W](D)$$
and
$$U \circ (V \circ W) = (U \circ V) \circ W$$

(*Note:* The reasoning used here applies quite generally to any "composition of transformations.")

## Exercises 2.4 (Page 72)

1. (a) $-(-n) = n$ for any integer $n$
   (b) $\dfrac{1}{\frac{1}{r}} = r$ for any positive rational number $r$
   (c) $\left[\begin{pmatrix} 1 & 2 & 3 & 4 \\ 4 & 3 & 1 & 2 \end{pmatrix}'\right]' = \begin{pmatrix} 1 & 2 & 3 & 4 \\ 4 & 3 & 1 & 2 \end{pmatrix}$

2. (a) $(a')' \circ [a' \circ a] = [(a')' \circ a'] \circ a$     $\circ$ is associative
          $(a')' \circ e = e \circ a$               $x' \circ x = e$ for all $x$
          $(a')' = a$                     $e \circ x = x \circ e = x$ for all $x$

   (b) By definition, the inverse of $a'$ is $(a')'$. Since $a \circ a' = a' \circ a = e$, the inverse of $a'$ is $a$. Hence both $(a')'$ and $a$ are inverses of $a'$. Since $a'$ has but one inverse, $(a')' = a$.

3. (a) $-(x+y) = -y + (-x)$
   (b) $\dfrac{1}{xy} = \dfrac{1}{y} \cdot \dfrac{1}{x}$

4. (a) (1) $(a \circ b)' \circ [(a \circ b) \circ (b' \circ a')]$
          $= [(a \circ b)' \circ (a \circ b)] \circ (b' \circ a')$     $\circ$ is associative
      (2) $(a \circ b)' \circ [\{(a \circ b) \circ b'\} \circ a']$       $\circ$ is associative and
          $= e \circ (b' \circ a')$                     $x' \circ x = e$ for all $x$
      (3) $(a \circ b)' \circ [\{a \circ (b \circ b')\} \circ a']$
          $= (b' \circ a')$                          Same as (2)
      (4) $(a \circ b)' \circ [\{a \circ e\} \circ a'] = b' \circ a$    $x \circ x' = e$ for all $x$
      (5) $(a \circ b)' \circ [a \circ a'] = b' \circ a'$        $x \circ e = x$ for all $x$
      (6) $(a \circ b)' \circ e = b' \circ a'$            $x \circ x' = e$ for all $x$
      (7)      $(a \circ b)' = b' \circ a'$             $x \circ e = x$ for all $x$

   (b) By definition, the inverse of $(a \circ b)$ is $(a \circ b)'$.
   Also $(a \circ b) \circ (b' \circ a')$
   $= [(a \circ b) \circ b'] \circ a'$             $\circ$ is associative
   $= [a \circ (b \circ b')] \circ a$             "
   $= [a \circ e] \circ a'$                    $x \circ x' = e$ for all $x$
   $= a \circ a'$                           $x \circ e = x$ for all $x$
   $= e$                                $x \circ x' = e$ for all $x$
   Similarly, $(b' \circ a') \circ (a \circ b) = e$. Hence $(b' \circ a')$ is an inverse of $a \circ b$. But $a \circ b$ has but one inverse. Hence $(a \circ b)' = b' \circ a'$.

7.  $(a \circ b) \circ (a \circ b) = (a \circ a) \circ (b \circ b)$     Given
    $a \circ [b \circ (a \circ b)] = a \circ [a \circ (b \circ b)]$     $\circ$ is associative
    $b \circ (a \circ b) = a \circ (b \circ b)$     Left cancellation
    $(b \circ a) \circ b = (a \circ b) \circ b$     $\circ$ is associative
    $b \circ a = a \circ b$     Right cancellation

11. (a) $\{\{1\}, \times \pmod 8)\}$, $\{\{1, 3\}, \times \pmod 8)\}$
    $\{\{1, 5\}, \pmod 8)\}$, $\{\{1, 7\}, \times \pmod 8)\}$
    as well as $\{\{1, 3, 5, 7\}, \pmod 8)\}$
    (b) $\{\{6\}, \times \pmod{10})\}$, $\{\{4, 6\}, \times \pmod{10})\}$
    $\{\{2, 4, 6, 8\}, \times \pmod{10})\}$
    (c) $\{\{e\}, \circ\}$, $\{\{e, r\}, \circ\}$, $\{\{e, s\}, \circ\}$
    $\{\{e, t\}, \circ\}$, $\{\{e, p, q\}, \circ\}$
    and $\{\{e, p, q, r, s, t\}, \circ\}$

13. For each element $x$ in a group we must have either $x = x'$ or $x \neq x'$. If $x \neq x'$ remove both $x$ and $x'$ from the group. This removes an even number of elements. The remaining elements must be even in number since the set has an even number of elements. But the remaining ones are those which are their own inverses, that is, $x = x'$.

## Exercises 2.5 (Page 81)

1.  (c) Let $a \leftrightarrow a^3$ be the one-to-one correspondence. Then
    $$xy \leftrightarrow (x \cdot y)^3 = x^3 \cdot y^3$$
    so that $xy \leftrightarrow x^3 \cdot y^3$.
    (d) Let $a \leftrightarrow -a$ be the one-to-one correspondence. Then
    $$x + y \leftrightarrow -(x+y) = (-x) + (-y)$$
    so that $x + y \leftrightarrow (-x) + (-y)$.
    (e) Let $a \leftrightarrow \dfrac{1}{a}$ be the one-to-one correspondence. Then
    $$x \cdot y \leftrightarrow \frac{1}{x \cdot y} = \frac{1}{x} \cdot \frac{1}{y}$$
    so that $x \cdot y \leftrightarrow \dfrac{1}{x} \cdot \dfrac{1}{y}$.
    (g) Let $a + b\sqrt{2} \leftrightarrow a - b\sqrt{2}$ be the one-to-one correspondence. Then
    $$(x + y\sqrt{2}) + (r + s\sqrt{2}) = (x + r) + (y + s)\sqrt{2}$$
    $$\leftrightarrow (x + r) - (y + s)\sqrt{2} = (x - y\sqrt{2}) + (r - s\sqrt{2})$$
    so that $(x + y\sqrt{2}) + (r + s\sqrt{2}) \leftrightarrow (x - y\sqrt{2}) + (r - s\sqrt{2})$.

(h) Let $(a+b\sqrt{2}) \leftrightarrow (a-b\sqrt{2})$ be the one-to-one correspondence. Then

$$(x+y\sqrt{2})(r+s\sqrt{2}) = (xr+2ys) + (xs+yr)\sqrt{2}$$
$$\leftrightarrow (xr+2ys) - (xs+yr)\sqrt{2} = (x-y\sqrt{2})(r-s\sqrt{2})$$

so that

$$(x+y\sqrt{2})(r+s\sqrt{2}) \leftrightarrow (x-y\sqrt{2})(r-s\sqrt{2}).$$

5. Let $a \leftrightarrow \bar{a}$ be an isomorphism for groups $\{S, \circ\}$ and $\{\bar{S}, \bar{\circ}\}$.
   (a) $x = x \circ e \leftrightarrow \bar{x} \bar{\circ} \bar{e}$ where $\bar{x}$ is the image of $x$ for all $x$. Hence $x \leftrightarrow \bar{x} \bar{\circ} \bar{e}$. But $x \leftrightarrow \bar{x}$ and the correspondence is one-to-one so that $\bar{x} = \bar{x} \bar{\circ} \bar{e}$. Similarly, $\bar{x} = \bar{e} \bar{\circ} \bar{x}$. This proves that $\bar{e}$ is the identity element for $\{\bar{S}, \bar{\circ}\}$. Since a group has but one identity element, we have that under this correspondence the identity $e$ is matched with the identity $\bar{e}$.

   (b) $\bar{e} = \overline{x \circ x'} = \bar{x} \bar{\circ} \bar{x}'$, $\quad \bar{e} = \overline{x' \circ x} = \bar{x}' \bar{\circ} \bar{x}$
   Since $\bar{e}$ is the identity for $\{\bar{S}, \bar{\circ}\}$, $\bar{x}'$ is the inverse of $\bar{x}$. Since $\bar{x}$ has but one inverse, $(\bar{x})' = \overline{(x')}$

   (c) $\bar{x} \bar{\circ} \bar{y} = \overline{x \circ y} = \overline{y \circ x} = \bar{y} \bar{\circ} \bar{x}$
   Hence $\bar{x} \bar{\circ} \bar{y} = \bar{y} \bar{\circ} \bar{x}$, proving the first part. Since the correspondence is an isomorphism between the two groups, the very same argument applies. We could also argue

   $$x \circ y \leftrightarrow \overline{x \circ y} = \bar{x} \bar{\circ} \bar{y} = \bar{y} \bar{\circ} \bar{x} \leftrightarrow y \circ x.$$

   Since the correspondence is one-to-one, $\bar{x} \bar{\circ} \bar{y}$ or $\bar{y} \bar{\circ} \bar{x}$ can be matched with but one element so that $x \circ y = y \circ x$.

8. (1) $x \leftrightarrow 2x$

   $$(x+y) \leftrightarrow 2(x+y) = 2x + 2y$$

   (2) $x \leftrightarrow -x$

   $$(x+y) \leftrightarrow -(x+y) = -x + (-y)$$

   In fact, $x \leftrightarrow ax$ for $a \neq 0$ will do.

9. (a) By direct substitution, as $r + s\sqrt{2}$ is a root,

   $$a(r+s\sqrt{2})^2 + b(r+s\sqrt{2}) + c = 0$$
   $$a(r^2 + 2rs\sqrt{2} + 2s^2) + b(r+s\sqrt{2}) + c = 0$$
   $$a(r^2 + 2s^2) + br + c + (2ars + bs)\sqrt{2} = 0$$

   But $a, b, r, s$, are integers. Hence

   $$a(r^2 + 2s^2) + br + c = 0 \quad \text{and} \quad 2ars + bs = 0$$

   If $s = 0$, then $r + s\sqrt{2}$ and $r - s\sqrt{2}$ are identical and there is nothing to prove. Suppose $s \neq 0$. Then since $s(2ar + b) = 0$, it follows that $2ar + b = 0$. This last equality will help us establish that $r - s\sqrt{2}$ is also a root.

Substituting $r - s\sqrt{2}$ for $x$, we obtain

$$ar(-s\sqrt{2})^2 + b(r - s\sqrt{2}) + c$$
$$= a(r^2 - 2rs\sqrt{2} + 2s^2) + b(r - s\sqrt{2}) + c$$
$$= a(r^2 + 2rs\sqrt{2} + 2s^2) - 4ars\sqrt{2} + b(r + s\sqrt{2}) - 2bs\sqrt{2} + c$$
$$= [a(r + s\sqrt{2})^2] + b(r + s\sqrt{2}) + c] - 2s\sqrt{2}(2ar + b) = 0$$
$$= 0 - 2s\sqrt{2}(0) = 0$$

Hence $r - s\sqrt{2}$ is also a root.

(b) Let $r + s\sqrt{2} \leftrightarrow r - s\sqrt{2}$ be the one-to-one correspondence producing the isomorphic mapping between the group

$$\{\{r + s\sqrt{2} \mid r \text{ and } s \text{ are integers}\}, +\}$$

and itself. That this is an isomorphic mapping was established in Exercise 6a. Hence

$$a(r + s\sqrt{2})^2 + b(r + s\sqrt{2}) + c$$
$$= ([ar^2 + 2as^2] + 2ars\sqrt{2}) + ([br + c] + s\sqrt{2})$$
$$\leftrightarrow [ar^2 + 2as^2] - 2ars\sqrt{2} + [br + c] - s\sqrt{2}$$
$$= [a(r - s\sqrt{2})^2] + b(r - s\sqrt{2}) + c$$

Since $r + s\sqrt{2}$ is a root of $ax^2 + bx + c = 0$, we have
$$a(r + s\sqrt{2})^2 + b(r + s\sqrt{2}) + c = 0.$$

Hence, substituting above, we get

$$0 \leftrightarrow a(r - s\sqrt{2})^2 + b(r - s\sqrt{2}) + c$$

However, we have shown that an identity gets mapped into an identity for an isomorphic mapping. Hence the image of 0 is the identity for addition. But there is but one identity for addition, namely 0, so that we must have

$$a(r - s\sqrt{2})^2 + b(r - s\sqrt{2}) + c = 0$$

as we had to prove.

12. (a) If a group has exactly two elements, its operation table must look like this:

| ∘ | e | a |
|---|---|---|
| e | e | a |
| a | a | e |

where $e$ is the identity and $a$ the other element.

For $a$ to have an inverse, it must follow that $a \circ a = e$. If $\{\{\bar{e}, b\}, \bar{\circ}\}$ is another group, then the isomorphism will be

$$e \leftrightarrow \bar{e}, \qquad a \leftrightarrow b, \qquad \circ \leftrightarrow \bar{\circ}$$

(b) If $\{\{e, a, b\}, \circ\}$ is a group, then we shall show that there is essentially but one operation table. Let us begin the table using the fact that $e$ is the identity. We first prove that no two elements in the same column or row of an operation table for a group can be the same. For suppose $a \circ x = a \circ y$. Then by left cancellation we would have $x = y$. Similarly, if $x \circ a = y \circ a$, then $x = y$ from right cancellation.

| $\circ$ | $e$ | $a$ | $b$ |
|---|---|---|---|
| $e$ | $e$ | $a$ | $b$ |
| $a$ | $a$ | $\square$ | |
| $b$ | $b$ | | |

The only entry possible for the frame $\square$ is "$e$" from our last remark.

| $\circ$ | $e$ | $a$ | $b$ |
|---|---|---|---|
| $e$ | $e$ | $a$ | $b$ |
| $a$ | $a$ | $x$ | $e$ |
| $b$ | $b$ | $y$ | $z$ |

But then from the same remark we must have $x = b$; then $y = e$, and $z = a$. Hence the table must be

| $\circ$ | $e$ | $a$ | $b$ |
|---|---|---|---|
| $e$ | $e$ | $a$ | $b$ |
| $a$ | $a$ | $b$ | $e$ |
| $b$ | $b$ | $e$ | $a$ |

13. Let $\{\{e, a, b, c\}, \circ\}$ be a group. We shall examine all possibilities for its operation table. We shall again use the fact that no element appears twice in any row or column.

(a)

| $\circ$ | $e$ | $a$ | $b$ | $c$ |
|---|---|---|---|---|
| $e$ | $e$ | $a$ | $b$ | $c$ |
| $a$ | $a$ | $\triangle e$ | $c$ | $z$ |
| $b$ | $b$ | $x$ | | |
| $c$ | $c$ | $y$ | | |

It follows that the frame $\square$ cannot have "$a$" or "$b$." Suppose we try "$c$." Then we could have either "$e$" or "$b$" for $\triangle$. Try "$e$" for $\triangle$. Then from the earlier remark it must follow that $x = c$, $y = b$, $z = b$. Hence the table now is

(b)

| $\circ$ | $e$ | $a$ | $b$ | $c$ |
|---|---|---|---|---|
| $e$ | $e$ | $a$ | $b$ | $c$ |
| $a$ | $a$ | $\triangle e$ | $c$ | $b$ |
| $b$ | $b$ | $c$ | $x$ | $y$ |
| $c$ | $c$ | $b$ | $z$ | $w$ |

If we take $x = e$, then $y = a$, $z = a$, $w = e$. If we take $x = a$, then $y = e$, $z = e$, $w = a$. In other words, we have either of the two tables (note that $x \neq b$ and $x \neq c$):

**Table I**

| ∘ | e | a | b | c |
|---|---|---|---|---|
| e | e | a | b | c |
| a | a | e | c | b |
| b | b | c | e | a |
| c | c | b | a | e |

or

**Table II**

| ∘ | e | a | b | c |
|---|---|---|---|---|
| e | e | a | b | c |
| a | a | e | c | b |
| b | b | c | a | e |
| c | c | b | e | a |

Tables I and II followed from using ▭$c$ and △$e$. Suppose we use ◯$b$. We then have in order (see Figure A): $x=e$, $y=e$, $z=c$, $u=a$, $v=b$, $w=e$, $t=a$, and the table becomes

| ∘ | e | a | b | c |
|---|---|---|---|---|
| e | e | a | b | c |
| a | a | b | c | x |
| b | b | z | w | t |
| c | c | y | u | v |

**Table III**

| ∘ | e | a | b | c |
|---|---|---|---|---|
| e | e | a | b | c |
| a | a | b | c | e |
| b | b | c | e | a |
| c | c | e | a | b |

Figure A

Tables I, II, and III followed from using "$c$" for ▭. We could have used "$e$" for ▭; then we could not have "$b$" for △, for then we would have to use "$c$" for ◯, which is not possible. Hence we must have "$c$" for △ and "$b$" for ◯. Now we are forced to have $x=e$ and $y=b$.

| ∘ | e | a | b | c |
|---|---|---|---|---|
| e | e | a | b | c |
| a | a | c | e | b |
| b | b | x |   |   |
| c | c | y |   |   |

Our table now is Figure B.

| ∘ | e | a | b | c |
|---|---|---|---|---|
| e | e | a | b | c |
| a | a | c | e | b |
| b | b | e | z | w |
| c | c | b | x | y |

Figure B

We are forced to use $x=a$, $y=e$, $z=c$, $w=a$ in Figure B yielding Table IV.

**Table IV**

| ∘ | e | a | b | c |
|---|---|---|---|---|
| e | e | a | b | c |
| a | a | c | e | b |
| b | b | e | c | a |
| c | c | b | a | e |

We observe that there are four tables. Only for one table, I, is $x \circ x = e$ for each of the four elements. For the other three the solution set for $x \circ x = e$ consists of exactly two elements. These three are tables for isomorphic groups under the following mapping:

$$\begin{array}{ccc} \text{II} & \text{III} & \text{IV} \\ e \leftrightarrow & e \leftrightarrow & e \\ a \leftrightarrow & b \leftrightarrow & c \\ b \leftrightarrow & a \leftrightarrow & a \\ c \leftrightarrow & c \leftrightarrow & b \end{array}$$

Moreover, we have the following additional isomorphic mappings:

IV and $\{\{0, 1, 2, 3\}, +(\text{mod } 4)\}$

$$\begin{array}{cc} e \leftrightarrow 0 & b \leftrightarrow 3 \\ a \leftrightarrow 1 & c \leftrightarrow 2 \end{array}$$

I and $\{\{1, 3, 5, 7\}, \times(\text{mod } 8)\}$

$$\begin{array}{cc} e \leftrightarrow 1 & b \leftrightarrow 5 \\ a \leftrightarrow 3 & c \leftrightarrow 7 \end{array}$$

## Review Exercises (Page 84)

1.

|   | (1) Closure | (2) Assoc. | (3) Identity | (4) Inverse | (5) Commutative |
|---|---|---|---|---|---|
| A. | Yes | Yes | Yes | No | No |
| B. | Yes | No $(aa)b \neq a(ab)$ | Yes | Yes | No |
| C. | Yes | Yes | Yes | Yes | Yes |
| D. | Yes | No $(aa)b \neq a(ab)$ | Yes | No | Yes |
| E. | Yes | No $(aa)b \neq a(ab)$ | No | No | No |
| F. | No | No | No | No | Yes |
| G. | Yes | No | No | No | Yes |
| H. | Yes | Yes | Yes | Yes | Yes |
| I. | Yes | Yes | Yes | Yes | Yes |
| J. | Yes | Yes | Yes (0) | No(−1) | Yes |

4. (a) We must show that if $x$ is in $Hb$, then $x$ is in $H$ and conversely. If $x$ is in $Hb$, then for some $h$ in $H$, $x = h \circ b$. But $e \circ b = b$ is in $H$. Hence $h \circ b$ is in $H$ so that $x$ is in $H$. Conversely, if $x$ is in $H$, we must show that $x$ is in $Hb$. As $b$ is in $H$, so is $b'$ in $H$. If $x$ is in $H$, so is $x = (x \circ b') \circ b$ in $H$. But as $x$ and $b'$ are in $H$, so is $x \circ b'$ in $H$. But $(x \circ b') \circ b$ is in $Hb$. Hence $x$ is in $H$.

(b) For every $x$ in $H$ make the association

$$x \leftrightarrow x \circ b$$

We must show that this association is one-to-one. If $y \leftrightarrow y \circ b$, then $y \circ b = x \circ b$ if and only if $x = y$. Hence no two elements of $H$ are associated with the same element of $Hb$. Moreover, every element of $Hb$ has a partner in $H$ by the given association. If $x$ is in $Hb$, then $x = h \circ b$ for some unique $h$ in $H$. The partner for $x$ is then $h$.

(c) Suppose $Hb$ and $Hc$ had a common element $x$. Then for some $h_1$ and $h_2$, $x = h_1 \circ b$ and $x = h_2 \circ c$. But then $h_1 \circ b = h_2 \circ c$ and $b = (h_1' \circ h_2) \circ c$. If $y$ is in $Hb$, then for some $h$, $y = h \circ b$. Hence

$$y = [h \circ (h_1' \circ h_2)] \circ c$$

so that $y$ is in $Hc$. Similarly,

$$c = (h_2' \circ h_1) \circ b$$

and if $z = h \circ c$,

$$z = [h \circ (h_2' \circ h_1)] \circ b$$

so that if $z$ is in $Hc$, $z$ is also in $Hb$. Hence, if $Hb$ and $Hc$ have one element in common, they are identical sets.

(d) Select $b_1$, not in $H$. Then considering $H = He$, $H$ and $Hb$ have no elements in common. Now select $b_2 \notin H \cup Hb_1$ if $G \neq H \cup Hb_1$. Hence $H$, $Hb_1$, $Hb_2$ have no elements in common by virtue of (c). If $G \neq H \cup Hb_1 \cup Hb_2$, then let $b_3 \notin H \cup Hb_1 \cup Hb_2$ and form $Hb_3$. If $G = H \cup Hb_1 \cup Hb_1 \cup Hb_3$, we are finished. Otherwise, continue in the same manner. As $G$ has a finite number of elements, this procedure will end and our objective will be achieved.

(e) As $G = H \cup Hb_1 \cup \cdots \cup Hb_n$ and $Hb_1$ has the same number of elements as $H$ (from (b)), $Hb_2$ has the same number of elements as $H$, etc., with no two subsets having an element in common. $G$ must have $n + 1$ times the number elements in $H$. This result is called the Lagrange Theorem.

(f) If $\{G, \circ\}$, $\{H, \bar{\circ}\}$ are two groups each having 5 elements, let $x \in G$ and $y \in H$, neither being an identity. Then $G$ can be generated by $x$ and $H$ generated by $y$. Thus:

$$G = \{x, x \circ x, x \circ x \circ x, x \circ x \circ x \circ x, x \circ x \circ x \circ x \circ x\}$$
$$H = \{y, y \bar{\circ} y, y \bar{\circ} y \bar{\circ} y, y \bar{\circ} y \bar{\circ} y \bar{\circ} y, y \bar{\circ} y \bar{\circ} y \bar{\circ} y \bar{\circ} y\}$$

Moreover, $x \circ x \circ x \circ x \circ x = e_G$ (the identity in $G$) and $y \bar{\circ} y \bar{\circ} y \bar{\circ} y \bar{\circ} y = e_H$ (the identity in $H$). If fewer factors gave the identity, then we would have a subgroup and the number of elements in this subgroup must divide 5 (the number of elements in $G$ and $H$). But 5 is prime and so they cannot have "proper" subgroups other than the group consisting simply of the identity element. An isomorphic mapping is

$$x \leftrightarrow y$$
$$x \circ x \leftrightarrow y \bar{\circ} y$$
$$x \circ x \circ x \leftrightarrow y \bar{\circ} y \bar{\circ} y$$
$$x \circ x \circ x \circ x \leftrightarrow y \bar{\circ} y \bar{\circ} y \bar{\circ} y$$
$$e_G \leftrightarrow e_H$$

6.    $a \circ a = a$        Given
      $a \circ a = e \circ a$      $e \circ x = x$ for all $x$
      $a = e$           Right cancellation

10. Form the direct product of $\{S, \circ\}$ and our nonabelian group $\{\{e, p, q, r, s, t\}, \circ\}$. (See Exercises 2.4, Exercise 9.)

11. Closure: All possible products are defined.
Associativity: Let $a, b, c \in \{1, -1\}$, $x^{-1} = x'$, $x^1 = x$

$$[(x, a) \cdot (y, b)] \cdot (z, c) \quad (x, a) \cdot [(y, b) \cdot (z, c)]$$
$$= (x \circ y^a, ab) \cdot (z, c) \quad = (x, a) \cdot (y \circ z^b, bc)$$
$$= ((x \circ y^a) \circ z^{ab}, abc) \quad = (x \circ (y \circ z^b)^a, abc)$$
$$= (x \circ (y^a \circ z^{ab}), abc) \quad = (x \circ (z^{ab} \circ y^a), abc) \text{ if } a = -1$$
$$\quad \quad = (x \circ (y^a \circ z^{ab}), abc) \text{ if } a = 1$$

But $\circ$ is commutative, proving $\cdot$ is associative.

Identity: $(x, a) \cdot (e, 1) = (x \circ e, a) = (x, a)$
$(e, 1) \cdot (x, a) = (e \circ x, a) = (x, a)$
$\therefore (e, 1)$ is the identity.

Inverse: $(x, 1) \circ (x', 1) = (x \circ x', 1) = (e, 1)$
$(x', 1) \circ (x, 1) = (x' \circ x, 1) = (e, 1)$
$\therefore (x, 1)' = (x', 1)$
$(x, -1) \circ (x, -1) = (x \circ x', 1) = (e, 1)$
$(x, -1)' = (x, -1)$

Noncommutative: Let $x \neq x'$ from our assumption on $\{S, \circ\}$. Then

$$(x, 1) \cdot (x, -1) = (x \circ x, -1), \quad (x, -1) \cdot (x, 1) = (x \circ x', -1) = (e, -1)$$

As $x \circ x \neq e$, we have $(x, 1) \cdot (x, -1) \neq (x, -1) \cdot (x, 1)$.

# CHAPTER 3

## Exercises 3.1 (Page 93)

1. The following sets of real numbers are well-ordered (the others are not): (c), (g), (j).

2. (a)

| | $a$ | $b$ | $q$ | $r$ |
|---|---|---|---|---|
| (1) | 65 | 7 | 9 | 2 |
| (2) | 70 | 7 | 10 | 0 |
| (3) | 135 | 7 | 19 | 2 |
| (4) | 15 | 7 | 2 | 1 |
| (5) | 4558 | 7 | 651 | 1 |

(b)

| | $a$ | $b$ | $q$ | $r$ |
|---|---|---|---|---|
| (1) | 65 | 17 | 3 | 14 |
| (2) | 70 | 17 | 4 | 2 |
| (3) | 135 | 17 | 7 | 16 |
| (4) | 15 | 17 | 0 | 15 |
| (5) | 4558 | 17 | 268 | 2 |

3. (a)

| | $a$ | $b$ | $q$ | $r$ |
|---|---|---|---|---|
| (1) | −65 | 7 | −10 | 5 |
| (2) | −70 | 7 | −10 | 0 |
| (3) | 135 | −7 | −19 | 2 |
| (4) | −15 | −7 | 3 | 6 |
| (5) | −4558 | 7 | −652 | 6 |

(b)

| | $a$ | $b$ | $q$ | $r$ |
|---|---|---|---|---|
| (1) | −65 | 17 | −4 | 3 |
| (2) | −70 | 17 | −5 | 15 |
| (3) | −135 | 17 | −8 | 1 |
| (4) | −15 | 17 | −1 | 2 |
| (5) | −4558 | 17 | −269 | 15 |

(c)

| | $a$ | $b$ | $q$ | $r$ |
|---|---|---|---|---|
| (1) | 58 | 6 | 9 | 4 |
| (2) | 58 | −6 | −9 | 4 |
| (3) | −58 | 6 | −10 | 2 |
| (4) | −58 | −6 | 10 | 2 |
| (5) | 187 | 13 | 14 | 5 |
| (6) | 187 | −13 | −14 | 5 |
| (7) | −187 | 13 | −15 | 8 |
| (8) | −187 | −13 | 15 | 8 |

## Exercises 3.2 (Page 96)

1. (a) 18: $\pm 1, \pm 2, \pm 3, \pm 6, \pm 9, \pm 18$
   (b) 30: $\pm 1, \pm 2, \pm 3, \pm 5, \pm 6, \pm 10, \pm 15, \pm 30$
   (c) 36: $\pm 1, \pm 2, \pm 3, \pm 4, \pm 6, \pm 9, \pm 12, \pm 18, \pm 36$ (d) 37: $\pm 1, \pm 37$
   (e) 48: $\pm 1, \pm 2, \pm 3, \pm 4, \pm 6, \pm 8, \pm 12, \pm 16, \pm 24, \pm 48$

2. (a) 4  (b) 2  (c) 1  (d) 48  (e) 4  (f) 19  (g) 23
   (h) 29

7. As $a$ is a divisor of $b$, for some integer $q$, $b = qa = a + (q-1)a$. As $a$ and $b$ are both positive integers, $q \geq 1$. Hence $(q-1)a \geq 0$ and $b \geq a$.

8. 2, 3, 5, 7, 11, 13, 17, 19, 23, 29, 31, 37, 41, 43, 47, 53, 59, 61, 67, 71, 73, 79, 83, 89, 97

9. (3, 5), (5, 7), (11, 13), (17, 19), (29, 31), (41, 43), (59, 61), (71, 73)

10. If the smallest divisor, say $d$, greater than 1 were not a prime, then it would be composite and there would be two positive integers $r$, $s$, each greater than 1 such that $d = rs$. But then $d$ would not be the smallest positive divisor greater than 1, for $r$ and $s$ are smaller than $d$.

12. Suppose $p$ were the greatest prime. Then consider the number $N = (1 \cdot 2 \cdot 3 \ldots \cdot p) + 1$. This number is greater than $p$ and therefore cannot be a prime according to our assumption. But $N$ is not divisible by $2, 3, 4, \ldots, p$, for in each case there is the remainder 1. Hence the smallest divisor of $N$ greater than 1 must be greater than $p$, and, according to previous Exercise 10, must be a prime. This contradicts the assumption on $p$.

## Exercises 3.3 (Page 100)

1. (a) 4   (b) 8   (c) 12   (d) 32   (e) 3
2. The GCD for three integers may be defined as the greatest integer dividing all three integers.
   (a) 4   (b) 2   (c) 4
3. 1
5. A counterexample would suffice here. Consider the numbers 6 and 8 having the GCD 2. However, the GCD of 6 and $6 + 3 \times 8$ is 6 and not 2.

## Exercises 3.4 (Page 109)

1. (a) $144 = 2 \cdot 2 \cdot 2 \cdot 2 \cdot 3 \cdot 3$   (b) $1000 = 2 \cdot 2 \cdot 2 \cdot 5 \cdot 5 \cdot 5$
   (c) $576 = 2 \cdot 2 \cdot 2 \cdot 2 \cdot 2 \cdot 2 \cdot 3 \cdot 3$   (d) $234 = 2 \cdot 3 \cdot 3 \cdot 13$
   (e) $324 = 2 \cdot 2 \cdot 3 \cdot 3 \cdot 3 \cdot 3$   (f) $423 = 3 \cdot 3 \cdot 47$

3. As $b$ and $c$ are relatively prime, there are integers $x, y$ such that $xb + yc = 1$. As $b$ and $c$ are divisors of $a$, there are integers $r$ and $s$ such that $a = rb$ and $a = sc$. We then have upon multiplying by $a$

$$axb + ayc = a$$

and replacing $a$ by $rb$ or $sc$ (on the left)

$$scxb + rbyc = a$$

or

$$(bc)sx + (bc)ry = a.$$

As $bc$ is a divisor of the left member of the equality, $bc$ must also divide the right member or $a$.

5. (a) By Theorem 3.5, since $b$ is relatively prime to $c$, then $b$ is relatively prime to $c \cdot c = c^2$ and hence also to $c^2 \cdot c = c^3$. Again, since $c^3$ is relatively prime to $b$, then $c^3$ is relatively prime to $b^2$.
   (b) Clearly, the argument in (a) can be repeated as often as needed to show that $b^r$ and $c^s$ are relatively prime for any positive integers $r$ and $s$.

6. (a) If $b^2$ and $c^3$ are relatively prime, then there are integers $x$ and $y$ such that $xb^2 + yc^3 = 1$ or also that $(xb)b + (yc^2)c = 1$.
   If $b$ and $c$ had a GCD, say $d > 1$, $d$ would divide 1 which is impossible.
   *Alternate proof*: If $b$ and $c$ were not relatively prime, then there would exist a common divisor $d > 1$ for $b$ and $c$. Since $d$ is a divisor of both $b$ and $c$, it must certainly be a divisor of both $b^2$ and $c^3$. This contradicts the hypothesis that $b^2$ and $c^3$ are relatively prime.

(b) Similar argument to alternate proof (a) but using $b^r$ and $c^s$ instead of $b^2$ and $c^3$.
*Alternate proof*: If $b^r$ and $c^s$ are relatively prime, then there are integers $x$ and $y$ such that $xb^r + yc^s = 1$ or also that

$$(xb^{r-1})b + (yc^{s-1})c = 1$$

If the GCD of $b$ and $c$ were $d > 1$, $d$ would divide 1 which is impossible.

7. (a) If $x \gtrless y$, then there are integers $a$, $b$ such that $ax + by = 1$. But then $y \gtrless x$.
   (b) Theorem 3.4: $a \gtrless b$ if and only if there exist integers $x$ and $y$ such that $ax + by = 1$.
   (c) If $ax + by = 1$, then from Theorem 3.4 it follows immediately that $a \gtrless b$, $a \gtrless y$, $x \gtrless b$, $x \gtrless y$.
   (d) If $ax + by = g$ where $g$ is the GCD of $a$ and $b$, then $\dfrac{a}{g}$ and $\dfrac{b}{g}$ are integers and $\left(\dfrac{a}{g}\right)x + \left(\dfrac{b}{g}\right)y = 1$. By Theorem 3.4 it follows that $x \gtrless y$.
   (e) If $a$ were not relatively prime to $c$, then they would have a common divisor $d > 1$. But then $d$ would divide $ab$ and $c$ and $ab \gtrless c$. Contradiction. Similarly for $b \gtrless c$.

10. (a) The method is more easily understood through an example. Consider the problem of finding the GCD of 5670 and 7650. Expressing each as a product of primes, we have

    $$5670 = 2 \cdot 3^4 \cdot 5 \cdot 7$$
    $$7650 = 2 \cdot 3^2 \cdot 5^2 \cdot 19$$

    Comparing factors we notice that the following are common divisors: 2, $3^2$, 5. Notice that each power of a prime is the smallest power common to both factorizations. The GCD is $2 \cdot 3^2 \cdot 5 = 90$.
    (b) The LCM is the product of the greatest powers of primes appearing in either factorization in (a): for example, the LCM of 5670 and 7650 is $2 \cdot 3^4 \cdot 5^2 \cdot 7 \cdot 19$.

11. 

|     |          | GCD | LCM  |
| --- | -------- | --- | ---- |
| (a) | (48, 63) | 3   | 1008 |
| (b) | (98, 144)| 2   | 7056 |
| (c) | (144, 50)| 2   | 3600 |
| (d) | (69, 92) | 23  | 276  |

12. Let the GCD of $a$ and $b$ be $g$. Then there are integers $x$, $y$ such that $a = gx$ and $b = gy$. Moreover, $x$ and $y$ are relatively prime, for otherwise $g$ would not be the GCD of $a$ and $b$. Hence any multiple of both $a$ and $b$ must contain as a factor $gxy$. That $gxy$ is the LCM can be proved by considering the prime factorizations of $g$, $x$, $y$ and the method in Exercise 10b for finding the LCM. Hence

    $$\text{LCM} \times \text{GCD} = gxy \cdot g = (gx)(gy) = ab$$

13. If $d$ is the GCD of $a$ and $b$, then for some integers $x$ and $y$, $a = dx$ and $b = dy$. If $x$ and $y$ were not relatively prime, then suppose their GCD were $g > 1$. Then for some integers $r$, $s$ we would have $x = gr$ and $y = gs$. But then $a = dx = dgr$ and $b = dy = dgs$ so that $dg$ would be a common divisor of $a$ and $b$. Since $dg > g$, we have a contradiction on $g$. Hence $x = \dfrac{a}{d}$ and $y = \dfrac{b}{d}$ are relatively prime.

Answers and Solutions to Selected Exercises 393

## Exercises 3.5 (Page 112)

1. The following are true (and the others are false): (a), (b), (c), (f), (g), (h), (j), (k), (l), (m), (n), (o), (p), (q), (r), (s), (t), (u), (v).

2. (a) 8 (b) 8 (c) 4 (d) 5 (e) 9 (f) 7 (g) 6 (h) 2
 (i) 11 (j) 12

3. (a) $\{3, 4\}$ (b) $\{2, 5\}$ (c) $\emptyset$ (d) $\emptyset$ (e) $\{2, 5\}$ (f) $\{1, 6\}$
 (g) $\{3, 4\}$

4. (a) 2 (b) 4 (c) 1 (d) 2 (e) 4 (f) 1 (g) 2 (h) 4

 $2^x \equiv 1 \pmod{7}$ if $x = 3n$ for some integer $n$.
 $2^x \equiv 2 \pmod{7}$ if $x = 3n + 1$ for some integer $n$.
 $2^x \equiv 4 \pmod{7}$ if $x = 3n + 2$ for some integer $n$.

5. (a) 2 (b) 1 (c) 1 (d) 1 (e) 2 (f) 4

## Exercises 3.6 (Page 119)

1. (a) True (b) False for every $a$ not a multiple of 3 (c) False for $m = 7$
 (d) True (e) False for $m = 3$, $a = 2$ (f) True (g) False for $m = 3$, $a = 2$
 (h) False for $m = 8$, $a = 3$ (i) False for $m = 7$, $a = 2$ (j) True
 (k) False for $a = 2$, $b = 1$, $m = 3$ (l) False; $2^{100} \equiv 2 \pmod{7}$
 (m) False; $3^{100} \equiv 4 \pmod{7}$

2. (a) $a = b + Km$ for some integer $K$.
 (1) $a^2 = (b + Km)^2 = b^2 + 2bKm + K^2m^2$
  $= b^2 + (2bK + K^2m)m$
  $a^2 \equiv b^2 \pmod{m}$
 (2) $a^3 = (b + Km)^3 = b^3 + 3b^2Km + 3bK^2m^2 + K^3m^3$
  $= b^3 + (3b^2K + 3bK^2m + K^3m^2)m$
  $\equiv b^3 \pmod{m}$
 (3) From (1) twice: $a^2 \equiv b^2 \pmod{m}$
  Hence $(a^2)^2 \equiv (b^2)^2 \pmod{m}$
  and $a^4 \equiv b^4 \pmod{m}$

 *Alternate proof:*
 (a) (1) Since $a \equiv b \pmod{m}$, it follows from Theorem 3.10c that

 $$a \cdot a \equiv b \cdot b \pmod{m}$$

 that is, $a^2 \equiv b^2 \pmod{m}$

 (2) Since $a \equiv b \pmod{m}$ and $a^2 \equiv b^2 \pmod{m}$, as just proved, it follows from Theorem 3.10c that

 $$a \cdot a^2 \equiv b \cdot b^2 \pmod{m}$$

 that is, $a^3 \equiv b^3 \pmod{m}$

 (3) Since $a \equiv b \pmod{m}$ and $a^3 \equiv b^3 \pmod{m}$, as just proved, it follows from Theorem 3.10c that

 $$a \cdot a^3 \equiv b \cdot b^3 \pmod{m}$$

 that is, $a^4 \equiv b^4 \pmod{m}$

2. (b) If $a \equiv b \pmod{m}$, then $a^n \equiv b^n \pmod{m}$ for every positive integer $n$.
Proof: We shall prove this by Mathematical Induction. (See Chapter 6, section 8.) Suppose that for any integer $K$, $a^K \equiv b^K \pmod{m}$. Then as $a \equiv b \pmod{m}$,

$$a^K \cdot a \equiv b^K \cdot b \pmod{m} \qquad \text{(Theorem 3.10c)}$$

and

$$a^{K+1} \equiv b^{K+1} \pmod{m}$$

However, we know that the congruence holds for $K = 1$. Hence it holds for 2, 3, ... and all the positive integers.

3. (a) If $a \equiv b \pmod{m}$, then $a = b + tm$ for an integer $t$. Then

$$Ka = K(b + tm) = Kb + (Kt)m$$

and so

$$Ka \equiv Kb \pmod{m}$$

(b) If $a \equiv b \pmod{Km}$, then $a = b + t(Km)$ for an integer $t$. Then

$$a = b + (tK)m \qquad \text{and} \qquad a \equiv b \pmod{m}$$

(c) If $Ka \equiv Kb \pmod{Km}$, then $Ka = Kb + t(Km)$ for some integer $K$. We are assuming $K \neq 0$. Hence

$$a = b + tm$$

and so

$$a \equiv b \pmod{m}$$

4. (a) (1) 4   (2) 6   (3) 10   (4) 2   (5) 4   (6) 4   (7) 2
       (8) 3   (9) 1   (10) 5

(b) Let $a \equiv r_1 \pmod{b}$, $a^2 \equiv r_2 \pmod{b}$, ..., $a^b \equiv r_b \pmod{b}$ where $r_1, r_2, \ldots, r_b$ are the smallest possible nonnegative integers. Hence each $r$ is in the set $\{0, 1, 2, 3, \ldots, b-1\}$. If all the $r$'s are different, then one of the $r$'s, say $r_n$, must be 1 because there are $b$ of them. If $r_n = 1$, then $a^n \equiv 1 \pmod{b}$ and we are finished. Suppose that two of the $r$'s were the same, say $r_x = r_y$. Then

$$a^x \equiv r_x \pmod{b}, \qquad a^y \equiv r_y \pmod{b}$$

Hence $a^x \equiv a^y \pmod{b}$. Since $x \neq y$, suppose $x > y$. Hence

$$a^x - a^y \equiv 0 \pmod{b}$$
$$a^y(a^{x-y} - 1) \equiv 0 \pmod{b}$$

As $b$ is relatively prime to $a$, it is relatively prime to $a^y$, so $b$ must divide $a^{x-y} - 1$ so that $a^{x-y} - 1 \equiv 0 \pmod{b}$ and $a^{x-y} \equiv 1 \pmod{b}$. Let $n = x - y$.

8. (a) $\begin{cases} x \equiv 5 \pmod{7} \\ y \equiv 3 \pmod{7} \end{cases}$   (b) $\begin{cases} x \equiv 4 \pmod{7} \\ y \equiv 2 \pmod{7} \end{cases}$   (c) $\begin{cases} x \equiv 5 \pmod{7} \\ y \equiv 5 \pmod{7} \end{cases}$

(d) $\begin{cases} x \equiv 6 \pmod{7} \\ y \equiv 1 \pmod{7} \end{cases}$

## Review Exercises (Page 121)

1. If $S = \{x \mid 2 < x < 3\}$, then $S$ has no least member (because if $2 < x < 3$, then
$$2 < \frac{2+x}{2} < \frac{x+x}{2} = x < 3).$$

**2.**

|       |           | GCD | LCM    |
|-------|-----------|-----|--------|
| (a)   | (243,732) | 3   | 59,292 |
| (b)   | (423,732) | 3   | 103,212|
| (c)   | (234,372) | 6   | 14,508 |
| (d)   | (227,367) | 1   | 83,309 |

**4.** (a) $\begin{cases} x \equiv 4 \pmod{7} \\ y \equiv 4 \pmod{7} \end{cases}$

**5.** (a) $6 \equiv \frac{2}{5} \pmod 7$, $2 \equiv \frac{3}{5} \pmod 7$, $5 \equiv \frac{6}{25} \pmod 7$

(b) $xy = 6 \cdot 2 = 12 \equiv 5 \pmod 7$  $\quad x + y = 6 + 2 = 8 \equiv 1 \pmod 7$
$z = 5 \equiv 5 \pmod 7$  $\quad \therefore x + y \equiv 1 \pmod 7$
$\therefore xy \equiv z \pmod 7$

# CHAPTER 4

## Exercises 4.1 (Page 129)

**3.** (a) $\{Z^+, +\}$ is not an abelian group because there is no identity element.
(b) $\{Q^+, +\}$ is not an abelian group because there is no identity element.
(c) $\{Z^*, +\}$ is not an abelian group because there is no identity element.
(d) $\{Q^*, +\}$ is not an abelian group because there is no identity element.
(e) $\{W_4, +\}$ is not a group because $+$ is not closed on $W_4$. Moreover, $x$ is not closed on $W_4$.
(f) $\{Z^-, +\}$ is not an abelian group
(g) × is not closed on $Z(\frac{1}{2})$
(h) × is not closed on $Z(\sqrt[3]{2})$
(i) × is not closed on $Q(\sqrt[3]{2})$
(j) × is not closed on $\sqrt{2}Z$
(k) × is not closed on $\sqrt{3}Z$
(l) × is not closed on $\sqrt{2}Q$
(m) × is not closed on $Q(\sqrt[3]{5})$
(n) × is not closed on $\pi Q$

## Exercises 4.2 (Page 140)

**1.** (a) $x = -4, y = 5, z = 2, w = -7$
(b) $x = 3, \quad y = 2, z = 4, w = -2$
(c) $x = -2, y = 1, u = 6, z = 3$

**2.** (a) $\begin{bmatrix} -3 & 3 \\ 3 & 3 \end{bmatrix}$ (b) $\begin{bmatrix} -3 & 3 \\ 3 & 3 \end{bmatrix}$ (c) $\begin{bmatrix} -9 & 8 \\ 20 & -16 \end{bmatrix}$ (d) $\begin{bmatrix} 5 & 1 \\ 4 & 12 \end{bmatrix}$
(e) $\begin{bmatrix} -4 & 9 \\ 24 & -4 \end{bmatrix}$ (f) $\begin{bmatrix} -31 & -10 \\ 17 & 6 \end{bmatrix}$ (g) $\begin{bmatrix} 9 & -2 \\ -8 & 8 \end{bmatrix}$ (h) $\begin{bmatrix} -22 & -12 \\ 9 & 14 \end{bmatrix}$

4. (a) $\begin{bmatrix} -9 & 16 \\ 1 & -6 \end{bmatrix}$  (b) $\begin{bmatrix} 21 & -8 \\ -17 & 6 \end{bmatrix}$  (c) $\begin{bmatrix} -54 & 48 \\ 120 & -96 \end{bmatrix}$

(d) $\begin{bmatrix} -14 & 42 \\ -7 & -14 \end{bmatrix}$  (e) $\begin{bmatrix} 21 & -70 \\ 35 & 98 \end{bmatrix}$  (f) $\begin{bmatrix} 39 & -54 \\ -3 & 48 \end{bmatrix}$

(g) $\begin{bmatrix} -16 & 2 \\ 18 & -4 \end{bmatrix}$  (h) $\begin{bmatrix} -94 & 40 \\ 88 & -192 \end{bmatrix}$

## Exercises 4.3 (Page 150)

1. (a) $-3$  (b) 3  (c) $-3$  (d) 1  (e) 2  (f) $n-3$  (g) $a$

8. Since $x \cdot x = x \Leftrightarrow x(x-1) = 0$, we seek consecutive integers whose product is a multiple of the modulus.
   (a) 0, 1, 4, 9  (b) 0, 1, 9, 10  (c) 0, 1, 9, 28

9. (a) 2, 3, 4, 6, 8, 9  (b) 2, 3, 4, 6, 8, 9, 10, 12, 14, 15, 16
   (c) 2, 3, 4, 6, 8, 9, 10, 12, 14, 15, 16, 18, 20, 22, 24, 26, 27, 28, 30, 32, 33, 34

10. (a) $\{0, 6\}$; $\{0, 6, 12\}$; $\{0, 6, 12\}$; $\{0, 6, 12, 18, 24\}$
    (b) $\{2, 4, 10\}$; $\{2, 16\}$; $\{2, 16, 20, 34\}$
    (c) $\{0, 4, 6, 10\}$; $\{0, 4\}$; $\{0, 4, 18, 22\}$

11. (a) $\{0, 6\}$  (b) $\{0, 6, 12\}$  (c) $\{0, 6, 12, 18, 24\}$

15. (a) $1^{-1} = 1$;  0 has no multiplicative inverse.
    (b) $1^{-1} = 1, 2^{-1} = 2$;  0 has no multiplicative inverse.
    (c) $1^{-1} = 1, 3^{-1} = 3, 5^{-1} = 5$;  0, 2, 4 have no multiplicative inverses.
    (d) $1^{-1} = 1, 3^{-1} = 7, 7^{-1} = 3, 9^{-1} = 9$;  0, 2, 4, 5, 6, 8 have no multiplicative inverses.
    (e) $1^{-1} = 1, (-1)^{-1} = -1$; all other integers have no multiplicative inverses.
    (f) No even integer has a multiplicative inverse.

    (g) (1) $\begin{bmatrix} x & 0 \\ 0 & x \end{bmatrix}^{-1} = \begin{bmatrix} \frac{1}{x} & 0 \\ 0 & \frac{1}{x} \end{bmatrix}$ for $x \neq 0$

    $\begin{bmatrix} 0 & 0 \\ 0 & 0 \end{bmatrix}$ has no multiplicative inverse.

    (2) $\begin{bmatrix} 1 & 0 \\ 0 & 1 \end{bmatrix}^{-1} = \begin{bmatrix} 1 & 0 \\ 0 & 1 \end{bmatrix}$    $\begin{bmatrix} -1 & 0 \\ 0 & -1 \end{bmatrix}^{-1} = \begin{bmatrix} -1 & 0 \\ 0 & -1 \end{bmatrix}$

    All other matrices have no multiplicative inverses.

16. $(-1)x = -(1 \cdot x)$     By Theorem 4.3
           $= -x$          Property of multiplicative identity 1: $1 \cdot z = z$ for all $z$

## Exercises 4.4 (Page 155)

1. (b) Let the one-to-one correspondence be $x \leftrightarrow \begin{bmatrix} x & x \\ 0 & 0 \end{bmatrix}$

$$x+y \leftrightarrow \begin{bmatrix} x+y & x+y \\ 0 & 0 \end{bmatrix} = \begin{bmatrix} x & x \\ 0 & 0 \end{bmatrix} + \begin{bmatrix} y & y \\ 0 & 0 \end{bmatrix}$$

$$x \cdot y \leftrightarrow \begin{bmatrix} xy & xy \\ 0 & 0 \end{bmatrix} = \begin{bmatrix} x & x \\ 0 & 0 \end{bmatrix} \cdot \begin{bmatrix} y & y \\ 0 & 0 \end{bmatrix}$$

(e) $(x+y\sqrt{2}) + (r+s\sqrt{2}) = (x+r) + (y+s)\sqrt{2} \leftrightarrow (x+r) - (y+s)\sqrt{2}$
$= (x - y\sqrt{2}) + (r - s\sqrt{2})$
$(x+y\sqrt{2})(r+s\sqrt{2}) = (xr + 2ys) + (xs+yr)\sqrt{2} \leftrightarrow (xr+2ys) - (xs+yr)\sqrt{2}$
$= (x - y\sqrt{2})(r - s\sqrt{2})$

2. Let the isomorphic correspondence be $x \leftrightarrow y = \bar{x}$.
   (a) Then $\bar{x} = \overline{x+0} = \bar{x} + \bar{0}$; $\bar{x} = \overline{0+x} = \bar{0} + \bar{x}$. Hence $\bar{0}$, the image of 0, is the additive identity of $R_2$.
   (b) $\bar{0} = \overline{x+(-x)} = \bar{x} + \overline{(-x)}$. But from (a) $\bar{0}$ is the additive identity in $R_2$. Hence the additive inverse of $\bar{x}$ is $-\bar{x}$, or $\overline{-x} = -(\bar{x})$.

3. (a) Suppose there were a one-to-one correspondence between $Q$ and $Z$. Then

$$\overline{1_Q} = 1_Z \quad \text{(Image of 1 in } Q \text{ is the 1 in } Z\text{.)}$$

$$\overline{2_Q} = \overline{1_Q + 1_Q} = \overline{1_Q} + \overline{1_Q} = 1_Z + 1_Z = 2_Z$$

$$1_Z = \overline{1_Q} = \overline{2_Q \times \frac{1}{2_Q}} = \overline{2_Q} \times \overline{\left(\frac{1}{2_Q}\right)}$$

$$\therefore 1_Z = 2_Z \times \overline{\left(\frac{1}{2_Q}\right)}$$

But $Z$ has no numbers which when multiplied by 2 give 1, so that $\overline{\left(\frac{1}{2_Q}\right)}$ is not defined and no such isomorphic correspondence exists.

## Review Exercises (Page 156)

4. If $\{Q, +, \cdot\}$ and $\{2Q, +, \cdot\}$ were isomorphic under $x \leftrightarrow 2x$, we would have

$$(ab) \leftrightarrow 2(ab)$$
$$\text{and } ab \leftrightarrow (2a)(2b)$$

But $2(ab) \neq (2a)(2b)$ if $ab \neq 0$.

7. $\{3, 7\}$

8. (a) $\begin{bmatrix} 7 & 8 \\ 6 & 7 \end{bmatrix}' = \begin{bmatrix} 7 & -8 \\ -6 & 7 \end{bmatrix}$

   (b) $\begin{bmatrix} a & a+1 \\ a-1 & a \end{bmatrix}' = \begin{bmatrix} a & -a-1 \\ -a+1 & a \end{bmatrix}$

(c) $\begin{bmatrix} 7 & 9 \\ 5 & 7 \end{bmatrix}' = \begin{bmatrix} \frac{7}{4} & -\frac{9}{4} \\ -\frac{5}{4} & \frac{7}{4} \end{bmatrix}$

(d) $\begin{bmatrix} a & a+2 \\ a-2 & a \end{bmatrix}' = \begin{bmatrix} \frac{a}{4} & \frac{-a-2}{4} \\ \frac{-a+2}{4} & \frac{a}{4} \end{bmatrix}$

10. (a) $\{T, +, \cdot\}$ is an ideal where $T$ may be

   $\{0, 2, 4, 6, 8, 10, 12, 14, 16, 18, 20, 22\}$
   $\{0, 3, 6, 9, 12, 15, 18, 21\}$
   $\{0, 4, 8, 12, 16, 20\}, \{0, 6, 12, 18\}$
   $\{0, 8, 16\}, \{0, 12\}, \{0\}$
   (b) $\{0, 9, 16\}$
   (c) $\{0, 6, 12, 18\}$

# CHAPTER 5

## Exercises 5.1 (Page 167)

2. (a) 2, 3, 4 do not have multiplicative inverses.
   (b) (1) There is no multiplicative identity.
       (2) No element has a multiplicative inverse.
   (c) There are numbers other than 0 that have no multiplicative inverse, for example, $2 + \sqrt{2}$.
   (d) (1) Multiplication is not closed. Thus $\sqrt[3]{2} \cdot \sqrt[3]{2} = \sqrt[3]{4}$, which is not in $Q(\sqrt{2})$.
       (2) Some multiplicative inverses do not exist. Thus $\frac{1}{\sqrt[3]{2}} = \frac{1}{2} \cdot \sqrt[3]{4}$, which is not in $Q(\sqrt{2})$.
   (e) Some elements have no multiplicative inverse, for example,
   $$\begin{bmatrix} 0 & 1 \\ 0 & 0 \end{bmatrix}$$
   (f) $a$ and $b$ have no multiplicative inverses.

3. (a) $\begin{bmatrix} a & b \\ c & d \end{bmatrix} \begin{bmatrix} r & s \\ t & u \end{bmatrix} = \begin{bmatrix} ar+bt & as+bu \\ cr+dt & cs+du \end{bmatrix}$

   But $ad - bc = 1$ and $ru - st = 1$. We must show that
   $$(ar+bt)(cs+du) - (as+bu)(cr+dt) = 1$$

Answers and Solutions to Selected Exercises 399

or $\quad arcs + ardu + btcs + btdu - [ascr + asdt + bucr + budt] = 1$

or $\quad (ardu - bucr) + (btcs - asdt) = 1$

or $\quad ru(ad - bc) + st(bc - ad) = 1$

or $\quad ru - st = 1$

(b) As multiplication of matrices is associative, this is surely the case for $M_1$.

(c) $\begin{bmatrix} 1 & 0 \\ 0 & 1 \end{bmatrix}$ is the multiplicative identity and is in $M_1$.

(d) $\begin{bmatrix} a & b \\ c & d \end{bmatrix} \begin{bmatrix} d & -b \\ -c & a \end{bmatrix} = \begin{bmatrix} ad - bc & 0 \\ 0 & ad - bc \end{bmatrix} = \begin{bmatrix} 1 & 0 \\ 0 & 1 \end{bmatrix}$

$\begin{bmatrix} d & -b \\ -c & a \end{bmatrix} \begin{bmatrix} a & b \\ c & d \end{bmatrix} = \begin{bmatrix} da - bc & 0 \\ 0 & da - bc \end{bmatrix} = \begin{bmatrix} 1 & 0 \\ 0 & 1 \end{bmatrix}$

(e) Yes

(f) No. A counterexample is

$\begin{bmatrix} 2 & 1 \\ 1 & 1 \end{bmatrix} \cdot \begin{bmatrix} 3 & 2 \\ 1 & 1 \end{bmatrix} = \begin{bmatrix} 7 & 5 \\ 4 & 3 \end{bmatrix}$

$\begin{bmatrix} 3 & 2 \\ 1 & 1 \end{bmatrix} \cdot \begin{bmatrix} 2 & 1 \\ 1 & 1 \end{bmatrix} = \begin{bmatrix} 8 & 5 \\ 3 & 2 \end{bmatrix}$ Not identical

(g) No. Addition is not closed. Thus

$\begin{bmatrix} 1 & 0 \\ 0 & 1 \end{bmatrix} + \begin{bmatrix} 1 & 0 \\ 0 & 1 \end{bmatrix} = \begin{bmatrix} 2 & 0 \\ 0 & 2 \end{bmatrix}$ and not in $M_1$

(h) (1) No (2) No

## Exercises 5.2 (Page 174)

1. (a) $(x - y) + y = (x + [-y]) + y \quad$ Definition of $x - y$
   $= x + ([-y] + y) \quad$ Addition is associative
   $= x + 0 \quad\quad\quad\quad -z + z = 0$ for all $z$
   $= x \quad\quad\quad\quad\quad z + 0 = z$ for all $z$

2. (a) $\dfrac{x}{y} \cdot y = \dfrac{x}{y} \cdot \dfrac{y}{1} = \dfrac{x \cdot y}{y \cdot 1} = \dfrac{x \cdot y}{1 \cdot y} = \dfrac{x}{1} \cdot \dfrac{y}{y} = x \cdot 1 = x$

3. (a) $(-3)5 = -(3 \cdot 5) = -(1) = 6$
   $(-3)5 = (4)5 = 6$

   (b) $(-3)(-5) = 3 \cdot 5 = 1$
   $(-3)(-5) = 4 \cdot 2 = 1$

   (c) $\dfrac{3}{5} = 3 \times \dfrac{1}{5} = 3 \times 3 = 2$

   $\dfrac{3}{5} = \dfrac{3 \times 3}{5 \times 3} = \dfrac{2}{1} = 2$

**400** Answers and Solutions to Selected Exercises

(d) $\dfrac{-3}{5} = (-3)\left(\dfrac{1}{5}\right) = (4)(3) = 5$

$\dfrac{-3}{5} = -\left(\dfrac{3}{5}\right) = -\left(3 \cdot \dfrac{1}{5}\right) = -(3 \cdot 3) = -2 = 5$

(e) $\dfrac{3}{-5} = \dfrac{3}{2} = 3\left(\dfrac{1}{2}\right) = 3(4) = 5$

$\dfrac{3}{-5} = \dfrac{-3}{5} = (-3)\left(\dfrac{1}{5}\right) = (4)(3) = 5$

(f) $\dfrac{-3}{-5} = \dfrac{4}{2} = 2$

$\dfrac{-3}{-5} = \dfrac{3}{5} = 3\left(\dfrac{1}{5}\right) = 3(3) = 2$

(g) $\dfrac{3}{5} + \dfrac{4}{5} = \dfrac{3+4}{5} = \dfrac{0}{5} = 0$

$\dfrac{3}{5} + \dfrac{4}{5} = 3\left(\dfrac{1}{5}\right) + 4\left(\dfrac{1}{5}\right) = 3(3) + 4(3) = 2 + 5 = 0$

(h) $\dfrac{3}{5} + \dfrac{1}{4} = \dfrac{3(4) + 1(5)}{5 \cdot 4} = \dfrac{5+5}{6} = \dfrac{3}{6} = \dfrac{1}{2} = 4$

$\dfrac{3}{5} + \dfrac{1}{4} = 3\left(\dfrac{1}{5}\right) + (2) = 3(3) + 2 = 2 + 2 = 4$

(i) $\dfrac{2}{5} \times \dfrac{3}{4} = \dfrac{2 \times 3}{5 \times 4} = \dfrac{6}{6} = 1$

$\dfrac{2}{5} \times \dfrac{3}{4} = 2\left(\dfrac{1}{5}\right) \times 3\left(\dfrac{1}{4}\right) = 2(3) \times 3(2) = 6 \cdot 6 = 1$

(j) $\dfrac{3}{5} \div \dfrac{2}{5} = \dfrac{3 \cdot 5}{5 \cdot 2} = \dfrac{1}{3} = 5$

$\dfrac{3}{5} \div \dfrac{2}{5} = 3\left(\dfrac{1}{5}\right) \div 2\left(\dfrac{1}{5}\right) = 3(3) \div 2(3) = 2 \div 6 = 2 \times \dfrac{1}{6} = 2 \times 6 = 5$

(k) $\dfrac{4}{5} \div 2 = 4\left(\dfrac{1}{5}\right) \times \left(\dfrac{1}{2}\right) = 4(3) \times (4) = 5 \times 4 = 6$

$\dfrac{4}{5} \div 2 = \dfrac{4}{5} \cdot \dfrac{1}{2} = \dfrac{4}{5 \times 2} = \dfrac{4}{3} = 4\left(\dfrac{1}{3}\right) = 4 \cdot 5 = 6$

(l) $\dfrac{6}{5} \div 3 = 6\left(\dfrac{1}{5}\right) \cdot \dfrac{1}{3} = 6(3) \cdot 5 = 4 \cdot 5 = 6$

$\dfrac{6}{5} \div 3 = \dfrac{6}{5} \times \dfrac{1}{3} = \dfrac{6 \times 1}{5 \times 3} = \dfrac{6}{1} = 6$

(m) $\dfrac{2}{5} \div \dfrac{3}{4} = \dfrac{2}{5} \cdot \dfrac{4}{3} = \dfrac{2 \cdot 4}{5 \cdot 3} = \dfrac{1}{1} = 1$

$\dfrac{2}{5} \div \dfrac{3}{4} = 2\left(\dfrac{1}{5}\right) \div 3\left(\dfrac{1}{4}\right) = 2(3) \div 3(2) = 6 \div 6 = 6 \times \dfrac{1}{6} = 1$

(o) $(3 - 2) + 5 = (3 + [-2]) + 5 = (3 + 5) + 5 = 1 = 5 = 6$

$(3 - 2) + 5 = (1 + 2 + [-2]) + 5 = 1 + 5 = 6$

(p) $\dfrac{3}{5} - \dfrac{2}{5} = \dfrac{3}{5} + \dfrac{-2}{5} = \dfrac{3}{5} + \dfrac{5}{5} = \dfrac{3+5}{5} = \dfrac{1}{5} = 3$

$\dfrac{3}{5} - \dfrac{2}{5} = 3\left(\dfrac{1}{5}\right) - 2\left(\dfrac{1}{5}\right) = 3(3) - 2(3) = 2 - 6 = 2 + (-6) = 2 + 1 = 3$

(q) $\dfrac{2}{3} - \dfrac{1}{3} = \dfrac{2}{3} + \dfrac{-1}{3} = \dfrac{2+[-1]}{3} = \dfrac{2+6}{3} = \dfrac{1}{3} = 5$

$\dfrac{2}{3} - \dfrac{1}{3} = 2\left(\dfrac{1}{3}\right) - (5) = 2(5) + (-5) = 3 + 2 = 5$

(r) $\dfrac{1}{3} - \dfrac{2}{3} = \dfrac{1}{3} + \dfrac{-2}{3} = \dfrac{1+[-2]}{3} = \dfrac{1+5}{3} = \dfrac{6}{3} = \dfrac{-1}{3} = -\dfrac{1}{3} = -5 = 2$

$\dfrac{1}{3} - \dfrac{2}{3} = 5 - 2\left(\dfrac{1}{3}\right) = 5 - 2(5) = 5 - 3 = 5 + (-3) = 5 + 4 = 2$

4. (a) {6}  (b) {6}  (c) {3}  (d) {1}  (e) {5}  (f) {5}  (g) {5}
   (h) {6}  (i)      (j) {2}  (j) {1, 6}  (k) {3, 4}  (l) { }  (m) {2, 5}
   (n) { }  (o) { }  (p) {0, 6}  (q) {6, 5}  (r) { }  (s) {2, 4}
   (t) {2, 6}  (u) { }  (v) { }  (w) { }  (x) {2, 5}  (y) {3, 4}
   (z) {1, 6}

5. (a) If $x \neq 0$, then $x$ has a multiplicative inverse $\dfrac{1}{x}$. Then

$$\dfrac{1}{x}(x \cdot x) = \dfrac{1}{x} \cdot 0$$

and

$$\left(\dfrac{1}{x} \cdot x\right) \cdot x = \dfrac{1}{x} \cdot 0$$

$$1 \cdot x = 0$$

$$x = 0$$

thus assuming $x \neq 0$ leads to a contradiction.

(b) No, because $x + x = 1 \cdot x + 1 \cdot x = (1 + 1) \cdot x$. Hence, if $1 + 1 = 0$, then $x + x = 0$ even if $x \neq 0$.

(c) in which case $x$ must be 0.

## Exercises 5.3 (Page 179)

1. (a) $x^2 + 1$; $x^2 + 2x + 1$      (b) $x^2 + 1$; $x^2 + x + 1$
   (c) $x^3 + x^2 + x + 1$; $x^3 + 1$      (d) $x^3 + x^2 + x + 1$; $x^3 + 1$
   (e) $x^4 + 1$; $x^4 + x^3 + x + 1$      (f) $x^4 + 1$; $x^4 + 2x^3 + 2x + 1$

2. (a) $x^2 + 2x + 1$    (b) $x^2 + 2x + 1$    (c) $x^3 + 3x^2 + 3x + 1$
   (d) $x^3 + x^2 + 3x + 3$    (e) $x^4 + 2x^2 + 1$    (f) $x^4 + 2x^2 + 1$

3. (a) $x^2 + xy + yx + y^2$    (b) $x^2 - xy - yx + y^2$
   (c) $x^3 + x^2y + xyx + yx^2 + xy^2 + yxy + y^2x + y^3$
   (d) $x^3 - x^2y - xyx - yx^2 + xy^2 + yxy + y^2x - y^3$    (e) $4x^2$    (f) $8x^3$

4. If $xy + yx = 0$ for all $x, y$ in the ring, then $xx + xx = 0$, that is, $x^2 + x^2 = 0$, $x^2 = -x^2$.
   (a) $x^2 + y^2$      (d) Same answers as (c)
   (b) $x^2 + y^2$      (e) 0
   (c) $x^3 + xy^2 + yx^2 + y^3$      (f) 0
   or $x^3 + x^2y + y^2x + y^3$
   or $x^3 + y^2x + yx^2 + y^3$
   or $x^3 + x^2y + xy^2 + y^3$

5. (a) (1) $a$    (2) 1    (3) 1    (4) 1    (5) $b, b$    (6) $b, b$

   (7) $\dfrac{1}{a} - \dfrac{1}{b} = b - a = b + a = 1$

   $\dfrac{1}{ab} = \dfrac{1}{1} = 1$

   (8) $\dfrac{1}{a} + \dfrac{1}{b} = b + a = 1$; $\dfrac{a+b}{ab} = \dfrac{1}{1} = 1$

   (9) $\dfrac{-a}{b} = \dfrac{a}{b} = a\left(\dfrac{1}{b}\right) = a \cdot a = b$

   $-\left(\dfrac{a}{b}\right) = -(a \cdot a) = -b = b$

   $-\dfrac{a}{b} = \dfrac{a}{b} = a\left(\dfrac{1}{b}\right) = a \cdot a = b$

(10) $\dfrac{1}{a} \div \dfrac{a}{b} = b \div (b) = 1$

$\dfrac{1}{a} \times \dfrac{b}{a} = b \times a = 1$

$\dfrac{1 \div a}{a \div b} = \dfrac{b}{b} = 1$

(b) (1) {0}  (2) {1}  (3) {b}  (4) {a}  (5) {1}  (6) {0, 1}
(7) {1, a}  (8) {a, b}  (9) { }  (10) {0, a}  (11) { }
(12) {1, a}  (13) {1, b}

6. (a) (1) 0  (2) e  (3) c  (4) e  (5) + and · are associative
(6) d  (7) f  (8) Multiplication distributes over addition

(f) (1) 1  (2) 1  (3) $x^2 + y^2$  (4) $x^2 + y^2$
(5) $x^3 + x^2y + xy^2 + y^3$  (6) $x^3 + x^2y + xy^2 + y^3$  (7) 1  (8) d
(9) a  (10) a  (11) a  (12) 1  (13) f

(g) (1) {0}  (2) {1}  (3) {d}  (4) {a}  (5) {e}  (6) {0, 1}
(7) {0, 1}  (8) {a, b}  (9) {a, b}  (10) {e}  (11) {c}
(12) {d}  (13) {d}  (14) {f}  (15) {f}  (16) {a, d}
(17) {b, d}  (18) { }

7. Suppose $a \cdot b = 0$. If $a = 0$, there is nothing to prove. If $a \neq 0$, then

$$\dfrac{1}{a}(a \cdot b) = \dfrac{1}{a}(0)$$

$$\left(\dfrac{1}{a} \cdot a\right) \cdot b = 0$$

$$1 \cdot b = 0$$

$$b = 0$$

Conversely, if $a = 0$ or $b = 0$, according to a basic ring theorem $ab = 0$ (Theorem 4.2).

## Exercises 5.4 (Page 190)

1. (a) {2, 5}, {2, 9}  (b) {1, 4}, {1, 8}  (c) {3, 4}, { }
(d) {2, 3}, { }  (e) {2, 5}, {3, 8}  (f) {6, 2}, {7, 1}  (g) {1, 3}, {7, 1}
(h) {5}, {3, 2}  (i) { }, {7, 9}  (j) {1}, {3, 4},  (k) {4, 5}, { }

2. (a) {2, −2}, {2, −2}  (b) {1, −3}, {1, −3}  (c) { }, { }
(d) { }, { }  (e) { }, $\left(\dfrac{\sqrt{15}}{3}, \dfrac{-\sqrt{15}}{3}\right)$  (f) { }, $\left(4 + \dfrac{\sqrt{15}}{3}, 4 - \dfrac{\sqrt{15}}{3}\right)$
(g) {1, −4}, {1, −4}  (h) { }, { }  (i) { }, { }  (j) { }, { }
(k) { }, { }

3. (a) The quadratic formula does not apply to a field for which $1 + 1 = 0$.

(b)
$$\begin{array}{c|c|c|c|c|c|c|c|c} & x & 0 & 1 & a & b & c & d & e & f \\ & x+1 & 1 & 0 & c & f & a & e & d & b \\ & x(x+1) & 0 & 0 & d & a & d & b & b & a \end{array}$$

Consider the quadratic $ux^2 + vx = w$, $uv \neq 0$. It is equivalent to

$$\left(\frac{ux}{v}\right)^2 + \left(\frac{u}{v}x\right) = \frac{uw}{v^2} \quad \text{or} \quad \frac{ux}{v}\left(\frac{ux}{v}+1\right) = \frac{uw}{v^2}$$

First use above table to determine $\frac{ux}{v} = z$. Then $x = \frac{zv}{u}$ $(u \neq 0)$, $z(z+1) = \frac{uw}{v^2}$

(1) $\{b, f\}$    (2) $\{d, e\}$    (3) $\{\ \}$

## Exercises 5.5 (Page 195)

1. (a) $\{\ \}$   (b) $\{\ \}$   (c) $\{1\}$   (d) $\{3\}$   (e) $\{7\}$   (f) $\{-4\}$
   (g) $\{-4\}$   (h) $\{-4\}$

2. (a) $\{0, 1\}$   (b) $\{0, 1, 2\}$   (c)–(h) $\{\ \}$

3. (a) $\{0, 1\}$   (b) $\{0, 1\}$   (c) $\{0, 3\}$   (d) $\{0, 5\}$   (e) $\{0, 9\}$
   (f) $\{0, -2\}$   (g) $\{0, -2\}$   (h) $\{0, -2\}$

4. (a) $\{1\}$   (b) $\{1\}$   (c) $\{4\}$   (d) $\{3, 5\}$   (e) $\{2, 6\}$   (f) $\{\ \}$
   (g) $\{4+\sqrt{15}, 4-\sqrt{15}\}$   (h) $\{4+\sqrt{15}, 4-\sqrt{15}\}$

5. (a) $\{0\}$   (b) $\{0, 2\}$   (c) $\{1, 4\}$   (d) $\{3, 5\}$   (e) $\{7\}$
   (f) $\{1.5, -4\}$   (g) $\{1.5, -4\}$   (h) $\{1.5, -4\}$

6. (a) $\{\ \}$   (b) $\{\ \}$   (c) $\{4\}$   (d) $\{0\}$   (e) $\{6\}$   (f) $\{2\frac{1}{3}\}$
   (g) $\{2\frac{1}{3}\}$   (h) $\{2\frac{1}{3}\}$

7. (a) $\{0\}$   (b) $\{0, 1\}$   (c)–(h) $\{3, 4\}$

8. (a) $\{0\}$   (b) $\{2\}$   (c) $\{0, 2\}$   (d)–(h) $\{2, 5\}$

9. (a) $\{0\}$   (b) $\{0\}$   (c) $\{0, 1\}$   (d) $\{0, 2\}$   (e) $\{0, 6, 7\}$   (f) $\{0\}$
   (g) $\{0\}$   (h) $\{0, 1+.5\sqrt{10}i, 1-.5\sqrt{10}i\}$

10. (a) $\{1\}$   (b) $\{1, 2\}$   (c) $\{0, 2, 3\}$   (d) $\{3, 4\}$   (e)–(g) $\{\ \}$
   (h) $\{\pm i, \pm\sqrt{5}i\}$

11. (a) $\{\ \}$   (b) $\{\ \}$   (c) $\{0, 1, 2, 3, 4\}$   (d) $\{4\}$   (e) $\{7\}$
   (f) $\{18\}$   (g) $\{18\}$   (h) $\{18\}$

12. (a) $\{0\}$   (b) $\{1, 2\}$   (c) $\{3\}$   (d) $\{4, 5\}$   (e) $\{6, 9\}$   (f) $\{\frac{1}{2}, -2\}$
   (g) $\{\frac{1}{2}, -2\}$   (h) $\{\frac{1}{2}, -2\}$

13. (a) $\{0\}$   (b) $\{2\}$   (c) $\{1, 4\}$   (d)–(h) $\{-1, 6\}$

14. (a) $\{0\}$   (b) $\{\ \}$   (c)–(h) $\{4\}$

15. (a) $\{\ \}$   (b) $\{\ \}$   (c) $\{2\}$   (d) $\{5\}$   (e) $\{4\}$   (f)–(h) $\{\frac{1}{3}\}$

16. (a) $\{0\}$   (b) $\{0, 1\}$   (c) $\{0, 1, 2, 3\}$   (d) $\{0, 1, 2, 3, 4, 5\}$
   (e) $\{0, 1, 2, 3, 4, 5, 6, 7, 8, 9\}$   (f)–(h) All permissible numbers except $-1$

17. (a) $\{\ \}$   (b)–(h) $\{1\}$

**18.** (a)–(h) { }

**19.** (a) { }   (b)–(h) {1}

**20.** (a)–(h) { }

**21.** (a)–(c) {1}   (d) {1, 2, 4}   (e) {1}   (f) {1}   (g) {1}

(h) $\left\{1, \dfrac{-1}{2} + \dfrac{\sqrt{3}\,i}{2}, \dfrac{-1}{2} - \dfrac{\sqrt{3}}{2}i\right\}$

## Review Exercises (Page 196)

1. $\{3, 8\}$; $\{W_3, + \pmod{3}, \cdot \pmod{3}\}$, $\{1\}$; $\{W_5, + \pmod{5}, \cdot \pmod{5}\}$
2. $\{0, 1\}$. Yes. Every field has the elements 0, 1.
3. $\{4, 8\}$; $\{W_5, + \pmod{5}, \cdot \pmod{5}\}$; solution is 3
4. $\{2, 10\}$. No. 2 is a solution for each field.
5. $\{1, 9\}$
10. $\{2\}$

## Exercises 6.1 (Page 206)

1. The subsets of $W_2 = \{0, 1\}$ which do not contain 0 are $\{1\}$, $\varnothing$.
   Can $P = \{1\}$? No, because $1 + 1 = 0$ which is not in $P$.
   Can $P = \varnothing$? No, because $P$ cannot be empty as Axiom $\mathcal{O}2$ requires that either 1 or $-1$ must belong to $P$. Note that $1 = -1$ for this field. Thus none of the subsets of $W_2$ qualifies to be $P$.

3. (a) If $1 + 1 = 0$, then $1 = -1$. From Axiom $\mathcal{O}2$ either 1 or $-1$ is in $P$. As $1 = -1$, both 1 and $-1$ are in $P$ so that $1 + (-1) = 0$ is in $P$, contradicting Axiom $\mathcal{O}1$.

4. If $x$ is an ordered field, then $x$ is in $P \cup N \cup \{0\}$. If $x \in P$, then $-x$ cannot also be in $P$, for $x + (-x) = 0$ would have to be in $P$, violating Axiom $\mathcal{O}1$. Moreover, if $-x = 0$, then $x = -0 = 0$, contradicting that $x \in P$. As $-x$ is in the field, the only remaining possibility is to have $-x$ in $N$.

## Exercises 6.2 (Page 211)

1. (a) If $-x = 0$, then $x + (-x) = x + 0$ and therefore $0 = x$, contradicting the assumption that $x \neq 0$.
   (b) Yes. The very same proof holds as in (a).
   (c) Let $e$ be the identity element of a group $\{S, \circ\}$. If $x \neq e$, then $x' \neq e$. If $x' = e$, then $x \circ x' = x \circ e$ and $e = x$, contradicting the assumption that $x \neq e$.

2. (a) We know that $1 \in P$ and $-1 \notin P$ and $-1 \neq 0$. Hence $-1 \in N$ so that $-1 < 0$.
   (b) If $x \in P$, then $x + x \in P$ or $x - (-x) \in P$ so that $x > -x$ or $-x < x$.

5. Let $P$ be the positive elements of $S$ that are also in $T$.
   (1) As 0 is not a positive element of $S$, 0 is not in $P$.
   (2) Let $x$ be any nonzero element of $T$; then $x$ is a nonzero element of $S$. If $x$ is positive in $S$, then $x$ is in $P$. If $x$ is negative in $S$, then $-x$ is positive in $S$ and so must be in $P$.

**Answers and Solutions to Selected Exercises**

(3) Let $x$ and $y$ be in $P$. Then $x$ and $y$ are positive elements of $S$ that are in $T$. As $\{S, +, \cdot\}$ is an ordered field, $x+y$ and $x \cdot y$ are positive elements in $S$. As $\{T, +, \cdot\}$ is a field, $x+y$ and $x \cdot y$ are in $T$. As $x+y$ and $x \cdot y$ are positive elements of $S$ that are also in $T$, it follows that $x+y$ and $x \cdot y$ are in $P$. Hence $P$ is closed under $+$ and $\cdot$.

## Exercises 6.3 (Page 219)

1. $\{x < 5\}$

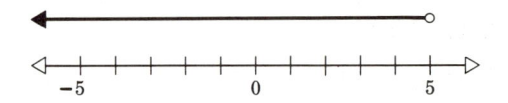

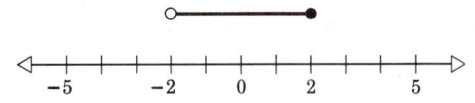

9. $\{-2 < x \leq 2\}$

15. (a) (i) Suppose $0 \leq x < y$. Let $p = y - x$. Then $x + p = y$ where $p > 0$.

   Then $(x+p)^3 = y^3$

   $x^3 + [3x^2p + 3xp^2 + p^3] = y^3$

   $x^3 < y^3$ because the bracketed quantity is positive.
   (*Note:* This result can also be established using Theorem 6.9c.)

   (ii) Suppose $x < 0 \leq y$. Then $x^3$ is negative and $y^3$ is nonnegative. Hence $x^3 < y^3$.

   (iii) Suppose $x < y < 0$. Then

   $$-x > -y > 0 \quad \text{or} \quad 0 < -y < -x.$$

   From (i) it follows that $(-y)^3 < (-x)^3$ or that $-y^3 < -x^3$ or $y^3 > x^3$ or $x^3 < y^3$.

   (b) If $x = y$, then $x^3 = y^3$, contradicting the given fact that $x^3 < y^3$. Hence $x \neq y$. If $y < x$, then from (a) $y^3 < x^3$, contradicting the given fact that $x^3 < y^3$. Thus $x < y$ is the only remaining possibility.

   (c) Suppose $x \neq y$; then $x \gtreqless y$ and hence by (a) $x^3 \gtreqless y^3$, contradicting $x^3 = y^3$.

## Exercises 6.4 (Page 226)

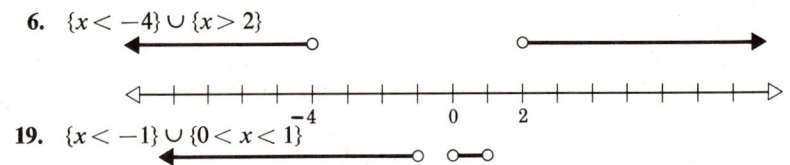

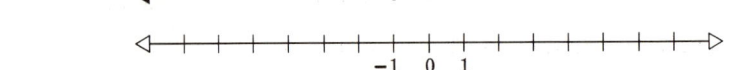

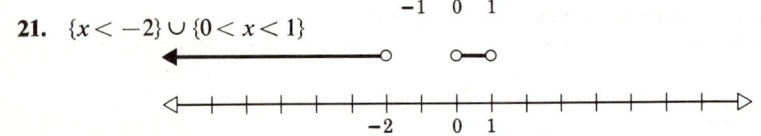

6. $\{x < -4\} \cup \{x > 2\}$

19. $\{x < -1\} \cup \{0 < x < 1\}$

21. $\{x < -2\} \cup \{0 < x < 1\}$

**30.**

**33.**

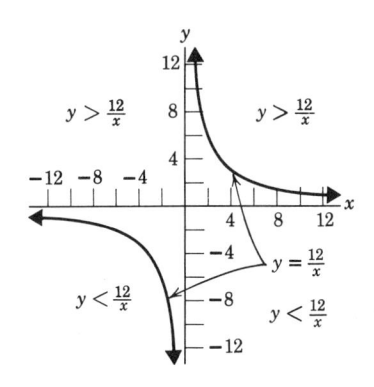

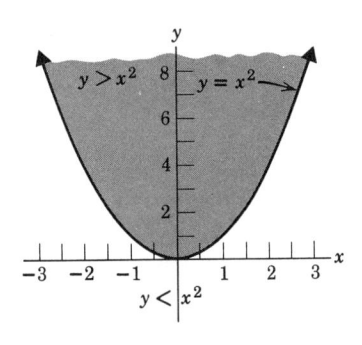

## Exercises 6.5 (Page 229)

**1.** As $x > 0$, $x + \dfrac{1}{x} \geq 2 \Leftrightarrow x^2 + 1 \geq 2x$

$$\Leftrightarrow x^2 - 2x + 1 \geq 0$$

$$\Leftrightarrow (x-1)^2 \geq 0$$

Equality holds if and only if $x = 1$. (*Alternate solution*) By Theorem 6.16a, if $x$ is positive, $(\sqrt{x})^2 + \left(\dfrac{1}{\sqrt{x}}\right)^2 \geq 2(\sqrt{x})\left(\dfrac{1}{\sqrt{x}}\right)$, that is, $x + \dfrac{1}{x} \geq 2$.

**4.** We must show $\sqrt{xy} > \dfrac{2xy}{x+y}$    for $x, y > 0$, $x \neq y$

We have $x + y \geq 2\sqrt{xy}$    from Theorem 6.16 for $\sqrt{x}, \sqrt{y}$

$$\therefore 1 \geq \dfrac{2\sqrt{xy}}{x+y} \quad \text{Theorem 6.9a; } z = \dfrac{1}{x+y}$$

and

$$\sqrt{xy} \geq \dfrac{2xy}{x+y} \quad \text{Theorem 6.9a; } z = \sqrt{xy}$$

**7.** (a) From $x > y$ and $z > w$ we have $x - y > 0$, $z - w > 0$ so that

$$(x-y)(z-w) > 0$$
$$xz - xw - yz + yw > 0$$

and

$$xz + yw > xw + yz$$

(b) $(x+z)(y+w) < (x+w)(y+z)$

$\Leftrightarrow xy + xw + zy + zw < xy + xz + wy + wz$

$\Leftrightarrow xw + zy < xz + wy$

$\Leftrightarrow xz + vw > xw + yz$ which is so from (a)

## Answers and Solutions to Selected Exercises

### Exercises 6.6 (Page 239)

4. $\{-2 < x < 12\}$   5. $\{-5, 4\}$   9. $\{-1 \leq x \leq 2\}$   14. $\{x \leq 0\}$   15. $\{5\}$
22. $\{\ \}$   23. $\{-1 - \sqrt{2.5} < x < -1 + \sqrt{2.5}\}$

31. (a)

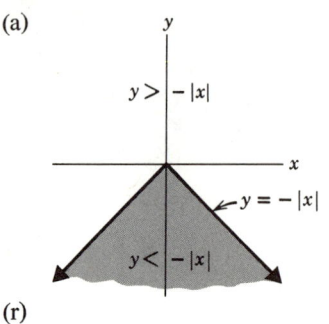

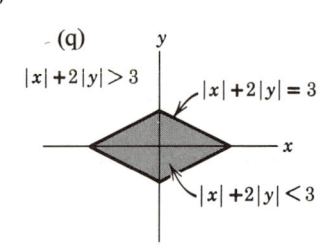

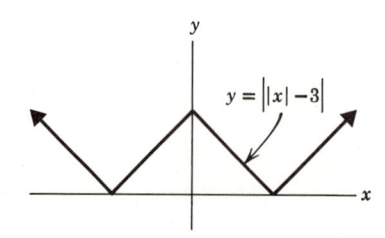

(q)

(r)

### Exercises 6.7 (Page 249)

1. The following are successor sets, and the others are not: (a), (b), (d), (f), (g), (k).

2. (a) $1 \in S_1 \cup S_2$ as $1 \in S_1$ and $1 \in S_2$. If $x \in S_1 \cup S_2$, then either $x \in S_1$ or $x \in S_2$. If $x \in S_1$, then $x + 1 \in S_1$ and so $x + 1 \in S_1 \cup S_2$. If $x \in S_2$, then $x + 1 \in S_2$ and so $x + 1 \in S_1 \cup S_2$.
   (b) Let $S$ be a union of successor sets. Then $1 \in S$ as $1 \in$ every successor set. If $x \in S$, then $x \in$ one of the successor sets, say $S_1$. Then $x + 1 \in S_1$ and so $x + 1 \in S$.

3. Suppose there were an integer $n$ between 0 and 1. We would then have $0 < n < 1$. Adding 1 would then yield $1 < n + 1 < 2$, implying that there is an integer between 1 and 2, thereby contradicting Theorem 6.32.

   Other proofs are possible, for example:
   Suppose there were at least one integer between 0 and 1. By the well ordering theorem (Th 6.36) there would be a *least* such integer $n$. Then since $0 < n < 1$ it would follow that $n \cdot 0 < n \cdot n < n \cdot 1$. Hence $0 < n^2 < n < 1$ showing that $n^2$ is less than $n$ and lies between 0 and 1. Contradiction.

8. The set of positive rational numbers does not have a least element.

### Exercises 6.8 (Page 253)

3. Let $A$ be the set of all positive integers $n$ for which the following equation holds:

$$1^2 + 2^2 + 3^2 + \cdots + n^2 = \frac{n(n+1)(2n+1)}{6}$$

If $n=1$, we have but one term for the left member and $1^2 = \dfrac{1(1+1)(2 \cdot 1 + 1)}{6}$ or $1 = \dfrac{1 \cdot 2 \cdot 3}{6}$ which is true, showing that $1 \in A$.

If positive integer $x$ is in $A$, then from the definition of $A$ we must have

$$1^2 + 2^2 + 3^2 + \cdots + x^2 = \dfrac{x(x+1)(2x+1)}{6}$$

If we add the next number $(x+1)^2$ to both members, we get

$$1^2 + 2^2 + 3^2 + \cdots + x^2 + (x+1)^2 = \dfrac{x(x+1)(2x+1)}{6} + (x+1)^2$$
$$= \tfrac{1}{6}(x+1)[x(2x+1) + 6(x+1)]$$
$$= \tfrac{1}{6}(x+1)[2x^2 + 7x + 6]$$
$$= \tfrac{1}{6}(x+1)(x+2)(2x+3)$$
$$= \tfrac{1}{6}(x+1)([x+1]+1)(2[x+1]+1)$$

which shows that $(x+1)$ must also be in $A$. Hence $A$ is a successor set. But $A \subseteq N$. It then follows by Mathematical Induction that $A = N$ and the initial equation holds for every positive integer.

8. Let $A$ be the set of all positive integers $n$ for which the following inequation holds:

$$2^n > n$$

We check $n=1$ in the above inequation getting $2 > 1$ which is true. Hence $1 \in A$. If positive integer $x$ is in $A$, then from the definition of $A$ we must have

$$2^x > x$$

If we multiply both members by the positive number 2, we get

$$2^x \cdot 2 > x \cdot 2$$

or

$$2^{x+1} > 2x$$

or

$$2^{x+1} > x + x$$

As $x$ is a positive integer, we have $x \geq 1$ so that $x + x \geq x + 1$. Hence

$$2^{x+1} > x + 1$$

showing that $(x+1)$ must also be in $A$. Hence $A$ is a successor set. But $A \subseteq N$. It then follows by Mathematical Induction that $A = N$ and that the initial inequation holds for every positive integer.

13. Let $A$ be the set of all positive integers $n$ for which the following becomes a true statement: $\dfrac{5^n - 2^n}{3}$ is an integer. If $n=1$, we get the true statement $\dfrac{5-2}{1}$ is an integer so that $1 \in A$.

If positive integer $x$ is in $A$, from the definition of $A$ the following must be true:
$$\frac{5^x - 2^x}{3} \text{ is an integer}$$
We would like to prove that for this $x$, $x + 1$ is also in $A$. Replacing $x$ by $x + 1$, we get
$$\frac{5^{x+1} - 2^{x+1}}{3} = \frac{5 \cdot 5^x - 2 \cdot 2^x}{3} = \frac{(3+2) \cdot 5^x - 2 \cdot 2^x}{3}$$
$$= \frac{3 \cdot 5^x + 2 \cdot 5^x - 2 \cdot 2^x}{3} = 5^x + 2\frac{(5^x - 2^x)}{3}$$
$$= \text{an integer} + 2 \times \text{(an integer)}$$
$$= \text{some integer}$$

Hence $(x + 1) \in A$ and $A$ is a successor set. As $A \subseteq N$ it follows by Mathematical Induction that $A = N$, and so $\frac{5^n - 2^n}{3}$ is an integer for every positive integer $n$.

26. Let $A$ be the set of all positive integers $n$ for which the following sentence is true:

The number of chords joining $n$ points of a circle in every way is $(\frac{1}{2})n(n-1)$.

For $n = 1$ we get the true statement that the number of chords joining one point of a circle in every way is
$$(\tfrac{1}{2})(1)(1 - 1) = 0, \quad \text{so } 1 \in A.$$
Let positive integer $x$ be in $A$. From the definition of $A$ we must have that the number of chords joining $x$ points of a circle in every way is $(\frac{1}{2})x(x-1)$. We would like to prove that for such an $x$, $x + 1$ is also in $A$. Let $A_1, A_2, \ldots, A_x, A_{x+1}$ be $x + 1$ points of a circle. Then the number of chords for the $x$ points $A_1, A_2, \ldots, A_x$ is $(\frac{1}{2})x(x-1)$. Introducing the point $A_{x+1}$ simply introduces the additional $x$ chords, $\overline{A_1 A_{x+1}}, \overline{A_2 A_{x+1}}, \ldots, \overline{A_x A_{x+1}}$, making the total number of chords for the $x + 1$ points $A_1, A_2, \ldots, A_x, A_{x+1}$.
$$(\tfrac{1}{2})x(x-1) + x = (\tfrac{1}{2})x^2 - (\tfrac{1}{2})x + x = (\tfrac{1}{2})x^2 + (\tfrac{1}{2})x$$
$$= (\tfrac{1}{2})(x+1)x = (\tfrac{1}{2})(x+1)([x+1]-1)$$
so $(x + 1) \in A$ and $A$ is a successor set. As $A \subseteq N$ it follows from Mathematical Induction that $A = N$ and that the initial sentence is true for every positive integer $n$.

29. Let $A$ be the set of all positive integers for which the following inequation is true:
$$(1.1)^{n+1} > 1 + .1(n + 1)$$
For $n = 1$ we get the true statement
$$(1.1)^2 > 1 + .1(2) \text{ because } 1.21 > 1 + .2$$
Hence $1 \in A$. Let positive integer $x$ be in $A$. From the definition of $A$ it follows that
$$(1.1)^{x+1} > 1 + .1(x + 1)$$
Multiplying both members by the positive number 1.1, we get
$$(1.1)^{[x+1]+1} > [1 + .1(x + 1)](1.1)$$
$$> 1.1 + .11x + .11$$
$$> 1 + .11x + .21$$
$$> (1 + .1x + .2) + (.01x + .01)$$
$$> 1 + .1(x + 2)$$
$$> 1 + .1([x + 1] + 1)$$

Answers and Solutions to Selected Exercises    411

so that $(x+1) \in A$, making $A$ a successor set. As $A \subseteq N$, it follows by Mathematical Induction that $A = N$ and the initial inequation is true for every positive integer $n$.

## Review Exercises (Page 255)

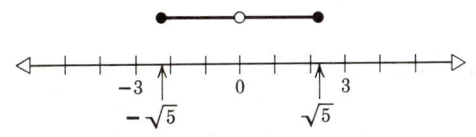

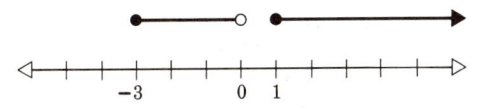

1. (b) $\{-\sqrt{5} \leq x < 0\} \cup \{0 < x \leq \sqrt{5}\}$

(e) $\{-3 \leq x < 0\} \cup \{1 \leq x\}$

(h) {All real numbers}

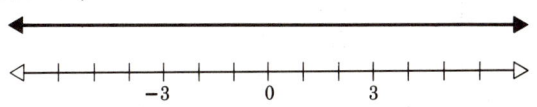

2. (a) $a^2 = -1 < 0$, contradicting Theorem 6.6: $a^2 \geq 0$

# CHAPTER 7

## Exercises 7.1 (Page 261)

1.  
|     | Upper Bound | Lower Bound |
| --- | --- | --- |
| (a) | Any $n \geq 10$ | Any $n \leq 9$ |
| (b) | Any $n \geq 1$ | Any $n \leq 0$ |
| (c) | Any $n \geq 1\frac{1}{9}$ | Any $n \leq 1$ |
| (d) | Any $n \geq 2\frac{2}{9}$ | Any $n \leq 2$ |
| (e) | Any $n \geq 1$ | Any $n \leq \frac{1}{2}$ |
| (f) | Any $n \geq 10$ | Any $n \leq -10$ |
| (g) | Any $n \geq 1$ | Any $n \leq -.1$ |
| (h) | Any $n \geq 1$ | Any $n \leq -1$ |
| (i) | Any $n \geq 2$ | None |
| (j) | Any $n \geq 4$ | Any $n \leq -4$ |
| (k) | Any $n \geq \sqrt{2}$ | Any $n \leq -\sqrt{2}$ |
| (l) | None | None |
| (m) | Any $n \geq \sqrt{3}$ | Any $n \leq -\sqrt{3}$ |
| (n) | None | None |
| (o) | Any $n \geq \frac{1}{9}$ | Any $n \leq -.1$ |
| (p) | None | Any $n \leq 1$ |
| (q) | Any $n \geq 2$ | Any $n \leq 1$ |
| (r) | Any $n \geq 2$ | Any $n \leq -\frac{3}{2}$ |
| (s) | Any $n \geq 2$ | Any $n \leq 1$ |
| (t) | Any $n \geq 1$ | Any $n \leq \frac{1}{2}$ |

(u) None    Any $n \leq 1$
(v) Any $n \geq 1$    Any $n \leq \frac{1}{2}$
(w) None    Any $n \leq 1$

2. (a) Let $c$ be a fixed element of $S$. Let $m$ be an upper bound for $A$ so that $x \leq m$ for every $x$ in $A$. Then $x + c \leq m + c$ for every $x$ in $A$ or for every $(x + c)$ in the set $A + c$ obtained by adding $c$ to every number in $A$. Hence $m + c$ is an upper bound for $A + c$.

## Exercises 7.2 (Page 266)

1.

|     | LUB            | GLB            |     | LUB  | GLB           |
|-----|----------------|----------------|-----|------|---------------|
| (a) | 10             | 9              | (j) | 4    | None          |
| (b) | 1              | 0              | (k) | $\frac{1}{4}$ | $\frac{1}{10}$ |
| (c) | $1\frac{1}{3}$ | 1              | (l) | None | 1             |
| (d) | $2\frac{2}{3}$ | 2              | (m) | 2    | 1             |
| (e) | 1              | $\frac{1}{2}$  | (n) | 2    | $-1$          |
| (f) | 10             | $-10$          | (o) | 1    | 0             |
| (g) | 1              | $-.1$          | (p) | 1    | $-\frac{1}{2}$ |
| (h) | 1              | $-1$           | (q) | 2    | 1             |
| (i) | 2              | None           | (r) | 1    | $\frac{1}{4}$ |

2. (a) Let $s$ be the fixed element of $S$ added to each element of $A$ to produce the set $A + s$. Let $b$ be the LUB of $A$. Then for every $x$ in $A$, $x \leq b$ and there is no smaller upper bound. Then $x + s \leq b + s$ for every $x$ in $A$. $b + s$ is the LUB for $A + s$, for if $c + s$ were a smaller upper bound, we would have $x + s \leq c + s < b + s$ and hence $x \leq c < b$, so that $c < b$ and $c$ would then turn out to be a lesser upper bound for $A$ than $b$.

## Exercises 7.3 (Page 275)

1. Suppose $2 = \dfrac{a^2}{b^2}$ where $a$ and $b$ are relatively prime; then $2b^2 = a^2$. The number of prime factors for $2b^2$ is an odd number, whereas the number of prime factors for $a^2$ is an even number. This is not possible from the uniqueness of prime factorization.

9. Suppose rational number $b > 0$ is such that $2 < b^2$. By Theorem 7.4 there is a positive rational number $r$ whose square is between 2 and $b^2$, that is,
$$2 < r^2 < b^2$$
We now show that $r < b$. If $r = b$, then $r^2 = b^2$. If $b < r$ and since both $b$ and $r$ are positive, we have $b^2 < r^2$ by Theorem 6.9. Hence $r < b$ is the only remaining possibility. Hence for every $b$ in $B$ there is a smaller number so that $B$ cannot have a least number.

## Review Exercises (Page 275)

1.     (a) LUB    (b) GLB
(1)    $\frac{1}{3}$    .3
(2)    .3    0
(3)    2.3    2
(4)    2.3    $-2.03$
(5)    1    $\frac{2}{3}$
(6)    None    1

2. Suppose $B$ had a least element, say $b$; then $3 < b^2$. By Theorem 7.4 there is a positive rational number $r$ such that $3 < r^2 < b^2$. As $r = b$ and $b < r$ are not possible, we must have $r < b$ as the only remaining possibility. Hence $b$ is not least, contradicting the assumption that $b$ was the least element in $B$.

5. (a) $\{x \mid 7 < x, x \in Q\}$  (b) $\{x \mid x < 7, x \in Q\}$  (c) $\{x \mid 7 < x^2, 0 < x, x \in Q\}$
   (d) $\{x \mid x^2 < 7, x \in Q\}$  (e) $\{x \mid 7 < x^2, 0 < x < 7, x \in Q\}$
   (f) $\{x \mid 2 < x^2 < 3, x \in Q\}$  (g) $\{x \mid 2 \leq x < 3, x \in Q\}$
   (h) $\{x \mid 2 < x \leq 3, x \in Q\}$

# CHAPTER 8

## Exercises 8.2 (Page 288)

3. (a) $\{U, +, \cdot\}$ is not a field.
   (b) $\{U, +, \cdot\}$ is not a field.
   (c) $\{U, +, \cdot\}$ is not a field.

5. (a) Yes, this is a vector space. $S$ is a field and $V$ is an abelian group.
   - ($v_1$) $s(\vec{v}_1 + \vec{v}_2) = s\vec{v}_1 + s\vec{v}_2$ checks readily, for the two values $s$ may have 0 or 1.
   - ($v_2$) $(s_1 + s_2)\vec{v} = s_1\vec{v} + s_2\vec{v}$. The only difficulty occurs when $s_1 = s_2 = 1$. But then we obtain $\vec{v} + \vec{v}$. Since addition is $+$(mod 2), all coefficients will be 0 so that $\vec{v} + \vec{v} = \vec{0}$, and this condition checks.
   - ($v_3$) $(s_1 s_2)\vec{v} = s_1(s_2\vec{v})$; again the only situation that needs checking is the case where $s_1 = s_2 = 1$, which checks immediately.
   - ($v_4$) $1 \cdot \vec{v} = \vec{v}$ clearly, for 1 is the identity for $\{0, 1\}$.
   
   (b) Yes, this is a vector space. $S$ is a field and $V$ is an abelian group.
   $S = \{A, +(\text{mod } 3), \cdot (\text{mod } 3)\}$   $V = \{A, +(\text{mod } 3)\}$ where $A = \{0, 1, 2\}$
   - ($v_1$) $s(\vec{v}_1 + \vec{v}_2) = s\vec{v}_1 + s\vec{v}_2$ $\Big\}$ since $S$ is a field, multiplication distributes over
   - ($v_2$) $(s_1 + s_2)\vec{v} = s_1\vec{v} + s_2\vec{v}$   addition.
   - ($v_3$) $(s_1 s_2)\vec{v} = s_1(s_2\vec{v})$   Multiplication is associative.
   - ($v_4$) $1 \cdot \vec{v} = \vec{v}$   1 is the multiplicative identity.
   
   (c) Yes, this is a vector space.
   $S = \{Q, +, \cdot\}$   $V = \{\sqrt{2}Q, +\}$   $S$ is a field and $V$ is an abelian group.
   - ($v_1$) $s(\vec{v}_1 + \vec{v}_2) = s(a\sqrt{2} + b\sqrt{2}) = s[(a+b)\sqrt{2}]$
     $= [s(a+b)]\sqrt{2} = (sa + sb)\sqrt{2}$
     $= (sa)\sqrt{2} + (sb)\sqrt{2}$
     $= s(a\sqrt{2}) + s(b\sqrt{2})$
     $= s\vec{v}_1 + s\vec{v}_2$
   - ($v_2$) $(s_1 + s_2)\vec{v} = (s_1 + s_2)[a\sqrt{2}]$
     $= [(s_1 + s_2)a]\sqrt{2} = (s_1 a + s_2 a)\sqrt{2}$
     $= (s_1 a)\sqrt{2} + (s_2 a)\sqrt{2}$
     $= s_1(a\sqrt{2}) + s_2(a\sqrt{2})$
     $= s_1\vec{v} + s_2\vec{v}$
   - ($v_3$) $(s_1 s_2)\vec{v} = s_1(s_2\vec{v})$   Multiplication is associative.
   - ($v_4$) $1 \cdot \vec{v} = \vec{v}$   1 is the multiplicative identity.

8. (a) Yes
   (b) No; $(1+x^2)+(-x^2)=1$. Degree is 0, not 2. Also it has no identity element.
   (c) No; $\{V, +\}$ has no identity element.

9. Yes, this a vector space. $\{R, +, \cdot\}$ is the field of scalars. We check that $\{V, +\}$ is an abelian group: $x_1 + 2y_1 = 0$ and $x_2 + 2y_2 = 0$ imply $(x_1 + x_2) + 2(y_1 + y_2) = 0$. Define addition for

$$\{V, +\} \text{ by } (x_1, y_1) + (x_2, y_2) = (x_1 + x_2, y_1 + y_2).$$

Also, if $x + 2y = 0$, then $ax + 2(ay) = 0$, so that $a(x, y) = (ax, ay) \in V$.
($v_1$) $a[(x_1, y_1) + (x_2, y_2)] = a(x_1 + x_2, y_1 + y_2) = (ax_1 + ax_2, ay_1 + ay_2)$
$\qquad = (ax_1, ay_1) + (ax_2, ay_2) = a(x_1, y_1) + a(x_2, y_2)$
($v_2$) $(a+b)(x, y) = (ax + bx, ay + by) = (ax, ay) + (bx, by) = a(x, y) + b(x, y)$
($v_3$) $(ab)(x, y) = (abx, aby)$
$\qquad = a[(bx, by)]$
$\qquad = a[b(x, y)]$
($v_4$) $1 \cdot (x, y) = (1 \cdot x, 1 \cdot y) = (x, y)$

## Exercises 8.3 (Page 296)

2. $a\vec{x} = a\vec{y}, a \neq 0$      Assumption

$\dfrac{1}{a}(a\vec{x}) = \dfrac{1}{a}(a\vec{x})$      "$\dfrac{1}{a}(a\vec{x})$" names but one vector

$\dfrac{1}{a}(a\vec{x}) = \dfrac{1}{a}(a\vec{y})$      Replacement Axiom

$\left(\dfrac{1}{a} \cdot a\right)\vec{x} = \left(\dfrac{1}{a} \cdot a\right)\vec{y}$      ($v_3$)

$1 \cdot \vec{x} = 1 \cdot \vec{y}$      $\dfrac{1}{a} \cdot a = 1$

$\vec{x} = \vec{y}$      ($v_4$)

Conversely,

$\vec{x} = \vec{y}$      Assumption
$a\vec{x} = a\vec{x}$      "$a\vec{x}$" names but one vector
$a\vec{x} = a\vec{y}$      Replacement Axiom

3. $a = b$      Assumption
$a\vec{x} = a\vec{x}$      "$a\vec{x}$" names but one vector
$a\vec{x} = b\vec{x}$      Replacement Axiom

Conversely

$a\vec{x} = b\vec{x}, \vec{x} \neq \vec{0}$      Assumption
$a\vec{x} + (-b\vec{x}) = \vec{0}$      $\{V, +\}$ is a group
$a\vec{x} + (-b)\vec{x} = \vec{0}$      Theorem 8.3
$(a-b)\vec{x} = \vec{0}$      ($v_2$)
$a - b = 0$      Theorem 8.2b
$a = b$      Addition of $b$ to both members

## Exercises 8.4 (Page 302)

1. Linearly independent: (b), (c), (e)
   Linearly dependent: (a), (d), (f), (g)

2. (a) $c_1 = 4$, $c_2 = -1$
   (b) $c_1 = 5$, $c_2 = -2$
   (c) $c_1 = 0$, $c_2 = 0$
   (d) $c_1 = -3a + 2b$, $c_2 = 2a - b$
   $\{(1, 2), (2, 3)\}$ is linearly independent, is a basis for $\{V_2, +\}$, and spans $\{V_2, +\}$.

7. (a) The following sets span $\{V_2, +\}$, and the remaining ones do not: (1), (2), (3), (5), (6), (7) provided $a \neq b$, (8) provided $a \neq -b$.
   (b) The following sets form a basis for $\{V_2, +\}$, and the remaining ones do not: (1), (7) provided $a \neq b$, (8) provided $a \neq -b$.

## Review Exercises (Page 303)

2. (a) This system satisfies all conditions for a vector space.
   (b) This system is *not* a vector space because the scalars do not form a field.
   (c) This system is *not* a vector space because property ($v_2$) fails as indicated by the following counterexample:

$$(1 + 1)(2, 3) = (2)(2, 3) = (2 \cdot 2(\bmod 3), 3) = (1, 3)$$

but

$$1(2, 3) + 1(2, 3) = (2 + 2(\bmod 3), 3 + 3(\bmod 4)) = (1, 2)$$

Hence $(1 + 1)(2, 3) \neq 1(2, 3) + 1(2, 3)$. (It is interesting to note that ($v_2$) is the only vector space property that fails.)

## Exercises 9.1 (Page 312)

4. By definition, if $\{\vec{v}_1, \vec{v}_2, \vec{v}_3\}$ is linearly dependent, there exist scalars $s_1$, $s_2$, $s_3$ *not all zero* such that

$$s_1 \vec{v}_1 + s_2 \vec{v}_2 + s_3 \vec{v}_3 = \vec{0}$$

If, say $s_1 \neq 0$, then this implies

$$\vec{v}_1 = -\frac{s_2}{s_1} \vec{v}_2 - \frac{s_3}{s_1} \vec{v}_3$$

expressing $\vec{v}_1$ in terms of the other two vectors $\vec{v}_2$ and $\vec{v}_3$. A similar argument holds if $s_2 \neq 0$ or $s_3 \neq 0$.

**5.** (a) $\begin{cases} 2x_1 - 3x_2 = 0 \\ 3x_1 + 4x_2 = 0 \end{cases} \Leftrightarrow \begin{cases} 1x_1 - \frac{3}{2}x_2 = 0 \\ 3x_1 + 4x_2 = 0 \end{cases}$

$\Leftrightarrow \begin{cases} 1x_1 + \frac{3}{2}x_2 = 0 \\ 0x_1 + \frac{17}{2}x_2 = 0 \end{cases}$

$\Leftrightarrow \begin{cases} 1x_1 + \frac{3}{2}x_2 = 0 \\ 0x_1 + 1x_2 = 0 \end{cases}$

$\Leftrightarrow \begin{cases} 1x_1 + 0x_2 = 0 \\ 0x_1 + 1x_2 = 0 \end{cases}$

Solution: $x_1 = 0$, $x_2 = 0$

**6.** (a) $x_1(2, 3) + x_2(3, 4) = 0 \Leftrightarrow \begin{cases} 2x_1 + 3x_2 = 0 \\ 3x_1 + 4x_2 = 0 \end{cases}$

(using Gauss-Jordan method) $\Leftrightarrow \begin{cases} 1x_1 + 0x_2 = 0 \\ 0x_1 + 1x_2 = 0 \end{cases}$

$\Leftrightarrow x_1 = 0$ and $x_2 = 0$

Hence the vectors are linearly independent.

(b) Linearly independent
(c) Linearly dependent
(d) Linearly independent
(e) Linearly independent

**7.** (a) $\begin{cases} 2x_1 - 3x_2 = 1 \\ 3x_1 + 4x_2 = 2 \end{cases} \Leftrightarrow \begin{cases} 1x_1 - \frac{3}{2}x_2 = \frac{1}{2} \\ 3x_1 + 4x_2 = 2 \end{cases}$

$\Leftrightarrow \begin{cases} 1x_1 - \frac{3}{2}x_2 = \frac{1}{2} \\ 0x_1 + \frac{17}{2}x_2 = \frac{1}{2} \end{cases}$

$\Leftrightarrow \begin{cases} 1x_1 - \frac{3}{2}x_2 = \frac{1}{2} \\ 0x_1 + 1x_2 = \frac{1}{17} \end{cases}$

Answers and Solutions to Selected Exercises 417

$$\Leftrightarrow \begin{cases} 1x_1 + 0x_2 = \dfrac{10}{17} \\ 0x_1 + 1x_2 = \dfrac{1}{17} \end{cases}$$

Solution: $x_1 = \dfrac{10}{17}$, $x_2 = \dfrac{1}{17}$

## Exercises 9.2 (Page 318)

1. $\begin{bmatrix} 1 & -3 & | & 0 \\ 2 & 5 & | & 0 \end{bmatrix}$

   $\begin{bmatrix} 1 & -3 & | & 0 \\ 0 & 11 & | & 0 \end{bmatrix}$

   $\begin{bmatrix} 1 & 0 & | & 0 \\ 0 & 1 & | & 0 \end{bmatrix}$

   Unique solution: $x_1 = 0$, $x_2 = 0$

8. $\begin{bmatrix} 1 & 1 & -1 & | & 0 \\ 1 & 4 & -3 & | & 0 \\ 4 & 1 & -2 & | & 0 \end{bmatrix}$

   $\begin{bmatrix} 1 & 1 & -1 & | & 0 \\ 0 & 3 & -2 & | & 0 \\ 0 & -3 & 2 & | & 0 \end{bmatrix}$

   $\begin{bmatrix} 1 & 0 & -\frac{1}{3} & | & 0 \\ 0 & 1 & -\frac{2}{3} & | & 0 \\ 0 & 0 & 0 & | & 0 \end{bmatrix}$

   Infinitely many solutions given by $\begin{cases} x_1 - \frac{1}{3}x_3 = 0 \\ x_2 - \frac{2}{3}x_3 = 0 \\ x_3 = \text{any real number} \end{cases}$

   or equivalently, $\begin{cases} x_1 = x_3 \\ x_2 = 2x_3 \\ x_3 = \text{any real number} \end{cases}$

9. $\begin{bmatrix} 1 & 1 & 1 & | & 6 \\ 1 & 2 & 3 & | & 14 \\ 2 & 1 & -1 & | & 1 \end{bmatrix}$

   $\begin{bmatrix} 1 & 1 & 1 & | & 6 \\ 0 & 1 & 2 & | & 8 \\ 0 & -1 & -3 & | & -11 \end{bmatrix}$

   $\begin{bmatrix} 1 & 0 & -1 & | & -2 \\ 0 & 1 & 2 & | & 8 \\ 0 & 0 & -1 & | & -3 \end{bmatrix}$

$$\begin{bmatrix} 1 & 0 & 0 & | & 1 \\ 0 & 1 & 0 & | & 2 \\ 0 & 0 & 1 & | & 3 \end{bmatrix}$$

Unique solution: $x_1 = 1$, $x_2 = 2$, $x_3 = 3$.

10. (a) $\begin{bmatrix} 4 & 6 \\ 6 & 8 \end{bmatrix}$

$\begin{bmatrix} 1 & \frac{3}{2} \\ 0 & -1 \end{bmatrix}$

$\begin{bmatrix} 1 & 0 \\ 0 & 1 \end{bmatrix}$ Unit matrix indicates that $\vec{v}_1$ and $\vec{v}_2$ are linearly independent.

11. (a) $\begin{bmatrix} 2 & 8 \\ 3 & a \end{bmatrix}$

$\begin{bmatrix} 1 & 4 \\ 0 & a-12 \end{bmatrix}$ This matrix will be singular if and only if $a - 12 = 0$, that is, if and only if $a = 12$.

(b) $\begin{bmatrix} 2 & 4+2a \\ 3 & 5+2a \end{bmatrix}$

$\begin{bmatrix} 1 & 2+a \\ 0 & -1-a \end{bmatrix}$ Singular matrix if and only if $-1 - a = 0$, that is, $a = -1$

(c) $\begin{bmatrix} 2 & 5 \\ a & 3a \end{bmatrix}$

$\begin{bmatrix} 1 & \frac{5}{2} \\ 0 & -\frac{a}{2} \end{bmatrix}$ Singular matrix if and only if $-\frac{a}{2} = 0$, that is, $a = 0$

(d) $\begin{bmatrix} 2 & 6 \\ a & 3a \end{bmatrix}$

$\begin{bmatrix} 1 & 3 \\ 0 & 0 \end{bmatrix}$ Singular matrix, regardless of the value of $a$.

Hence $\vec{v}_1$ and $\vec{v}_2$ are linearly dependent for all real values of $a$.

(e) $\begin{bmatrix} 2 & 0 \\ a & 3 \end{bmatrix}$

$\begin{bmatrix} 1 & 0 \\ 0 & 3 \end{bmatrix}$ This matrix is never singular, regardless of the value of $a$.

Hence there are *no* values of $a$ that will make $\vec{v}_1$ and $\vec{v}_2$ linearly dependent.

## Exercises 9.3 (Page 329)

9.

|   | x | y | z |   |
|---|---|---|---|---|
| ① | 2 | $-1$ | 2 |
| 3 | $-1$ | 0 | 5 | Initial tableau
| 1 | 9 | $-4$ | 3 |

| 1 | 2 | $-1$ | 2 |
| 0 | ⑦ (with $-$) | 3 | $-1$ | Tableau after first iteration
| 0 | 7 | $-3$ | 1 |

| 1 | 0 | $-\frac{1}{7}$ | $\frac{12}{7}$ |
| 0 | 1 | $-\frac{3}{7}$ | $\frac{1}{7}$ | Tableau after second iteration
| 0 | 0 | 0 | 0 |

The original system of three equations is now seen to be equivalent to the following system of two equations:

$$\begin{cases} x - \frac{1}{7}z = \frac{12}{7} \\ y - \frac{3}{7}z = \frac{1}{7} \end{cases}$$

or equivalently,

$$\begin{cases} x = \frac{1}{7}z + \frac{12}{7} \\ y = \frac{3}{7}z + \frac{1}{7} \end{cases}$$

Since $z$ can be any real number, there are *infinitely* many solutions. This is due to the fact that the original system was "redundant." In fact, the third equation can be obtained by multiplying the first equation by 4 and subtracting the second equation from this result.

## Review Exercises (Page 330)

1. (a) $x = -17$, $y = 11$
   (b) $x = 2$, $y = 2$, $z = 1$
   (c) No solution
   (d) $x = -\frac{2}{5}z + 2\frac{2}{5}$, $y = \frac{3}{5}z + 1\frac{2}{5}$, $z =$ any number

2. (a) $2\vec{v}_1 - \vec{v}_2 + \vec{v}_3 = \vec{0}$, hence linearly dependent
   (b) Linearly independent
   (c) Linearly independent

3. (a) $a = -4\frac{1}{2}$  (b) $a = 4\frac{1}{2}$  (c) $a = 2$  (d) $a = -3\frac{1}{3}$  (e) $a = 0$
   (f) $a = 0$  (g) $a = 1$ or $a = 2$

# Answers and Solutions to Selected Exercises

## Exercises 10.1 (Page 341)

1. (a) $\begin{bmatrix} 1 & 1 \\ 1 & -1 \end{bmatrix} \begin{bmatrix} x \\ y \end{bmatrix} = \begin{bmatrix} 5 \\ 1 \end{bmatrix}$

   (c) $\begin{bmatrix} 1 & 1 & 1 & 1 \\ 2 & 1 & -3 & -1 \end{bmatrix} \begin{bmatrix} x \\ y \\ z \\ w \end{bmatrix} = \begin{bmatrix} 9 \\ -2 \end{bmatrix}$

2. (b) $\begin{bmatrix} 6 & 4 & 7 \\ 13 & 2 & 6 \\ -2 & 4 & 5 \end{bmatrix}$

   (c) $\begin{bmatrix} 3 & 10 \\ 4 & 10 \end{bmatrix}$

   (b) and (c) illustrate the noncommutativity of matrix multiplication.

   (d) $\begin{bmatrix} 0 & 0 \\ 0 & 0 \end{bmatrix}$

   This shows that a product of two nonzero matrices can yield the zero matrix.

## Exercises 10.2 (Page 347)

1. (a) $\begin{bmatrix} 1 \\ 2 \\ 3 \end{bmatrix}$   (d) $[1]$   (t) $[66]$   (v) $\begin{bmatrix} 2 & 5 & 8 \\ 4 & 10 & 16 \\ 6 & 15 & 24 \end{bmatrix}$

5. (c) $\sum_{i=1}^{4}(i+1) = 2+3+4+5 = 14$

   (h) $\sum_{j=1}^{2}\sum_{i=1}^{3}(i+j) = \sum_{j=1}^{2}[(1+j)+(2+j)+(3+j)]$
   $= \sum_{j=1}^{2}(6+3j) = (6+3)+(6+6) = 21$

   (p) $\sum_{j=1}^{2}\sum_{i=1}^{3}(ij^2) = \sum_{j=1}^{2}(1j^2+2j^2+3j^2) = \sum_{j=1}^{2}(6j^2) = 6(1)^2+6(2)^2 = 30$

6. (a) $\sum_{i=1}^{4} x^i = x^1+x^2+x^3+x^4$   (c) $\sum_{i=1}^{4} ix_i = 1x_1+2x_2+3x_3+4x_4$

7. (a) $\begin{bmatrix} 2 & 3 & 4 & 5 \\ 3 & 4 & 5 & 6 \\ 4 & 5 & 6 & 7 \end{bmatrix}$

## Exercises 10.3 (Page 359)

2. $AB = I$
   $\therefore (AB)C = IC = C$
   $\therefore A(BC) = C$
   $\therefore AI = C$
   $\therefore A = C$

Answers and Solutions to Selected Exercises    421

## Exercises 10.4 (Page 366)

**2.** (a)

| 7 5 | 1 0 |
|---|---|
| 4 3 | 0 1 |
| 1 $\frac{5}{7}$ | $\frac{1}{7}$ 0 |
| 0 $\frac{1}{7}$ | $-\frac{4}{7}$ 1 |
| 1 0 | 3 −5 |
| 0 1 | −4 7 |

(b) $\begin{bmatrix} \frac{5}{4} & -3 \\ -\frac{3}{4} & 2 \end{bmatrix}$     (c) $\begin{bmatrix} 3 & -2 \\ 11 & -7 \end{bmatrix}$

Inverse of $\begin{bmatrix} 7 & 5 \\ 4 & 3 \end{bmatrix} = \begin{bmatrix} 3 & -5 \\ -4 & 7 \end{bmatrix}$

**3.** (a) No inverse

**4.** (a) $\begin{bmatrix} 3 & -7 \\ -2 & 5 \end{bmatrix}\begin{bmatrix} 5 & 7 \\ 2 & 3 \end{bmatrix}\begin{bmatrix} x \\ y \end{bmatrix} = \begin{bmatrix} 3 & -7 \\ -2 & 5 \end{bmatrix}\begin{bmatrix} 2 \\ 3 \end{bmatrix}$

$\begin{bmatrix} 1 & 0 \\ 0 & 1 \end{bmatrix}\begin{bmatrix} x \\ y \end{bmatrix} = \begin{bmatrix} -15 \\ 11 \end{bmatrix}$

$\begin{bmatrix} x \\ y \end{bmatrix} = \begin{bmatrix} -15 \\ 11 \end{bmatrix}$

that is, $x = -15, y = 11$

**11.** $A^2 = \begin{bmatrix} 5 & 7 \\ 2 & 3 \end{bmatrix}\begin{bmatrix} 5 & 7 \\ 2 & 3 \end{bmatrix} = \begin{bmatrix} 39 & 56 \\ 16 & 23 \end{bmatrix}$ and $-8A = (-8)\begin{bmatrix} 5 & 7 \\ 2 & 3 \end{bmatrix} = \begin{bmatrix} -40 & -56 \\ -16 & -24 \end{bmatrix}$

Hence

$$A^2 - 8A + I = \begin{bmatrix} 39 & 56 \\ 16 & 23 \end{bmatrix} + \begin{bmatrix} -40 & -56 \\ -16 & -24 \end{bmatrix} + \begin{bmatrix} 1 & 0 \\ 0 & 1 \end{bmatrix} = \begin{bmatrix} 0 & 0 \\ 0 & 0 \end{bmatrix}$$

From this we obtain

$$I = 8A - A^2 = (8I - A)A,$$

showing that  $8I - A = A^{-1}$, that is,

$$A^{-1} = \begin{bmatrix} 8 & 0 \\ 0 & 8 \end{bmatrix} - \begin{bmatrix} 5 & 7 \\ 2 & 3 \end{bmatrix} = \begin{bmatrix} 3 & -7 \\ -2 & 5 \end{bmatrix}$$

**12.** $B^2 = \begin{bmatrix} 4 & 5 \\ 2 & 3 \end{bmatrix}\begin{bmatrix} 4 & 5 \\ 2 & 3 \end{bmatrix} = \begin{bmatrix} 26 & 35 \\ 14 & 19 \end{bmatrix}$

$$rB^2 + sB + tI = \begin{bmatrix} 26r + 4s + t & 35r + 5s \\ 14r + 2s & 19r + 3s + t \end{bmatrix}$$

Hence we must determine $r, s, t$ so that

$$\begin{cases} 26r + 4s + t = 0 \\ 35r + 5s \phantom{+ t} = 0 \\ 14r + 2s \phantom{+ t} = 0 \\ 19r + 3s + t = 0 \end{cases}$$

The second and third of these equations are equivalent and yield

$$s = -7r$$

Substituting in the first and fourth equations yields

$$26r - 28r + t = 0$$
$$19r - 21r + t = 0$$

Each of these is equivalent to

$$t = 2r$$

Hence $rB^2 + sB + tI = r(B^2 - 7B + 2I) = 0$ for all real $r$, that is, the desired equation is

$$B^2 - 7B + 2I = 0$$

Using this we obtain

$$I = \tfrac{1}{2}(7B - B^2) = (\tfrac{7}{2}I - \tfrac{1}{2}B)B$$

Hence $B^{-1} = \tfrac{7}{2}I - \tfrac{1}{2}B$

$$= \begin{bmatrix} \tfrac{7}{2} & 0 \\ 0 & \tfrac{7}{2} \end{bmatrix} - \begin{bmatrix} 2 & \tfrac{5}{2} \\ 1 & \tfrac{3}{2} \end{bmatrix}$$

$$= \begin{bmatrix} \tfrac{3}{2} & -\tfrac{5}{2} \\ -1 & 2 \end{bmatrix}$$

## Review Exercises (Page 369)

1. (a)(2) $\begin{bmatrix} 1 & 1 & 2 \\ 4 & 3 & -1 \\ -3 & 2 & 4 \end{bmatrix}$

# Index

Abel, 54, 125
Abelian (commutative) group, 86, 125
    nonabelian (noncommutative) group, 86
Absolute value, 94, 230
Addition of matrices, 132-133, 351-352
Algebraically incomplete, 184
Anticommutative, 158
Archimedean property, 273
Arithmetic mean, 228
Artin, Emil, 307
Associative operation, in $S$, 34
    on $S$, 34
Associative triple, 32
Associativity, 32, 57
Automorphism, 83
Axioms, for a field, 166
    for a group, 58-59, 68, 73-74
    for a ring, 127
    of order, 204

Basis, 301
Bell, E. T., 333
Betweenness, 209
Binary operation, 5, 6, 7
    in a set, 8
    on a set, 7
Boolean ring, 157
Bounded set, 261
Bourbaki, N., 161
Boyer, Carl, 278

Cancellation Theorem, 70, 148
    left cancellation, 70
    right cancellation, 70
Cartesian product, 343
Cauchy, 54

Cauchy-Schwarz inequality, 229, 230
Cayley, Arthur, 131
Clock arithmetic, 10
Closure, 24, 57
Coefficient matrix, 313, 334
Column vector, 342
Command Group, 76
Common multiple, 108
Commutative operation, in $S$, 30
    on $S$, 30
Commutative rings, 138
Commutativity, 26, 57
    for a binary operation, 29
    for a group, 57
Complete ordered field, 272
Completeness property, 268
Completing the square, 185
Complex numbers, 184, 203
Composite integers, 96
Composite numbers, 96
Composites, 96
Conformable matrices, for addition, 352
    for multiplication, 340, 345
Congruence (modulo an integer), 110, 111
Cyclic Group, 77, 81, 84

Diagonal system, 323
Direct product, 73
Directed segment, 282
Distributes, 126, 127
Distributive, 91
Distributivity Axiom, in a ring, 144
Division Theorem, 89ff, 92
Divisor, 91, 95
Divisor of zero, 144, 145
    proper divisor of zero, 145

423

Eddington, Sir Arthur, 53
Euclidean algorithm, 97ff
Eudoxus, condition of, 273
Exponentiation, 28

Factor, 95
Function, 344
Fundamental Theorem of Arithmetic, 101, 106

Galois, Évariste, 53
Gauss, 54, 322
Gauss-Jordan Complete Elimination Procedure, 320
Gauss-Jordan Method, 307ff
Gaussian integers, 131
Geometric mean, 228, 230
Greater than ($>$), 208
Greatest common divisor (GCD), 97
Greatest lower bound (GLB), 265
Group, 53, 58
  abelian, 59
  command, 76
  commutative, 59
  cyclic, 76
  Klein, 84, 139
  nonabelian, 86
  noncommutative, 62, 86
Group properties, 54ff
  weaker definition, 73
Groupoid, 69

Halmos, P. R., 294
Hardy, G. H., 125
Harmonic mean, 228, 230
Henkin, Leon, 198
Homogeneous system, 330
Hyperbolic integers, 157

Ideal of a ring, 157
Idempotent element, 142
Identity element, 36, 57
Identity mapping, 83
Identity matrix, 146
Image of a product, 78
Indices, 343
Induction, Mathematical, 242ff
Inequations (inequalities), 212ff
Inner product, 345
Integers, 4, 56, 89, 164, 248
  Division Theorem, 93
Integral domain, 149
Inverse elements, pair of, 43
Inverse of an inverse, 71
  of a product, 71

Inverses, 40, 42, 57, 171
Irrational, 184
Isomorphic, 78
Isomorphic groups, 77
Isomorphic systems, 74ff
Iteration, 324

Jordan, Wilhelm, 322

Klein Group, 84, 139
Kronecker, Leopold, 89

Lagrange's Theorem, 85
Least common multiple (LCM), 108, 109
Least residue, 122
Least upper bound (LUB), 263, 264
Less than ($<$), 208
Linear combination, 297
Linearly dependent, 297, 299
Linearly independent, 299
Liouville, 54
Lower bound, 260

Mathematical system, 4
Matrix, 131ff, 312, 344
  addition of, 132-133, 351-352
  $m \times n$, 344
  multiplication of, 133, 346
  scalar, 141, 153, 164
  singular, 315
  two by two, 131ff
  unit, 314
Max (maximum), 38
Midpointing, 8
Min (minimum), 38
Modular arithmetic, 10
Multiplication, left, 69
  right, 69
Multiplication of matrices, 133, 346

Natural numbers, 242
Negative, 206
Negative integers, of an ordered field, 242, 248
Neutral element, 36
Nilpotent element, 142
Nonabelian group, 86
Noncommutative group, 62, 86
Noncommutative ring, 138
Number theory, 89ff

One-to-one correspondence, 76
Operation, 5, 6
Ordered field, 203ff
Ordered pair, 7

# Index

Pair of inverse elements, 43
Parabolic integers, 157
Partition, 206
Peano's Axioms, 248
Perfect squares, 178, 182
Permutation group, 57-58
Pivot, 308, 309
Positive, 204, 205
Positive integers, of an ordered field, 242, 248
Prime integers, 95
Prime numbers (primes), 95
   twin primes, 96
Principle of Mathematical Induction, 244
Product, image of, 78
Product of matrices, 133, 346
Pythagoras, 2
Pythagorean triplets, 49-50

Quadratic equation, 185
Quadratic formula, 2, 182, 185
Quotient, 91, 169

Rational numbers, 4, 56, 164, 184, 203, 248
Real numbers, 184, 203
Reciprocal, 147, 169
Relatively prime, 104
Remainder, 91
Replacement, Axiom of, 39
Residue, 112
Ring, 127
   noncommutative, 138
Rings, 125ff
Row vector, 342
Russell, Bertrand, 203

Scalar, 285, 287
Scalar matrices, 141, 153, 164
Shakespeare, William, 259
Singular matrix, 315
Spanning set, 301
Spans, 301

Special properties of 0 and 1, 170
Square root, 182
Subfield, 249
Subgroup, 67, 73, 85, 126
Subring, 165
Successor, 243
Successor subset, 243

Tableau, 323, 326
Theorems, for a group, 68ff
Transitivity, 209
Triangle inequality, 233-234
Triangular system, 322
Trichotomy, 209
Tucker, Albert, 322
Twin primes, 96
Two by two matrices, 131ff

Unary operation, 5
Unique Factorization Theorem, 106
Unique solution, 47-48, 70
Uniqueness, of identity element, 38, 69, 170
   of inverse element, 46, 69, 170
Unit element, 146
Upper bound, 260

Vector, 282, 287
   column, 334
   geometric, 282
   row, 342
   space, 279, 287
Vector space, 279, 287
   axioms for, 287

$W_3$, $W_7$, 13
$W_n$, 14
Weiner, Norbert, 294
Well-Ordering Axiom, 92
Well-Ordering Theorem, 247
Whitehead, A. N., 1
Whole numbers, 4, 89